Expanding Access to Science and Technology

Note to the Reader from the UNU

With science and technology having an increasing impact on our social and natural environments as well as on the conduct of the scientific community itself, the UNU considered it important to create a forum for leading experts to contemplate the directions in which specific advances are leading us. To meet this end, a seminar series on "Frontiers of Science and Technology" was established in 1991. The first symposium under this series looked at the role of the emerging theory of chaos in discussing "The Impact of Chaos on Science and Society."

The symposium on "Expanding Access to Science and Technology – The Role of Information Technologies" was the second symposium under the "Frontiers" series. New information technologies are fundamentally restructuring traditional ways of access to science and technology. However, with the growing amount of scientific information and the current rate of technological innovation there is a trend to greater disparities among nations in terms of information access and retrieval, and the ability to apply existing data to social and economic development. Based on the symposium, the current publication assesses the new technological potential in providing access to, retrieval, exchange, and handling of information. In this context intelligent information access and its impact on information retrieval and transfer as well as opportunities of developing countries in utilizing the new information technologies are of major concern.

Expanding Access to Science and Technology: The Role of Information Technologies

Proceedings of the Second International Symposium on the Frontiers of Science and Technology Held in Kyoto, Japan, 12–14 May 1992

Edited by Ines Wesley-Tanaskovic, Jacques Tocatlian, and Kenneth H. Roberts

The United Nations University
Tokyo, Japan

United Nations University Press
The United Nations University, 53-70, Jingumae 5-chome, Shibuya-ku,
Tokyo 150, Japan
Tel.: (03) 3499-2811   Fax: (03) 3499-2828
Telex: J25442   Cable: UNATUNIV Tokyo

Typeset by Asco Trade Typesetting Limited, Hong Kong
Printed by Permanent Typesetting and Printing Co. Ltd., Hong Kong
Cover design by Takashi Suzuki

UNUP-844
ISBN 92-808-0844-3
United Nations Sales No. E.94.III.A.1
04700 P

Contents

Preface

*Ines Wesley-Tanaskovic, Jacques Tocatlian, and
Kenneth H. Roberts*

Recognizing that science and technology have an increasing impact on our social and natural environments, the United Nations University decided to have a forum for international experts to contemplate the directions in which certain advances are leading society. To this end a series of international symposia on "Frontiers of Science and Technology" are being carried out in cooperation with major Japanese universities.

The first symposium in this series was held in 1991, in cooperation with the University of Tokyo, to assess the impact of chaos on science and society, reflecting the growing body of knowledge on chaotic behaviour in a variety of scientific disciplines.

The second symposium was held in Kyoto in 1992 in cooperation with Kyoto University, and its aim was to evaluate the potential of the new information technologies to improve information handling, retrieval, and exchange and, in general, to expand access to science and technology.

It is recognized that new information technologies are fundamentally restructuring traditional ways of providing access to science and technology. However, despite the current rate of technological innovation, there is a trend to greater disparities among nations in terms of access to information and the ability to apply knowledge for social and economic development. The opportunities of developing countries to use the new information technologies were a major concern of the second symposium.

The Kyoto symposium reviewed the experiences and strategies of major international information programmes over the last 25 years that aimed at improving the capabilities of countries, and recognized that a significant infrastructural gap and lack of a systematic approach persist. Therefore, the symposium took as its central focus the requirements for planning future infrastructure development in a more systematic way by fully taking into account the new technologial potential.

The symposium also examined technological experiences in database and data bank construction and use, communication networks, and the problems encountered by the developing countries in acquiring, adapting, and using the new information technologies.

One general area attracted particular attention: intelligent information access and its impact on information retrieval and transfer. Intelligent information access is a central area for research and development in information and computer sciences. It is expected that the advances in human-computer interaction and particularly in interactive technologies of natural language processing will open new perspectives. Access through natural language will become more creative for seeing in the computer an agent with a broad spectrum of communication means with the human user. Language must be integrated with other channels to attain multimodality for information access. The resulting hypermedia systems based on cognitive theories of human information processing open up horizons.

The paradigm of "sub-languages" that are domain-specific promises to offer an imaginative solution to the problems of user-computer interaction: the user as a processor of information. This approach encounters perceptual and processing limitations as well as applications to the user interface, such as input devices, displays, and dialogue design. "Flexible information processing" is another paradigm that is considered essential for the advancement of information technology applications in the real-world environment and that was examined at the symposium in terms of its future prospects.

A panel discussion at the symposium's end evoked recommendations for international cooperation in expanding access to science and technology.

Opening Address

Heitor Gurgulino de Souza
Rector, the United Nations University

Welcome ladies and gentlemen, distinguished guests, friends to the international symposium on "Access to Science and Technology: The Role of Information Technologies." This conference is the second in a series of Japan-based events carried out by the United Nations University (UNU) under its programme on the Frontiers of Science and Technology.

The United Nations University is a young institution with a University Centre established in Tokyo in 1975 following a recommendation in 1969 by U Thant, the then Secretary-General of the United Nations. It was designed to be truly international and devoted to peace and progress. For mobilizing international scholarly resources, however, a completely new kind of academic institution was required. Accordingly, the work of the UNU is not limited to its University Centre here in Japan, but it pursues its objectives as well through worldwide networks of scholars and through centres and programmes in Helsinki, Maastricht, Nairobi, Caracas, and Macau. Professor Charles Cooper, Director of our UNU Institute for New Technologies in Maastricht, and Dr. Zhou Chao Chen, representing our International Institute for Software Technology that is currently being established in Macau, are with us at this conference.

The Charter of the United Nations University calls for "due attention to the social sciences and the humanities as well as natural sciences, pure and applied." It expresses a particular concern that the UNU counteract the isolation of scholars and scientists from the third world. For this symposium we have invited computer scientists, mathematicians, economists, and social scientists to assess how the potential offered by new information technologies may be applied to improving access to science and how it may be used on behalf of the developing world.

The UNU's research programmes address global issues and their implications. Thus, global environmental change, third world development, cross-national implications of science and technology, as well as governance of growing mega-cities are among the objects of our attention. The increased awareness of the complexities underlying these issues requires multidisciplinary and new scientific approaches, sophisticated models, new ways of thinking and perceiving our natural and social environments. In this respect, information technologies have redefined the frontiers of science and technology over and over again for the past three decades, thus gradually shaping the path towards what is commonly referred to as "the information society."

Today we are experiencing a shift from single-purpose terminals to intelligent multi-purpose input-output devices, from limited to abundant storage capacity, and from dedicated to multi-purpose networks. The rate of diffusion of technological innovation is determined, though, by economic, social, political, and cultural factors. These are equally profound in developed and developing societies. However, the obstacles encountered by the developing countries in achieving technological innovation are high, and we must overcome a situation in which some parts of the world are connected to and other parts are disconnected from the information infrastructure, a world where the industrialized countries are "on-line" and the developing countries are "off-line." For the latter, the crucial issue is not just technological innovation but how to use it for improvements in the social and economic situation of their people. Only then will we be able to talk of truly "global technologies."

In organizing this conference I am grateful for the cooperation we received from the University of Kyoto, and my thanks go particularly to Professor Hiroo Imura, President of the University, and to its former President, Professor Yasunori Nishijima, who has been involved in this project since its very early planning stages.

Furthermore, my thanks go to Professor Huzihiro Araki of the Research Institute for Mathematical Sciences, whose support was essential for convening this symposium.

I am also grateful for the help we received from our international advisers, Dr. Jacques Tocatlian and Dr. Ines Wesley-Tanaskovic. Dr. Wesley-Tanaskovic has been in charge of our microprocessors and informatics training programmes for many years. Both of them will have the challenging task of synthesizing the diverse approaches from different scientific disciplines represented here this week for a future book publication. Finally, I would like to invite all of you to visit the special exhibition that will be set up by Fujitsu Corporation tomorrow, and in this context I want to thank Fujitsu also for its support of this symposium.

Before giving the floor to Professor Hiroo Imura, I thank all participants and observers for coming here today and I want to wish you a successful meeting. I look forward to the results and I hope to gain some new insights myself by talking to many of you personally during the reception given tonight here at the Kyoto International Community House.

Opening Address

Hiroo Imura
President, Kyoto University

Good morning Dr. de Souza, Dr. Araki, ladies and gentlemen. On behalf of Kyoto University, one of the organizers of this symposium, I have the pleasure of extending a cordial welcome to all participants to the second International Symposium on the Frontiers of Science and Technology. I have to express my sincere thanks especially to all speakers who have come a long distance. As already mentioned by Dr. de Souza, the first symposium, which was held in Tokyo last year, was focused on chaos in science. This year the subject is information. In the field of science and technology, very rapid accumulation of information is causing a variety of problems. The increase of numbers of scientific papers is not linear but exponential, and this has caused increases of review journals, computerized retrieval systems, and many other sophisticated methods. The progress of information transfer, however, has caused disparities not only among countries as already discussed by Dr. de Souza, but also among individuals.

Let me discuss this a little more by taking the human body as a metaphor. The human body consists of about 60 trillion cells, 10,000 times greater than the population in the world. It is a society of cells. Among such an enormous number of cells, there are many ways of information transmission. The nervous system is like a telephone or telefax and sends information through the nerve fibres. It is a very rapid and efficient way of information transmission but wires and special instruments are required. In order to facilitate information transmission by this system, you have to increase wires and instruments. Another important system is the endocrinal system. The endocrinal system distributes information as chemical substances, that is, hormones, by the bloodstream throughout the body. The cells that have a special antenna, or receptor, can catch specific information. There are several ways to facilitate information transfer in this system. One is to increase the number of receptors; another way is to change information in such a way as to enable it to be caught by different receptors. These functions and modifications are analogous to the development of information techniques to be discussed in this symposium.

In the past centuries, liberty, equality, and humanity have been considered most important for humankind. In such a global age as today, we must add one more important thing: communication. In the field of science and technology, communication of new information is, needless to say, quite important. In this symposium, the importance of transmission for science, economy, humankind, new technologies for information retrieval and transfer, and the need for international cooperation, especially for the developing countries, will be discussed.

We have to improve the nature of information and to increase the receptive mechanisms in order to achieve our goals. I hope that this three-day symposium will be a great success. Thank you very much.

Session 1

Access to Science and Technology and the Information Revolution

Chairperson: Huzihiro Araki

Introduction: Access to Science for the Benefit of Mankind

Sir John Kendrew

I think that this audience will find no difficulty with the idea that science is for the benefit of mankind. But I have to remind you that not everyone would agree with this opinion; there are some who think that science has had a negative influence on the spiritual values of mankind, and indeed in my own country a certain Dr. Appleyard has just published a book [1] that, I understand, strongly maintains this point of view. But as I say, I believe it will not be necessary to argue this point with those present here today.

Communication is an integral part of science and of the mechanism for increasing human knowledge. The lonely genius remains frustrated and useless without communication. In mathematics the classical example was Ramanujan, a poor and uneducated Indian whose quality was only revealed when he wrote a letter containing some of his results to the Cambridge mathematician Professor Hardy, who recognized his genius and brought him to England to work. Without this letter most of Ramanujan's remarkable theorems would probably have been lost to the human race.

Communication is of basic importance at three levels. The first is communication within science itself. I think we would all agree that free communication is essential for progress in science, even though there have been problems at the frontier with technology because in the world of commerce you cannot reveal everything you are doing. These frontier problems have been exacerbated in recent times; in many fields, my own of molecular biology in particular, the old openness has to some extent disappeared. Some of my younger colleagues do not want to talk in public about their work, because they think that if they speak about their own "secret," somebody else will exploit it and make money out of it that they could have made themselves.

When I began research in the 1950s I don't think any of us ever imagined that molecular biology would one day have any practical value; we thought of it only in terms of increasing human knowledge for its own sake; but now of course it has given rise to the major industry of biotechnology and fundamental advances in medical treatment. Of course, this is only a partial view: science is also for the benefit of mankind and this is indeed reflected in the title of my present talk. But the process of transmission of knowledge from the academic world, with its tradition of openness, to the world of technology and commerce does present difficulties that have still not been resolved. In my view they are very serious problems, though they do not fall within the remit of the present meeting.

The second level at which communication is of basic importance is especially in the developing world, as the Rector has already mentioned. And in this context I am not thinking only of what, as a western European, I characterize as the South, but also of the East; many of the problems that exist, for example in Africa, are present also in the countries of Eastern Europe following, and in spite of, the big political changes there. In the developing world, in this wider sense, there is a tremendous thirst for knowledge and at the same time a severe lack of journals, of books, and of means of rapid communication at low cost. The phrase "at low cost" is tremendously important in countries where financial resources are extremely limited; and here, I think, we enter a field where the experts present at this meeting can make enormous contributions. Improvements here can help, can be of really practical assistance, in beginning to reduce the gap between the affluent nations from which most of us here come and the poor nations of the world – a gap that, alas, in spite of all that has been done so far, is still increasing rather than getting smaller. In the most literal sense the future of the human race depends on changing the increase into a decrease, and the new technologies to be discussed at this meeting can make a very important contribution.

My third problem of communication is one that more particularly concerns the affluent countries. I refer again to the development of an anti-science movement, indeed in some circles an anti-intellectual movement. It becomes increasingly difficult to persuade young people to enter the profession of science. In the United States, for example, young foreign scientists are increasingly being imported to fill the gap. And even in Japan, where a few years ago problems of this kind seemed to be non-existent, we read that there are now difficulties in recruitment to the profession. At bottom, the problem is one of communication: young people simply do not understand the importance or the excitement of science. They need better education, and so indeed do their parents who give them their values. And beyond them, ministers and administrators need to know more of science. These are often very competent and well-trained people, but their education was generally not in science; and this is why in most countries governments do not provide adequate support for science and do not understand that science is at the basis of the civilization and health that they enjoy, and of the improvements for which they hope in the future.

So we have a whole set of world problems that demand better communication for their solution. In this meeting we are to discuss the technical means for communication – only a part of the problems, but certainly a part sufficiently important to justify the United Nations University's choice of the topic for this symposium.

Historically the first of the mechanisms for communication was the human voice; in eighteenth century Europe, scientists used to travel a great deal and discuss their results verbally; and this is a mechanism very important today on a larger scale, with the proliferation of international meetings; the only problem here is to provide funds and facilities for younger scientists of all countries, including the developing ones, to attend them. It's all very well to have professors moving around to talk to one another, but it is perhaps even more important for young people to do so. At least some of the political obstacles to travel in Europe have been removed, but the financial ones remain there and in many other parts of the world.

Next came the written word, the journals with their scientific papers and the newspapers with their mission of educating and informing the public. Here I believe we have a relatively new difficulty. In a recent issue of *Nature* you will find a very interesting article by Professor Donald Hayes of the Department of Sociology at Cornell entitled "The Growing Inaccessibility of Science" [2]. It is an account of a piece of statistical research examining the language used over a period of years in various types of publication. He allots an index of intelligibility to the language used by various types of publication; for an internationally read newspaper the index is 0; technical articles in scientific journals are in the range +40 to +60; fiction is −20; casual conversation between adults about −40; and at the bottom of his scale, farm workers talking to cows are about −60. He then examines popular science journals; until 1947 *Nature* was near 0, since when its index has risen decade by decade, until today it is about +30; *Science* began in 1883 at −8.5 and took off from a near-zero level around 1960; today it is +28; *Scientific American* remained near zero until 1970 and then increased; when its index reached +15, there was a decline of over 125,000 subscribers; when later its index dropped back to +10, there was a coincident increase in subscriptions. These figures mean that even for professional scientists much of the literature is unintelligible except to those actually working in the same field – including, by the way, most of the contributions to the present symposium! Experiments like Basic English have been abandoned, and we see no solution in sight. Is this adequate communication? How can we expect to communicate with citizens at large when we cannot even communicate with other scientists?

One of my professors was Lawrence Bragg, who held the view that young scientists should not read too many journals; if they did, he believed, they would probably discover that the experiment they had thought of doing had been done already; if they did not, they would do their experiment with an open mind and might discover something new. He also used to say that if you did read a journal, the important thing was flipping over the pages so that your eye might be caught by something entirely unexpected. I don't think he would have approved of abstract journals or *Current Contents*, or of the computerized database searches that are commonplace today.

Another of my professors was Desmond Bernal, and I have been thinking a great deal of him in these last days because just 44 years ago the Royal Society held its Scientific Information Conference [3] with a purpose not dissimilar to that of our meeting today, discussing as it did some of the then contemporary technical advances like punched cards and microfilm. I was present myself when Bernal proposed his scheme whereby journals should not be published in their present form but only as a set of titles and abstracts, so that you would write in to the editorial office and request an offprint of papers that interested you rather than subscribing to the whole journal. Well, of course the idea got nowhere, but it has been echoed in recent times with the promotion of publications like *Current Contents*. Of course in those first post-war years, one spent an immense time sending off postcards to authors asking for offprints; and a little later on, when photocopiers became common, in copying papers page by page – another technical advance of great importance.

Now we have in our hands a still more important technical advance, the computer; and this has given us access to databases and abstracts, and simple pieces of software like word processors have enormously facilitated the preparation and revision of manuscripts for press and the handling of them by the printer. Now things move still further; I went to a meeting last September in the United States where I heard an account of a new journal in the medical field that will be entirely computerized; that is to say, you will read it by calling down papers onto your computer screen, and then copying anything you wish to retain: still another step in the direction of that "unrealistic" 1948 proposal by Desmond Bernal.

Of course another important development in this field is electronic mail. I personally use it every day and I only wish I could give a better report of its efficiency. The last of the Rector's symposia was about chaos, and this would aptly describe what often goes on in electronic mail: dozens of different networks with different types of address, not always linking with one another; the same address has to be read backwards or forwards depending on whether you are in the United States or in the United Kingdom. In spite of months of effort, I have been able to establish only one-way communication between myself and a certain important international organization located in Belgium. It's not a technical mess, but it is certainly an administrative mess, in strong contrast with the telefax system, which is well standardized and in general provides excellent, trouble-free service. I mention electronic mail because I believe it is particularly important for isolated laboratories and in particular for scientists in the developing world (where the problems are not of course just administrative but also technical, since the system depends on adequate telephone links). If electronic mail works, and you are trying to do an experiment described in the literature, you can ask one of the authors for help and get the answer back in just a few minutes.

Communication in science is a means to an end, and the technical advances we are going to discuss are means to that means. But they are tremendously important, not only to those of us who are fortunate enough to work in the advanced countries of the world, but even more so to those in developing countries who cannot hope to improve their position without the provision of these technical means at prices they can afford.

REFERENCES

1. Appleyard, B. (1992). *Understanding the Present: Science and Soul of Modern Man*. London: Picador.
2. Hayes, D.P. (1992). "The growing inaccessibility of science." *Nature* 356: 739.
3. The Royal Society (1948). *The Royal Society Scientific Information Conference* (21 June–2 July 1948). The Royal Society.

Keynote Presentation: The Impact of Information Technology on the Access to Science

David R. Lide

ABSTRACT

Access to scientific information is crucial to continued scientific advance and to technological progress. After discussion of the diversity of information requirements, this paper takes up mechanisms for the organization of data. The importance and advantages of computer and telecommunications technology in access to scientific data are described, and a brief overview is given of the availability of numerical and factual databases in different areas of science. The paper concludes with a consideration of data as an international commodity and of prospects for future developments.

1. Introduction

It is no coincidence that the birth of modern science followed shortly after the introduction of the printing press, more reliable sea transportation, and the other technological innovations that brought Europe out of the medieval world. Science could not have developed in the way it did without the capability of scientists to communicate their results, ideas, and speculations to each other. While a few profound advances in science have resulted from the insights of a single individual working in isolation, closer analysis always shows that those insights rested upon a body of information developed by other scientists, often over a long period of time, and made available to their colleagues through one or another mechanism of information transfer. The importance of access to scientific information – experimental results, interpretations, and theories – looms even greater when one looks at the translation of basic scientific advances into useful technology that improves the lot of mankind. Advances in our understanding of nature would have little impact on humanity at large if the knowledge remained confined to the laboratory or university. An effective mechanism for communicating that knowledge to scientists and engineers is crucial to the development of new and better technology that benefits us all.

This paper will give an overview of the ways in which modern information technology is affecting the access to scientific and technical information. The development of digital computers and high-speed communication networks has already had profound effects on information storage, retrieval, and dissemination. Nevertheless, we are probably still in the early stages of this electronic revolution. Even in highly developed countries, only a small fraction of scientists make significant use of the information technology now available. We can expect changes in the next 40 years just as dramatic as those we have seen in the 40 years since digital computers entered our lives.

2. Diversity of Information Requirements

Access to technical information is crucial to all phases of the scientific process. However, a scientist's information needs can vary from simple items like laboratory instrument manuals and catalogues to megabytes of data telemetered from a space probe millions of miles away. In discussing the effect of modern information technology on data access, it is helpful to break the subject down into several categories of information requirements; for example:
 − access by research scientists to raw data obtained elsewhere
 − access to the archival scientific literature
 − access to reliable factual data
 − access by government officials and the public at large to scientific findings that affect the general welfare
These aspects will be discussed in turn and important differences noted.

In certain fields of science, the common pattern is for many scientists, often in different countries, to collaborate on analysing the data obtained in a single large facility. One example is high-energy physics, where a few extremely expensive particle accelerators provide data for a worldwide community of theoretical physicists. Another is space science, where the data gathered by satellites and space probes are distributed to many investigators for interpretation. In such areas, networks are already in place that allow access to massive amounts of data by dozens (sometimes hundreds) of researchers who are collaborating on a project. Electronic bulletin boards and computer conferences allow the participants to try out new ideas and obtain their colleagues' reactions virtually in real time. This has introduced a new dimension to scientific collaboration, especially at the international level. While those involved in "big science" led the way, many others who are working on more modest research problems have adopted the same approach. We can expect a rapid growth of this type of research collaboration as low-cost, high-capacity networks are introduced throughout the world. The possibilities for bringing third-world scientists into collaborations of this kind are particularly intriguing.

The second type of access is to the archival scientific literature. As this literature has grown over the last generation, it has become increasingly difficult for scientists to follow their fields of interest. The introduction of on-line searching of abstract files 25 years ago has been a major factor in alleviating this problem. This facility is now available in every major field of science, making it possible to search millions of papers in a very short time and retrieve citations to pertinent documents. The next step in this evolution will be to make the full text of scientific papers accessible electronically. Experiments of this type have already started. The American Chemical Society provides on-line access to its journals, but without the graphical items. European publishers in the biomedical area have established the ADONIS program, which provides current journals to libraries in CD-ROM form. A purely electronic journal, *Current Clinical Trials*, has been started by the American Association for the Advancement of Science to publish papers on testing of new drugs. A new "paper" is accessible via an on-line network within 24 hours of its acceptance by the editor. Thus, the momentum is building for a transition from the traditional printed journal, which has served as the archival record of science for the last 300 years, to a new pattern of electronic dissemination.

Access to numerical data and other forms of factual information presents a different set of considerations. This type of information has traditionally been published in handbooks and compilations; it represents a distillation (ideally, including a measure of critical analysis) of the data reported in the archival literature. Such data are needed at every level, from basic research to engineering. Great strides have been made in the use of computer technology for accessing this kind of data; a more detailed discussion appears later in the paper.

The final topic deals with the needs of public officials and private citizens. In the early days of computers, certain visionaries predicted that every government official would soon be able to access the full information base of science, leading to a better understanding and wiser decisions. Expensive demonstration systems have even been built for this purpose. However, the reality has not quite met the promise. There are too many opportunities for a non-technical person to misinterpret or misuse the results that he can instantly access via an electronic system. Quick access is not so important as a balanced, intelligent analysis of the information. In this arena, the human mind is still far ahead of the computer.

It should not be inferred that these dramatic advances in the techniques for accessing scientific information have occurred without problems. In fact, the introduction of electronic technology has put a great deal of stress on our traditional information mechanisms. There is a widespread perception in the scientific community that electronic access is too expensive. On the other hand, many organizations in the information business have found that the revenue from new electronic services does not make up for the ensuing loss of income from their traditional printed products. The scientific societies that publish scholarly journals in their fields are deeply concerned about the economics of the new media. Many questions of copyright and protection of intellectual property in the electronic age remain unsolved. The institution of peer review, which has been so important in maintaining the integrity of scientific publications, is threatened by the computer bulletin boards offering rapid but unscreened access to new research. New mechanisms will be needed to assure that scientists receive proper credit for their intellectual contributions, since the origin of individual pieces of data is sometimes lost when they are incorporated in large databases. Like all upheavals, the information revolution is causing its share of disruptions in the behaviour patterns and culture of the scientific community.

3. Numeric and Factual Databases

One class of scientific information where the new technology promises to have major impact is the hard factual data, usually numeric in nature, which form the lifeblood of science and are an essential ingredient in the transfer of scientific knowledge into useful technology. At its most simplistic level, a scientific datum involves an object and an attribute; examples are the boiling-point of benzene, the radius of the earth, and the gestation period of the elephant. Of course, it is usually necessary to specify certain auxiliary parameters if the meaning of the datum is to be completely clear. Thus the boiling-point of a liquid is ambiguous unless the pressure is specified, and the value of the radius of the earth differs between the equator and the poles. It is often necessary to associate a complex set of "metadata" with each data point in order to make that datum useful.

The preservation and dissemination of such data have been recognized as crucial since the beginning of modern science. In our own time, the cost to society of acquiring scientific data has risen enormously, not only in regard to data produced by expensive devices such as satellites and particle accelerators, but even for routine measurements done in the laboratory. It is therefore incumbent on us to assure that scientists and engineers have ready access to all the existing data that might expedite their work.

The particular field of science has a large bearing on the nature of the data encountered in that discipline. In particular, data tend to have different characteristics in the physical sciences, geosciences, and biosciences. The table is taken from a report [8] prepared by CODATA, the ICSU Committee on Data for Science and Technology. This table indicates the various ways to categorize data and gives examples in the three broad areas of science. While the details of the classification are not relevant to this paper, it is important to note that most physical science data are independent of location and time; in principle, the measurements can be repeated at a different place and time with the same result. Furthermore, much of the data in physics and chemistry can be analysed in terms of well-established quantitative

Varieties of categories of data

	Categories of data	Chemistry/physics	Geo-/astro-sciences	Biosciences
a_1	Data that can be measured repeatedly	Most data	Geol. structures, rocks Accel. due to gravity Fixed stars	Most data
a_2	Data that can be measured only once		Volcanic eruptions Solar flares, novae	Rare specimens Fossils
b_1	Location-independent	Most data	Minerals Global tectonics	Most data, excluding extraterrestrial
b_2	Location-dependent		Rocks, fossils Astronomical data Meteorological data	Rare specimens Fossils
c_1	Primary observational or experimental data	Optical spectra Crystallographic F-values	Seismographic records Weather charts	Physiological data (e.g., respiration rates, blood volumes, etc.) Biochemical data (e.g., composition of tissues and organs)
c_2	Combinations of primary data with the aid of a theoretical model	Fundamental constants Crystal structures	Fossil zoning Temp. distribution in sun	Genetic code Body surface area Model of vascular bed Dimensions of tracheo-bronchial tree
c_3	Data derived by theoretical calculation	Molecular properties calculated by quantum mechanics	Solar eclipses predicted by celestial mechanics	Prediction of phenotypic expression from genotypes

d_1	Determinable data	Most macroscopic data	Elements of planetary orbits	Gene loci Chromosome numbers
d_2	Stochastic data	Polymer data Structure-sensitive properties	Soil and rock composition Solar flares Frequency of visible meteors per unit interval	Most data
e_1	Quantitative data	Most data	Seismic data Meteorological data	Physiological data Biochemical data
e_2	Semi-quantitative data	Mohs hardness scale	Wind force scale	
e_3	Qualitative data	Chemical struc. formulae Properties of nuclides	Rock classification Classification of stellar spectra Fossil shapes	Amino acid sequences Taxonomic classification of organisms
f_1	Data presented as numerical values		Meteorological data	Physiological data Biochemical data
f_2	Data presented as graphs or models	Phase diagrams Stereoscopic molecular diagrams Molecular models	Geological maps Weather maps Sky mapping at a particular radio frequency (e.g., 21 cm)	Metabolic pathways Electrocardiograms Electroencephalograms
f_3	Symbolic data		Lithology in bore hole data	

Note: A given group of data can be categorized simultaneously by several "facets" a, b, c, etc.; for instance, the nature of meteorological data characterized as a_2, b_2, c_2, d_2, e_1, and f_1 (or f_2).

theories. It is therefore possible to cross-check data against theory and compare them with data on other materials. This provides a means of evaluating a set of data to establish its level of quality and, in some cases, to represent a large amount of data in a concise mathematical form.

In the geosciences, on the other hand, much of the pertinent data is location dependent and some (such as the data associated with earthquakes or solar flares) come from non-repeatable observations. Bioscience data, at least of the classical variety, are dominated by the variability of living organisms. Thus one must specify not only the central value of some characteristic, but also the range of values found, and sometimes even the form of statistical distribution. Because of these discipline-dependent factors, the design of a data storage and dissemination system must be approached very carefully, taking into consideration the inherent nature of the data and the way they are going to be used.

4. Evaluation and Quality Control

The quality of experimental or observational data may vary widely, depending on the care taken by the scientist who did the research. Furthermore, most measurements depend on some form of calibration, which can change over the years. The risk has long been recognized, especially in the physical sciences, of assuming a piece of data taken from the literature is valid without further checking. A distinct methodology of data analysis and evaluation has evolved, leading to compilations of "evaluated data" that can be used with confidence by the general scientific community. The details of the methodology vary with the type of data but it usually includes a careful study of the way the measurement was made (as described by the author); application of various corrections needed because of changes in temperature scale, fundamental constants, and the like; and comparison with applicable theory. Ideally, this evaluation procedure is applied systematically to a large body of data, so that any discrepant numbers are more visible.

This approach to quality control is not so easily applied in the geosciences and biosciences because of the different nature of the data, as already discussed. The most important consideration is to establish quality control before the experiment or observation is made. Thus the calibration of the instruments should be carefully documented and a valid statistical design established. Nevertheless, an independent peer review after the results are published often turns up errors and inconsistencies.

5. Traditional Access Mechanisms

Until the present generation, most scientific data were stored as ink on paper, in the form of tables of numbers or graphs. These data can be accessed in the archival research literature that is preserved in major libraries. Some journals maintain depositories, often as microfilm, where authors can put additional data too voluminous to print in a journal article. However, retrieving data from the primary literature is not an easy task, even with the help of abstracting services. Most abstracting and indexing services are oriented more to concepts, ideas, and theories than to the data content of a paper.

In the physical sciences, the need to aggregate and organize the data in the primary literature became apparent more than a century ago. The great German handbooks, such as Beilstein, Gmelin, and Landoldt-Bornstein, were started at that time in order to give scientists easier access to data. This represented a great advance, and the handbooks still function today. Another important step occurred in the 1920s, when the *International Critical Tables* were published [13]. This project introduced the idea of critical evaluation and selection of the best data, rather than simply recording all data found in the literature. More recently, other publication outlets for evaluated data in physics and chemistry have appeared. The *Journal of Physical and Chemical Reference Data* was started in 1972 as a joint project of the American Chemical Society, American Institute of Physics, and the National Bureau of Standards. This journal publishes papers with recommended data based upon an evaluation of all pertinent values found in the literature; the method of evaluation and criteria for selecting the data are fully documented. Somewhat similar publication series have been started in Germany and in the former Soviet Union. International organizations such as CODATA, IUPAC (the International Union of Pure and Applied Chemistry), and IUCr (the International Union of Crystallography) have also published many high-quality data books. Such efforts not only make data easier to locate but also assure that the user gets the most reliable values.

Much data in the geo- and biosciences is also preserved in the primary literature and in handbooks and compilations. In addition, certain types of data have traditionally been kept at the site of the measurements or in special depositories. Museums and culture collections are important repositories for biological data. Astronomical observatories have collections of photographic plates of stellar observations going back many years. The system of World Data Centers was set up by the ICSU at the time of the 1957 International Geophysical Year to preserve data from various geophysical observations, including earthquakes, solar flares, and tidal waves. An important feature of the World Data Centers is that records are duplicated at several sites throughout the world, in order to protect against loss of data through a disaster at one of the centres.

Patterns for storing and making accessible scientific data, even before the computer age, were therefore quite diverse. Many directories have been prepared to help scientists locate data, especially data in fields outside their own specialties. Two recent efforts of this type can be mentioned. CODATA has produced a CODATA Referral Database in computerized form that contains descriptions of data centres and depositories throughout the world [3]. It does not supply factual data but is intended to guide a user to organizations that can possibly provide the data needed; many of these data sources are particularly oriented to developing countries. The International Council of Scientific and Technical Information (ICSTI) has recently published a directory of numerical databases [11] that is also a useful guide.

6. Electronic Access to Scientific Data

The potential benefits of storing numerical scientific data in computerized formats were recognized in the early days of digital computers. Geophysics was one of the first fields to take advantage of electronic data storage, driven by the enormous quantity of data obtained from instruments on satellites. Distribution of data on magnetic tapes began during the International Geophysical Year. Other disciplines followed somewhat later. In the field of physics, for example, the International Atomic Energy Agency began to exchange tapes of neutron cross-section data in the 1960s. Databases of crystallographic data and certain types of spectroscopic data were established by 1970.

One of the most important advances in the 1960s was the introduction of computer storage and retrieval of chemical structures. This was led by the Chemical Abstracts Service, which was faced with handling records on the millions of the chemical compounds reported in the literature. Chemical structures represent a rather special type of data, essentially a record of the connections between atoms in the molecule. With this database available, it became possible to check a newly reported compound against the database and determine whether it actually was new. Furthermore, one can search the database and retrieve structures having specified features that are associated with particular chemical or biological activity. This has become a powerful tool for the development of new drugs.

In spite of these successes, the introduction of computer storage and retrieval of scientific data on a broad scale did not begin until the 1980s. Two technological developments were mainly responsible. One was the introduction of the personal computer, which enabled individual scientists to gain access to data sources, avoiding the cost and frustrations of dealing with a mainframe computer. The other was the growth of low-cost telecommunications networks. Scientists now have many options for communicating with remote computers from their own desk or laboratory.

Of course, the availability of spectacular technology does not assure benefits to the scientific community. Indeed, some of the early demonstrations of computerized data retrieval were in the nature of a *tour de force* and did not offer any real improvement over the traditional means of data access. It is important to look realistically at the potential advantages of electronic data dissemination in comparison with traditional methods. The principal advantages are:

1. More efficient access: the search software allows more detailed indexing than is practical with printed works; expert systems provide a new dimension in leading the user to the data he needs.
2. Linkage of related databases: different databases, even in remote physical locations, can be searched and manipulated simultaneously.
3. Multidimensional searches: desired combinations of properties (attributes) can be specified and materials selected that satisfy those requirements. This is very difficult to do with print media.
4. Direct data transfer: retrieved data can be transformed directly to computer programs for further manipulation, thus avoiding additional effort and possible errors.
5. Simplicity of updating: data files can be updated quickly and inexpensively, in contrast to printed reference works.

Electronic access to scientific data is developing along two parallel, and to some extent competing, paths. The first is the on-line approach, where the data banks are maintained at one geographical location (or perhaps on several linked mainframes) and users access this computer over a telecommunications network. The second approach is to distribute individual databases of floppy disks or CD-ROM to each user for installation into his own PC or workstation. Both modes have their advantages and disadvantages. Networks generally offer more powerful software and provide the capability of accessing many different databases from one entry point. Files can be kept up to date and errors corrected in an orderly fashion. However, the cost tends to be high and many people feel a psychological barrier in using an on-line network when they are being charged for every minute they are connected to the system. When databases are provided on diskettes or other transportable media, the user can load the file into his PC and then work with it at leisure, without incurring any further cost. In this mode, innovative uses of databases would seem more likely. However, updates and revisions must be physically sent to each user, who must make a conscious effort to replace the old data with the new.

It seems likely that both modes of dissemination will go through further evolution, without one mode being completely displaced by the other. The outcome will depend on economic and sociological factors, as well as strictly technical ones.

The following is a brief overview of computerized numerical databases that are currently available in different areas of science. Since the number of databases is very large and new issues appear frequently, no claim can be made that this survey is comprehensive. However, an effort will be made to give a flavour of what is now available.

6.1 Physical Sciences and Engineering

Numerical databases in physics, chemistry, materials science, and related engineering fields range from very large, multi-megabyte data collections to low-priced diskettes intended for educational purposes. Some representative examples are given below, categorized by the scientific subject matter they cover.

6.1.1 Spectroscopy

Most spectroscopic databases are intended for use in identifying unknown chemical substances. Therefore, they tend to be very large. The most widely used are in mass spectroscopy, which is a common analytical technique for identifying chemicals in industrial process control, environmental monitoring, and many other areas. The NIST/EPA/NIH Mass Spectral Database, available from the National Institute of Standards and Technology in the United States, contains data on more than 60,000 compounds. The version of this database for personal computers is distributed on diskettes and CD-ROM; it includes search software that permits a user to match peaks in an unknown sample with peaks in the database and thereby identify the chemical compound (fig. 1). The Wiley Register of Mass Spectral Data is a similar database, with over 150,000 substances. Several carbon-13 NMR databases exist, some designed for installation in NMR spectrometers and some accessible on-line. The largest one, containing about 70,000 entries, is maintained by the Fachinformationzentrum (FIZ) in Karlsruhe, Germany. It can be accessed on-line through STN International. Many collections of infrared spectral data are available in electronic form. Again, some are installed in instruments and others are on-line. The Handbook of Data on Organic Chemistry (HODOC) database combines spectroscopic and physical property data in one database covering over 25,000 organic compounds. It is maintained by CRC Press and can be accessed on-line through STN International; a CD-ROM version is planned (figs. 2, 3, 4, 5).

Increasing attention is being given to integrated data banks that allow searching of several types of spectra simultaneously. This is potentially a powerful technique for establishing positive identification of chemical substances that may be present in complex mixtures. In Japan, the National Chemical Laboratory for Industry has developed a large Spectral Database System of this type, and in Germany the Chemical Concepts organization is working on a similar system called "SPECINFO."

Quality control in such large databases is not an easy task. Certain automated procedures have been developed [9], but a trained spectroscopist is often needed to resolve discrepancies between spectra obtained from different sources.

6.1.2 Crystallography

Like spectroscopy, crystallography is a field of science characterized by the need to handle very large amounts of data. So, it is not surprising that crystallographers were among the first to adapt digital computer technology to their needs. The Cambridge Crystallographic Data Centre in the United Kingdom maintains a data bank on the crystal structures of organic compounds that contains over 80,000 substances. This database stores the three-

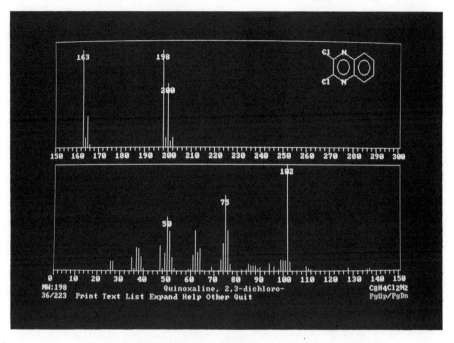

Figure 1 The NIST/EPA/NIH Mass Spectral Database

Figure 2 The CRC Press Database on Properties

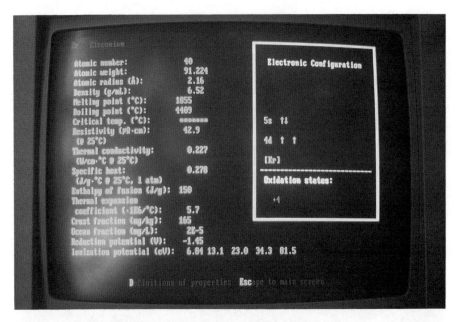

Figure 3 The CRC Press Database on Properties

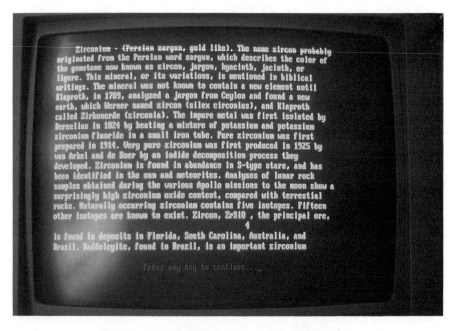

Figure 4 The CRC Press Database on Properties

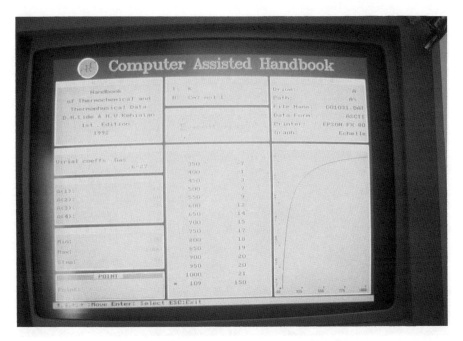

Figure 5 The CRC Computer-assisted Handbook

dimensional coordinates of all the atoms in each molecule, which allows the molecular structure to be displayed in graphical form. One of the most important applications of the database is in drug design, where it aids in pinpointing complex chemical compounds whose three-dimensional structure includes features that are likely to produce the desired biological activity. The Cambridge Centre licenses the data bank to pharmaceutical and chemical companies and makes it available to academic research groups through affiliates in all major countries. It can also be searched on-line through the CAN/SND system in Canada.

Another important crystallographic database is the NIST Crystal Data, which contains data on over 150,000 inorganic and organic crystals (figs. 6, 7, 8). This database does not contain the full structure, but only the lattice constants that are used to identify crystalline materials. Two other databases of this type are the Powder Diffraction File and the NIST/Sandia/ICDD Electron Diffraction Database. All of these are distributed by the International Centre for Diffraction Data.

6.1.3 Physical Properties of Chemical Substances

Two German data collections of great historical importance have recently become available in computerized versions. These are the Beilstein Database, covering a wide range of properties of organic compounds, and the Gmelin Database, which serves a similar function for inorganic chem-

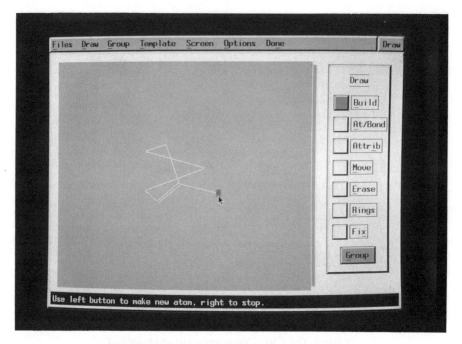

Figure 6 The NIST Structure and Properties Database

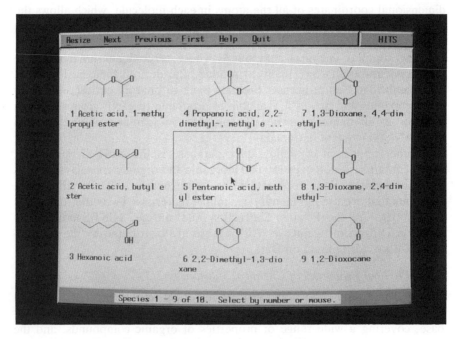

Figure 7 The NIST Structure and Properties Database

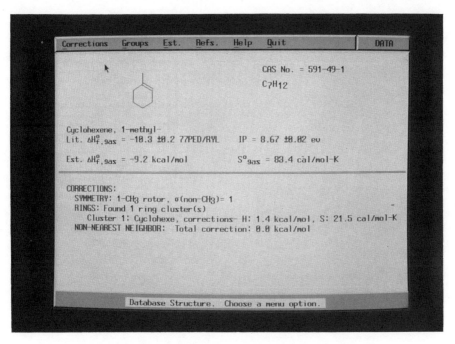

Figure 8 The NIST Structure and Properties Database

istry. Both can be accessed on STN International and other networks. The HODOC Database, mentioned earlier, is a much smaller database of organic compounds, but combines physical property with spectroscopic information.

6.1.4 Thermodynamic Properties

Data on thermodynamic and thermophysical properties find wide use in both industry and basic research. Many databases in this category are available. The Design Institute for Physical Property Data (DIPPR) has created a data bank of about 1,200 chemical substances of highest importance to industry. The DIPPR Database includes programs that calculate about 20 properties as a function of temperature. It is an example of a system that combines a collection of numerical constants with powerful computational software that generates data for the exact conditions requested by the user. It is available in magnetic tape and diskette formats and is on-line on STN. Another data bank designed for chemical industry needs is DECHEMA, developed in Germany; it also includes data on mixtures. The Thermodynamic Research Center (TRC) at Texas A&M University offers a large database on STN; they also distribute a PC diskette that provides vapour pressure data on about 5,000 substances. The NIST Chemical Thermodynamics Database contains standard-state thermodynamic properties for about 15,000 inorganic substances, and the database version of the JANAF Thermo-

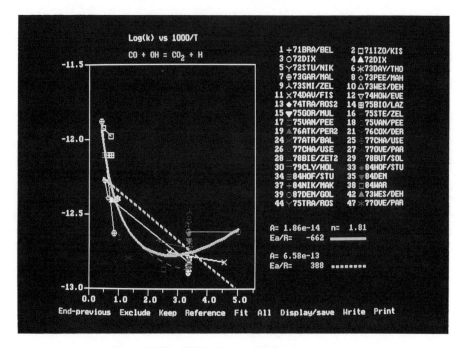

Figure 9 The NIST Chemical Kinetics Database

chemical Tables covers high temperature data on 1,800 substances. Both are distributed by NIST in magnetic tape form and are available on STN. Other NIST databases cover pure fluid properties, hydrocarbon mixtures, refrigerants, water and steam, molten salts, and thermodynamic property estimation.

6.1.5 Chemical Kinetics

Data on rates of chemical reactions are very important for environmental modelling and other applications. The NIST Chemical Kinetics Database contains data on over 5,000 gas-phase reactions (figs. 9, 10). This PC database displays data in both tabular and graphical form and permits the user to choose different models for fitting the data.

6.1.6 Nuclear and Particle Physics

The International Atomic Energy Agency (IAEA) in Vienna has led the international efforts to create databases of neutron cross-sections and other nuclear properties that are essential to the design of nuclear power reactors. These databases are distributed as magnetic tapes by the IAEA, as well as by national organizations such as the Nuclear Data Center of the Japan Atomic Energy Research Institute and the National Nuclear Data Center at Brookhaven National Laboratory in the United States. Data on fundamental particles and other data for high energy physics may be obtained

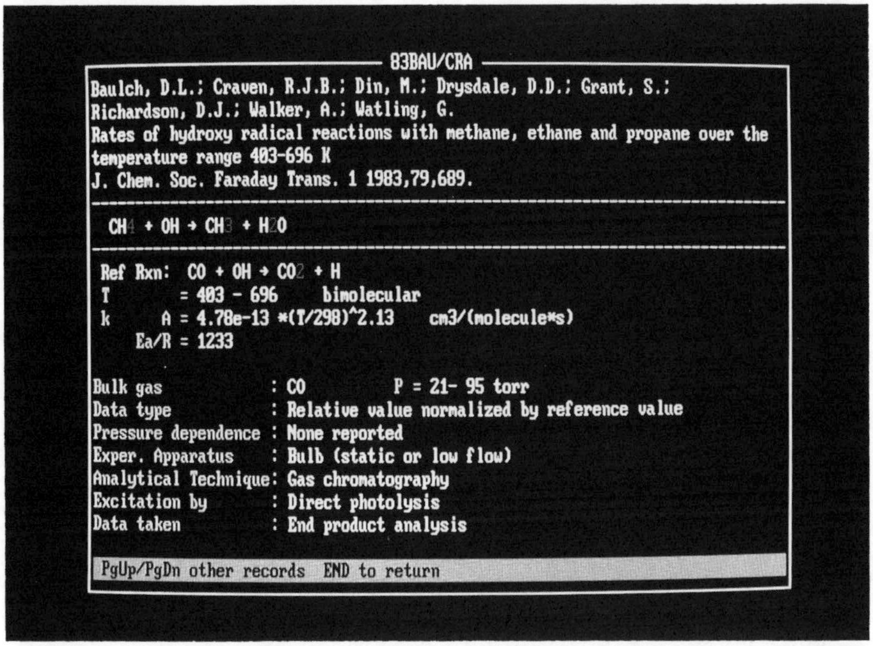

Figure 10 The NIST Chemical Kinetics Database

on magnetic tape from the Particle Data Center at Lawrence Berkeley Laboratory.

6.1.7 Engineering Materials

An extensive effort has been organized in the last few years to create databases on properties of materials of engineering interest. These emphasize structural materials such as metals, ceramics, polymers, and composites, but also include special types such as electronic materials. The effort has been led by the Materials Properties Data Network (MPDN), a non-profit organization established in the United States for the purpose of encouraging the production of databases and the development of an on-line delivery system. The MPDN is now operating under the Chemical Abstracts Service, and its databases form a separate module of STN International. Databases now available to the public on MPDN cover aluminium and copper alloys, steels, plastics, structural ceramics, and advanced materials for aerospace design. A sophisticated menu-driven interface helps to lead users to the right database, and results can be displayed in any desired set of units.

Many other groups are developing materials databases for specialized applications. Groups in France have been especially active in this area [1]. The NIST in the United States distributes PC databases on corrosion performance, phase diagrams of ceramics, and tribology. CODATA has an active Task Group on Materials Database Management that serves as a coordinating mechanism for these various efforts [7].

6.2 Biosciences

The systematic collection of information on the taxonomy and behaviour of biological species has a long and distinguished history. Numerous efforts are in progress to convert this type of information from paper to computerized form. The International Union of Biological Sciences (IUBS) has a Commission on Plant Taxonomic Databases that is attempting to establish standards for data exchange and promote cooperation between the databases. It is now studying how the community of taxonomic institutions and database developers might design and organize a global database system that would cover all the world's plants. This is clearly a long-range task, but the first steps of agreeing on nomenclature and adopting data exchange formats are in progress. Similar efforts have been started in zoology; an example is the MEDIFAUNE Databank, which covers the fauna of the Mediterranean region [4].

It is a daunting challenge to computerize massive amounts of data, some of it going back several centuries. One of the major problems is the variability in names of plant and animal species. The intelligent design of a database (or a series of linked databases) requires absolute conformity to an agreed-upon terminology. In an effort to alleviate this problem, CODATA has established a Commission on Standardized Terminology for Access to Biological Databanks [2]. The plan is for this group to coordinate the efforts of the ICSU unions and other international bodies to reach agreement on standardized terminology for use in data banks.

The situation is different in the realm of cellular and molecular biology. Since these are much newer fields, such a large backlog of data does not exist, and it is easier to establish mechanisms that facilitate the transfer of new data from the laboratory to data banks in a systematic fashion. In microbiology, a detailed coding scheme was published in 1988 under the auspices of CODATA and the International Union of Microbiological Sciences (IUMS) [12]. This scheme provides standardized codes for all the data elements likely to be of interest in a data bank on micro-organisms. At the molecular biology level, CODATA established a Task Group on Protein Sequence Databanks [6] in 1984. This group has promoted collaboration among the organizations throughout the world that maintain protein sequence data banks. A standard interchange format has been adopted that permits groups to exchange sequence data easily, even though each group has its own computer hardware and software.

The greatest challenge in the area of biological data banks is DNA sequences, especially the human genome. In the brief period since techniques for determining DNA sequences in chromosomes were first introduced, an immense amount of data has been accumulated. Even so, it is small compared to the 3 billion base pairs of the human genome. Major programmes have been started in several countries to map the human genome and eventually determine the full DNA sequence. Fortunately, the problem of storing all this data in data banks has been faced before the sequencing began on a large scale. Genome data banks have been established at the National Center for Biotechnology Information, part of the National Library of Medicine in the United States, the European Molecular Biology Laboratory in Germany, Los Alamos Scientific Laboratory (GENBANK), and elsewhere. Mechanisms are in place for these groups to maintain common standards and formats.

From this discussion, it is evident that the development of publicly accessible computerized data banks is at an earlier stage in the biosciences than in the physical sciences. Nevertheless, a considerable amount of biological data can already be accessed. The Microbial Strain Data Network (MSDN) [10] was established about five years ago with support from UNEP, CODATA, and other organizations. The MSDN serves as a gateway to a wide range of reference sources on micro-organisms and cell strains, some in computerized form and others not. The CODATA/IUIS Hybridoma and Monoclonal Antibody Databank provides data on about 20,000 hybridomas. Both protein and DNA sequence data are distributed in magnetic tape form by the groups that maintain the various databases. Mention should also be made of the Protein Data Bank at Brookhaven National Laboratory in the United States, which distributes tapes with data on the three-dimensional structures of protein molecules.

6.3 Geosciences

Some areas of the geosciences have a long-established pattern of compiling and organizing observational data. Geology and astronomy are examples of fields where records go back many centuries. Fields such as oceanography and meteorology also feature a considerable amount of historical data, but modern measurement techniques have led to a major expansion in the quantity of data that must be managed. Finally, remote sensing measurements from satellites and space probes are now causing a staggering data explosion.

This great size and diversity of geoscience data have given rise to data centres that maintain a multiplicity of parameter-specific data banks and provide a dissemination service to the scientific community. The system of World Data Centers coordinated by the ICSU Panel on World Data Centers links about 40 centres, each supported by its own national government. The centres exchange information and maintain some duplicate records to prevent loss in case of natural disasters. While many of the older records are in the form of paper, microfilm, and photographs, an increasing part of the holdings of these centres is in computerized form.

The scope of the World Data Centers is now described as "geophysical, solar, and environmental." Among the many topics covered are seismology, volcanology, geomagnetism, cosmic rays, solar emissions, tsunamis, etc. The individual centres disseminate computerized data to the scientific community in the form of magnetic tapes and optical disks. A summary of the services available may be found in the ICSU publication *Guide to the World Data Center System* [5].

Another ICSU activity is represented by the Federation of Astronomical and Geophysical Data Services (FAGS), which links 10 permanent data services conducted by several scientific unions. The participating centres analyse observational data from all parts of the world, checking for quality and consistency. Subjects covered include geomagnetic indexes, sunspot observations, gravity variations, and precise data on the rotation of the earth. Also included is the stellar data centre in Strasbourg, France, which maintains extensive computerized records on the positions and other features of stars.

Research on the atmosphere and oceans is another area where large quantities of data are being generated. This comes from satellite observations as well as land-based and ship-based measurements. The World Ocean Circulation Experiment (WOCE) and the Tropical Ocean Global Atmosphere (TOGA) are examples of scientific programmes that have the objective of synthesizing these data through the use of global models. Data exchange in oceanography is coordinated by the International Oceanographic Commission (IOC), while the World Meteorological Organization (WMO) serves a similar function in meteorology.

Many centres in different countries serve as distribution points for geoscience data. Taking the United States as an example, meteorological data are handled by the National Climatic Data Center in Asheville, NC; space science data by the National Space Science Data Center in Greenbelt, Md.; data on the oceans by the National Oceanographic Data Center in Washington, D.C.; various geophysical and geological records by the National Geophysical Data Center in Boulder, Colo.; and so on. Each of these centres maintains tens of thousands of magnetic tapes and provides copies on request. Much concern has been expressed about the longevity of these tapes, and studies are in progress on ways to assure that scientists will have access to the data a century from now.

In view of this bewildering assortment of geoscience data sources, most users need a great deal of help in locating what they want. In an effort to put some order into the situation, the National Space Science Data Center has developed a Global Change Master Directory. The Master Directory, which can be assessed on-line, utilizes software that switches the user to actual data records, which may exist in many different computer centres throughout the world. If a direct connection is not possible, the user is told how to contact other centres that may help him. The National Oceanographic and Atmospheric Administration (NOAA) is using the same software to provide access to some of its data.

Finally, a very interesting project designed to provide geoscience data at low cost should be mentioned. This is the Global Change Database Project (GCDP), developed by the ICSU Panel on World Data Centers with the

aid of the US National Geophysics Data Center, which is intended to provide data at medium or low spacial resolution in forms that can be easily
used by individual scientists. The pilot phase produced a diskette displaying
vegetation index data for Africa. Plans call for other types of data to be
added, such as topography, soil type, land use, ecosystem classification, and
climate summaries. A Global Ecosystem Database in CD-ROM format is
now being tested. Emphasis is being given to research needs in developing
countries, and the pilot diskettes have already been used in training workshops under the UNEP/GRID programme.

7. Data as an International Commodity

It is evident from this brief summary that scientific data banks are not constrained by national boundaries. Very few significant data banks have been
developed solely by scientists within a single country. Most have involved international collaboration, sometimes informal but often under the auspices
of an ICSU body or some other international organization. The task of collecting, evaluating, and organizing data in order to create a useful data bank
is expensive and time-consuming. International cooperation can greatly enhance the efficiency of this process.

The international character of data banks is even more apparent when dissemination is considered. In many existing on-line networks, the user is un-
aware of the geographical location of the database he is accessing. The technology for providing instantaneous access to data banks in any part of the
world has been perfected. The problems at this stage are primarily economic, legal, and political ones. For example, different national policies on
financing of data banks can be a real barrier to a collaboration that makes
great sense scientifically. Some governments heavily subsidize the creation
and maintenance of scientific data banks, on the condition that access be
provided at minimal cost to the user, while other governments may provide
seed money but require that the data banks, once established, become self-
supporting. The resulting disparity in user fees makes collaboration and
data exchange difficult. Political leaders often have little appreciation for
the way in which new information technology has made certain policies
obsolete.

8. The Future

This paper has given an overview of the way in which computer and telecommunication technology is changing the modes of accessing scientific and
technical information. Particular emphasis has been given to numerical and
factual data banks, although the general trends apply equally to other types
of information. It is a safe prediction that the level of electronic access will

grow over the next decade, since the momentum is strong. However, the rate of growth will be dependent on our success in overcoming some of the barriers to the use of the new information technology that still exist.

Cost forms one of the most significant barriers, and we can hope that efficiencies will be introduced that will reduce the cost of creating and maintaining computerized data banks. In addition, policy and attitudinal changes will be needed. For example, the free use of books in a scientific library is taken for granted, but most libraries still change a user or his research project for accessing an on-line data bank. Changes in financial policies of universities and other public institutions could do much to encourage greater use of the available technologies.

Deficiencies in the interface between the computer and the human user form another type of barrier. In spite of frequent discussion of "user-friendly interfaces," the fact is that most data banks are still clumsy to use. The need to learn – and remember – a special set of commands or protocols for each system is a real barrier to the widespread use of data banks. Graphics capabilities are still limited for most personal computer systems; subscripts, special characters, and different type fonts are not generally available. Thus, the display of retrieved data appears primitive compared to a well-composed printed text. Furthermore, in using the new technologies to store and retrieve information, we have probably been influenced too much by the familiar paper-based formats, just as the early automobiles resembled horse-drawn carriages. More imaginative thinking about the retrieval and display of data from computer databases will undoubtedly lead to much more useful and effective information systems.

More innovative approaches to guiding the user to the specific data he wants will also help expand the utility of data banks. While some attempts have been made to develop expert systems that analyse a user's needs, a more serious effort is required. Human translation of a query from someone who is seeking a source of data is often a slow and uncertain process; it is to be hoped that expert systems can be developed that are more efficient. Directories such as the NASA Global Change Master Directory will also help overcome this barrier. What hopefully will evolve is a widely accepted system for keeping track of electronic databases that is analogous in function to the tools available to librarians to locate books.

One essential requirement for effective use of modern technology is free flow of information across national boundaries. The current situation is much better in this regard than at certain times in the past, when restrictions were imposed by various countries on military, ideological, or political grounds. Perhaps the main concern in this decade is that countries may attempt to restrict the export of technical information on grounds of economic competitiveness. Such restrictions would have a debilitating effect on access to data for scientific purposes.

In summary, the new information technology that had its birth three decades ago has proved its effectiveness for accessing many types of scientific and technical information. We are now entering a stage where extensions and refinements should make it possible for all scientists to obtain quickly and easily the information needed in their work. Wise policies on the part of national governments can greatly expedite this process.

REFERENCES

1. "Banquees de Donnees Materiaux" (1988). *Metaux Corrosion Industrie* 63 (753–754).
2. Blaine, L. (1990). "CODATA Commission on Terminology and Nomenclature of Biology." In P.S. Glaeser, ed. *Scientific and Technical Data in a New Era.* New York: Hemisphere Publishing Corp.
3. CODATA Referral Data Base. Available from CODATA Secretariat, 51 blvd. de Montmorency, 75016 Paris, France.
4. Fredj, G., and M. Meinard (1990). "Biological Databanks: A Preliminary Survey." In P.S. Glaeser, ed. *Scientific and Technical Data in a New Era.* New York: Hemisphere Publishing Corp.
5. *Guide to the World Data Center System* (1987). ICSU, 51 blvd. de Montmorency, 75016 Paris, France.
6. Keil, B., ed. (1987). "Directory of Protein and Nucleic Acid Sequence Data Sources." *CODATA Bulletin* 65.
7. Koslov, A., and E.F. Westrum, eds. (1990). "The Provision of Materials Property Data via Computerized Databanks." *CODATA Bulletin* 22 (2).
8. Kotani, M., ed. (1975). "Study of the Problems of Accessibility and Dissemination of Data for Science and Technology." *CODATA Bulletin* 16.
9. Lias, S.G. (1989). "Numeric Databases for Chemical Analysis." *J. Res. Nat. Inst. Stnds. Tech.* 94: 25–35.
10. Microbial Strain Data Network. Information may be obtained from MSDN, 307 Huntingdon Road, Cambridge CB3 OJX, UK.
11. *Numeric Databases: A Directory.* International Council for Scientific and Technical Information, 51 blvd. de Montmorency, 75016 Paris, France.
12. Ragosa, M., M.I. Krichevskey, and R.R. Colwell (1984). *Coding Microbiological Data for Computer.* New York: Springer-Verlag.
13. Washburn, E., ed. (1921–32). *International Critical Tables.* New York: McGraw-Hill.

Session 2A

Experiences with International Cooperation and the Developing Countries

Chairperson: Carlos Correa

A Critical Evaluation of Experiences and Strategies

Jacques Tocatlian

ABSTRACT

After providing an overview of the contributions of various types of bodies participating in international cooperation – (1) non-governmental organizations and professional associations, (2) organizations of national character, and (3) intergovernmental organizations, with those of the United Nations system in a separate category – the paper discusses three intergovernmental conferences that considered the question of improving access to scientific and technical information. Particular attention is given to an examination of the strategies that governed the establishment and execution of the sets of actions and programmes that emerged. Finally, attention is given to special difficulties in access encountered by the developing countries.

1. Introduction

During my last three of four years at Unesco, I had the privilege to be concerned with the project for the Revival of the Ancient Library of Alexandria [25]. I cannot help but recall that this great library of classical antiquity was in many ways the first information and learning centre with an international dimension. Its policy was to collect manuscripts in different languages from every source, translate the texts, prepare them for use by bibliographic control, and make them available to the scientific community of the time. It was in the intellectual environment of this information centre that scholars came to explore, exchange information, invent, and study astronomy, physics, mathematics, geometry, anatomy, biology, geography, literature, philosophy, and engineering. The numerous findings and inventions born under the roof of the ancient Bibliotheca Alexandrina are clear evidence of the close relationship that has always existed between information and the advancement of science and the role that scientific and technical information (STI) plays in the discovery of new frontiers.

From the time of the international library of antiquity to the international information systems, networks, and services of today, history is rich with examples of international cooperation pointing the way for the information world of tomorrow.

The most advanced countries have indeed already entered the Information Age – a creature of information technology that itself results from the marriage of computers and telecommunications, hardware and software, information systems and services. We are told that "by the year 2000, to all intents and purposes, information technology will be able to create a nearly information-transparent world, while fiber optics will carry libraries of information to anyone, anywhere, who pushes a button!" [9]. It seems that three revolutionary technological changes will be required to bring about affordable individual access to global on-line information: efficient large-scale database construction and maintenance, high speed digital transmission networks, and highly precise intelligent searchware. "As these technological revolutions appear over the next several decades, they will result in a worldwide information system that will have a major impact on the entire information industry" [11]. And, indeed, on society itself, since they will affect the way people work, the way they act and organize themselves, and even the way they think.

We are gathered in Kyoto today to explore the role of information technology in facilitating access to science and technology. The objective of the symposium is to assess the potential of scientific and technological developments for enhancing the capacity to handle, transfer, exchange, and access information. Towards the conclusion of the symposium a panel will discuss and recommend new modalities of international cooperation for the future.

It seems useful to consider past experiences in international cooperation. A shared knowledge of past efforts and a better understanding of the strategies used, their impact and limitations, will help prepare the future. This paper is not a comparative review of international information systems and programmes, nor is it an evaluation of performances and results. Although the information needs of the developing countries permeate the whole presentation, it cannot be considered a review on the subject. The panel in Session 2B, "Achievements and Limitations in International Cooperation As Seen by the Developing Countries," is complementary since it will provide a perception that the developing countries have of international assistance, international programmes, and other schemes and systems set up under the banner of "international cooperation."

The paper first describes the various patterns of international cooperation and then analyses three experiences and strategies resulting from high-level intergovernmental conferences. In the three cases sovereign states discussed the question of improving access to STI. Their recommendations and the sets of actions that emerged provide matter for a critical evaluation of the strategies selected for international cooperation.

From the outset it should be emphasized that the international support systems involved in international cooperation, whether governmental or non-governmental, bilateral or multilateral, can hope to play only a catalytic role in assisting national efforts. Decisions on the nature of involvement in new technological areas, the kind of infrastructures to create, and the areas for priority action are all primarily the responsibility of the developing countries concerned [32].

International cooperation among nations is founded on the belief that everyone stands to gain from the benefits of sustainable growth, prevention of deterioration of the national environment, and satisfaction of people's basic needs – including access to information.

As we approach the end of the millennium, we observe that the developing countries are recognizing the value of self-generative efforts to orient their internal development strategies as an essential precondition to engaging in international cooperation. At the same time, the international community should conceive new ways of organizing international cooperative efforts that will take the real needs of the developing countries into greater account [32].

A first set of questions comes to mind: What bodies are concerned with international cooperation in the field of information? What are the prevailing patterns? What are the driving forces behind such cooperation? What are the strong points and weaknesses of these different patterns? What are the implications for the developing countries?

2. Patterns of International Cooperation

The literature contains several papers that cover some specific aspects of international cooperation in information. In general, the literature on this subject is descriptive rather than analytical or critical. Some review articles do provide a useful general overview [e.g. 1, 4, 10, 12, 14, 16, 21, 22].

We may attempt to elucidate the subject under three main headings:
(1) Professional associations and international non-governmental organizations (NGOs)
(2) National systems, agencies, institutions, and foundations
(3) Intergovernmental organizations

But before doing so, it may be useful to establish working definitions. In this paper the term "information" is generally used in a generic sense, irrespective of the sources, form of presentation, or transfer medium used. The term "data" denotes groups of numerical and statistical facts. The term "information system" is also used in a generic sense to denote libraries, documentation and information services, data banks, etc., as well as networks of these components.

2.1 Professional Associations and International Non-governmental Organizations (NGOs)

International cooperation has long been an essential characteristic of both the scientific and the information communities.

Since the early days scientists have developed a tradition of interchange of information and data. This tradition has survived the challenges of distant communication, wars, and totalitarian regimes. Of course, political, military, and industrial interests prevent a totally free exchange of information, but cooperation remains an intrinsic element in the advancement of science – a fact well illustrated at the international level in the work of the ICSU, the International Council of Scientific Unions.

Parallel to the trend towards cooperation among scientists is, of course, that in the information professions. The International Federation of Library Associations and Institutions (IFLA) and the International Federation for Information and Documentation (FID) are known internationally for their achievements in this area. The FID will soon celebrate its hundredth anniversary and is the senior NGO in the information field. To name but a few of the numerous non-governmental bodies that have programmes for fostering worldwide cooperation in information transfer, we may also cite the International Federation for Information Processing (IFIP), the International Council for Scientific and Technical Information (ICSTI), the ICSU Committee on Data for Science and Technology (CODATA), the International Council on Archives (ICA), and the International Organisation for Standardization (ISO).

It should be recalled that many national professional associations in the industrialized countries also have international cooperative programmes and, therefore, participate in international cooperation.

The work of these international associations and non-governmental organizations is essential and their contribution in international cooperation is fundamental. They are non-profit organizations, the driving force of which is the advancement of their respective professions in the service of society. They usually assemble, on a voluntary basis, top specialists in their fields and implement an impressive variety of useful activities and projects. However, there may never be resources enough to allow all their plans to be carried out. A critical study of these organizations [34] showed as early as 1973 that their proliferation to help solve information problems created an information problem of its own! Their number is continuing to grow, especially in Europe, underlining an urgent need for overall coordination. A recent *World Guide* [5] lists over 600 associations in the field of library, archives, and information science around the globe, of which 76 are international.

Most NGOs are concerned with the problems of the developing countries, and many have branches in or members from the various regions of the world. However, all in all, the participation and influence of the members from the third world remain weak – in many cases because of the high cost of travelling involved. Consequently most of the NGOs and professional associations having international missions and programmes are still primarily oriented around Europe and North America.

2.2 National Systems, Agencies, Institutions, and Foundations

Under this second pattern of cooperation we have grouped structures of national character – created, funded, and governed essentially at the national level – such as national information systems and services, national development agencies, and institutions and national foundations that undertake some form of international cooperation but are neither professional associations, NGOs, nor intergovernmental organizations.

In the last 30 years, agreements signed by information systems from different countries, in the same or in different regions, have grown in number and proliferated rapidly. These may provide for the operation of joint information services; the sharing in information systems' input and output; the creation of databases; the setting up of information networks; the distribution of information products; the development of common tools; the exchange of indexed literature; and the training of personnel; as well as, in general, for the sharing of workload.

Such cooperative work patterns exist among publishers and editors, abstracting and indexing services, information systems and services – at every step of the information-transfer chain, from the producer of information to the final user [22].

The impetus for this type of international cooperation or sharing of resources varies. It is usually economic, aiming essentially at reducing product costs, increasing timeliness and reliability, improving access, and extending the usefulness of recorded information. The driving force is often commercial, information being considered a commodity. Economic and time pressures are forcing organizations to share rather than duplicate information and resources. In most cases international cooperation among institutions and services is achieved rapidly when the economic advantages of doing so become clear. In the next section we shall see that many regional cooperative schemes, systems, and networks, linking a number of national institutions in the developing countries, are sponsored by intergovernmental organizations.

Under this second pattern of international cooperation many national development agencies or institutions carry out bilateral assistance to the developing countries. Most industrialized countries have a national agency, ministry, or programme within a governmental structure devoted to international cooperation, especially with the developing countries. We may cite for the purpose of illustration JAICA, the Japanese International Cooperation Agency, and the Japanese Ministry of Foreign Affairs; NORAD, the Norwegian Agency for Development Cooperation; SIDA, the Swedish International Development Agency; DANIDA, the Danish International Development Agency; The British Council; The Direzione Generale Cooperazione Allo Sviluppo, Italian Ministry of Foreign Affairs; the French

Ministries of Cooperation and of Foreign Affairs; the USAID, the US Agency for International Development; and the BMZ, the German Ministry for Economic Cooperation.

The impetus for such international cooperation, often referred to as "bilateral assistance," is essentially political in nature; it aims most often at helping friendly countries. The recipient countries may sometimes feel that this type of cooperation has more "strings attached" than does the more "neutral" cooperation with the NGOs or the intergovernmental organizations. However, the relatively larger funds invested per project by these development agencies provide incentive. In addition, there is often an element of project evaluation with strong possibilities of follow-up and phasing of the project until the national authorities can absorb its management.

A very interesting tendency can be observed among some national development agencies to shift from offering purely bilateral to the so-called "multi-bilateral" assistance. In this framework individual development projects, while financed by a donor agency, are entrusted to a specialized agency of the UN system such as Unesco, FAO, or UNIDO, for execution. In such cases, the national development agency enters into a funds-in-trust agreement with the executing agency. This trend indicates that some donor agencies recognize the competence of the specialized agencies in their respective fields and the difficulty they themselves have in dealing with projects in a very wide array of technical fields. The "multi-bilateral" approach is also preferred in sensitive areas such as communication, because, as said earlier, the association with the UN is perceived as "neutral" and void of political interests.

One agency that deserves to be singled out on two accounts is Canada's International Development Research Centre (IDRC). First, although the IDRC is funded by the government, like the other development agencies mentioned above, it is autonomous in its policies and activities. Its Board of Governors is international and reflects the non-partisan, multicultural nature of the organization. It assists developing countries in creating their own long-term solutions to pressing development problems. The second important fact is that the IDRC has designated information as one of its major sectors of activity.

This is not generally the case. Development agencies may sometimes include an information-related component within a larger development project, but seldom assign priority to a project dealing exclusively with the establishment of an information system or network in a developing country. Access to scientific and technical information is generally not viewed as a need at the same level as food, health, education, etc., when priorities for assistance to the developing world are assigned.

This is also true of the numerous foundations that in the industrialized world support a wide variety of activities in many different areas, including music, restoration of art, study grants, etc. There again information-related activities tend to find little favour except in the form of support to publications.

2.3 Intergovernmental Organizations

While there is ample justification for bilateral modes of cooperation among countries and for their preferred orientation, it is generally recognized that there are some problems that are universally significant and appropriate for multilateral efforts, and that require concerted political will. Access to STI falls within this category. We shall consider this issue under two broad classes of intergovernmental organizations: (1) Intergovernmental organizations outside the UN system and (2) the intergovernmental organizations belonging to the UN system.

2.3.1 Intergovernmental Organizations outside the UN System

Since the end of the Second World War, we have witnessed the emergence of regional groupings of countries for cooperative purposes in the areas of politics, economics, and development. In many cases, the countries have recognized regional cooperation in information as a necessary basis for their cooperation in other fields. This has been the case, to various degrees, for instance, for the Arab League, the Organisation for Economic Co-operation and Development (OECD), the Organization of American States (OAS), the Association of South East Asian Nations (ASEAN), the European Commission (EC), and Les Sommets de la Francophonie (summits of countries using French as one of their languages), which created BIEF, a data bank of francophone countries that constitutes an example of successful North-South cooperation.

A regional approach to the development of cooperative information systems appeals to the countries concerned

in that it focuses attention on their specific needs and fits into a framework of other co-operative programs in the social, economic, and cultural fields. The possibility of sharing their resources for the development of national information infrastructures and for the improvement of their capacity to utilize international information systems, services and programs is also a positive feature of regional co-operation. . . . Similarly, the sharing of resources among countries of the same region results in greater effectiveness, particularly with regard to the training of information personnel, the use of telecommunications, the elaboration and application of norms and standards, and the improvement of access to information sources. [16]

We shall cite as an example of successful regional cooperation the EURONET – the European On-Line Information Network. Homet [8] described European policy in mass communication and telecommunication in the 1970s and pointed out the domination, rarely challenged, of the national postal, telephone, and telegraph agencies (PTTs). The engineering predilection for a single, standardized telecommunications system prevailed over arguments in favour of innovation. In fact, when the Commission of European Communities decided to establish EURONET, no public European network existed, but thought had been given at some point to building a private network with limited PTT involvement. By mid-1975 it became evident that such a network would make economic sense only if it were eventually extended for computer services and community-wide information. The technical network of EURONET became a sub-network of a public PTT network. It created for the first time a distance-independent tariff. Following the agreement with the national telecommunications authorities, EURONET was replaced by interconnected national networks in 1985.

In parallel, the European Commission encouraged the creation of European databases and their use across the Community. International collaboration was stimulated and many projects were supported by the Commission. By the end of 1988, Europe was offering more than 900 databases on 88 computer hosts. The direct information access network in Europe became known as EURONET DIANE. Many users were dissatisfied with the variety of retrieval languages that had to be used, so the Commission encouraged the use of the Common Command Language (CCL). Now, the European Commission Host Organization (ECHO), a non-commercial organization, offers access to unique databases and data banks that are not available on other on-line host services. ECHO is also a Community instrument for the development of the information services market and the promotion of new technologies [3].

This is an example of concrete and successful international cooperation within a regional group of countries. Two important ingredients were basic to the success: political will and adequate funding. All technical problems could be solved in due time.

2.3.2 Intergovernmental Organizations of the UN System

Most organizations within the UN system have developed information systems to support their internal needs as well as international information systems in their fields oriented toward member states. We are concerned here with the latter, since they contribute to improved access to science and technology. Large amounts of substantive information are gathered and disseminated by these systems. Improving the accessibility of the UN information resources has been, since its creation in 1983, one of the main objectives of ACCIS, the Advisory Committee for the Coordination of Information Systems. Earlier, the IOB, the Inter-Organization Board for Information Systems, had had this role. *The Directory of United Nations Databases and Information Services* [33] produced by ACCIS is a guide to 872 computerized databases and information systems and services.

These cover a wide variety of subjects, including natural resources and the environment, agriculture, industry, health, population, human settlements, science and technology, and education. In spite of their shortcomings, they have had a very positive impact. It is doubtful that such international development could have occurred solely under the auspices of national governments or private firms and in the absence of the "UN family." The legitimacy of UN systems has generated in the developing countries an interest and activity in the information field that would not have come about in their absence [7]. In fact, a developing country, in order to participate in and take advantage of the UN information systems has to develop a minimum information base, that is to train personnel, collect nationally produced documentation as input into the international systems, organize the diffusion to national users of the information made available through the UN information systems, and in many cases obtain high-level decisions and develop national policies. In all cases, a certain infrastructural development, including the use of information technology, is necessary. A statement by C. Keren, who reviewed the literature [10] deserves repetition: "Information activities in less developed countries would probably never have reached their present state without the active support of international organizations, such as Unesco, UNIDO, FAO and IAEA; professional organizations, such as FID and IFLA; and funding organizations, such as IDRC."

On the problematic side, one may say that countries without the necessary information base have not been able to benefit from these systems. The information provided by many of them is in the form of bibliographic citations, which are of limited usefulness if access to the primary literature cannot be provided. The information retrieved from these systems in large quantities often needs to be organized, evaluated, and digested by qualified personnel with a good grasp of the scientific subject covered before it can be utilized by the end user. In addition, these systems are costly and their operating budgets at the international level are usually. low. This often impedes the granting of substantial assistance to the developing countries.

A few selective examples will serve to illustrate the wide array of information services offered within the UN family:

(a) The United Nations Bibliographic Information System (UNBIS) of the Dag Hammarskjöld Library is an on-line bibliographic and factual information system covering the publications and documents of the United Nations. About 25 per cent of the citations concern STI.

(b) The United Nations Centre on Transnational Corporations (UNCTC) has developed the Corporate Profile System (CPS) dealing with the activities of transnational corporations in developing countries and the issue of technology development (transfer of technology, the role of transborder data flow, impact of new micro-electronic technologies). Over 60 per cent of users of the system are from the developing countries.

(c) The Economic Commission for Africa (ECA) operates a regional multi-disciplinary bibliographic information system called the Pan-African Documentation and Information System (PADIS), containing references to information on African economic, social, scientific, and technological development.

(d) The United Nations Industrial Development Organization (UNIDO) operates two major information systems, the Industrial and Technological Information Bank (INTIB) and the Technological Information

Exchange System (TIES), combining referral, retrospective search, and the provision of consolidated and repackaged information. The first, INTIB, covers bibliographic information generated by UNIDO as well as information on institutions and technology suppliers. TIES provides information on the terms and conditions of technology contracts. UNIDO also assists countries in establishing information services for industry.

(e) The United Nations Environment Programme (UNEP) operates the on-line International Referral System for Sources of Environmental Information (INFOTERRA), which directs users to sources of information in a wide range of scientific and technological topics pertaining to the environment. UNEP also runs the International Register of Potentially Toxic Chemicals (IRPTC).

(f) The United Nations University (UNU) operates a bibliographic database entitled Abstracts of Selected Solar Energy Technology (ASSET), covering solar, wind, and bioconversion energy.

(g) The two major information systems of the Food and Agriculture Organization (FAO) are the International Information System for Agricultural Sciences and Technology (AGRIS) and the Current Agricultural Research Information System (CARIS). AGRIS is a decentralized cooperative bibliographic network of centres in charge of collecting, processing, and disseminating information on published agricultural literature. CARIS is a referral system on ongoing research in the field. FAO also provides technical assistance for strengthening national information services in agriculture in the developing countries.

(h) The International Labour Organisation (ILO) operates the International Labour Documentation (LABORDOC) – a global bibliographic database covering industrial relations, technological changes, labour laws, employment, etc.

(i) The World Health Organization (WHO) is a highly decentralized organization comprising a headquarters in Geneva, six regional offices, and a number of programme coordinating units. Information systems exist at all levels on a large number of specific medical subjects.

(j) The World Intellectual Property Organization (WIPO) operates an International Patent Documentation Centre (INPADOC) that provides information on technological solutions as described in patent documents.

(k) The International Nuclear Information System (INIS) of the International Atomic Energy Agency (IAEA) covers the substantive literature of nuclear science and its peaceful applications. It is organized on the same pattern as AGRIS. It provides for decentralized input, centralized processing, decentralized access, and utilization of information.

(l) The World Weather Watch (WWW) provided by the World Meteorological Organization (WMO) is considered among the most successful and truly global cooperative information networks.

(m) Unesco, the United Nations Educational, Scientific, and Cultural Organization, carried out in 1991 through its clearing-house an inventory of its information services [30] and lists 69 operational databases as well as 15 under development (46 per cent referral, 43 per cent bibliographic, and 10 per cent numerical). Among the numerical ones, we may cite the *Statistical Yearbook*, which provides access to over 2.1 million statistics from over 200 countries and territories concerning population, education, science, culture, communication, and information.

But Unesco is unique within the UN family since it not only provides information services in its areas of competence, as do the other UN agencies, but it also covers "information" as a subject and has developed programmes in this field known as UNISIST and the General Information Programme (PGI). Through these, Unesco has been concerned with improved access to STI and has provided a conceptual framework for the establishment of national, regional, and international information systems and services, including technical assistance to the developing countries. We have seen that other agencies, such as FAO and UNIDO, also provide technical assistance to member states to create national structures. Because of their scope, magnitude, "horizontal" nature, and evolution, UNISIST and PGI deserve particular attention and will be the subject of separate sections in this paper.

Another separate section will be devoted to the United Nations Conference on Science and Technology for Development (UNCSTD), the recommendations of which are implemented by the Intergovernmental Committee on Science and Technology for Development with the support of the United Nations Centre on Science and Technology for Development. It will be remembered that UNCSTD also gave particular attention to the problem of access to STI.

The analysis of Unesco's programmes in STI, UNISIST and the PGI, their evolution, changes in emphasis, difficulties and achievements, as well as the analysis of UNCSTD's original ambitions and later problems of implementation, will provide a basis for evaluating past experiences and strategies. These programmes have offered, at various periods, high-level international forums for the expression by all countries of their information needs and requirements, their wishes and priorities, and their views regarding international cooperation in the field of information. The lessons to be drawn are important and should help in designing future programmes and aid reflection on new modalities of international cooperation.

Emphasis will be placed on strategies, choices, and approaches rather than on activities, projects, and modalities of action. Regarding modalities, we may simply recall that international programmes of cooperation have attempted to reach their objectives by a range of actions, including the convening of intergovernmental conferences, congresses, and symposia; meetings of working groups and committees; publication and diffusion of guidelines, studies, surveys, and technical documents; promotion of norms, methods, and standards; demonstration of new technologies through pilot projects; organization of training programmes and workshops; granting of fellowships, equipment, software; provision of experts and consultants; etc. In view of the emergence of new information technologies, we may ask whether: there are modalities, other than those mentioned above, to be experimented with. Will the new information technologies offer new ways of communication and exchange that will allow innovation and the exploration of new horizons in international cooperation?

3. Selected Experiences and Strategies

3.1 UNISIST I

In the 1960s the attention given to "Big Science" was paralleled by an uncoordinated development of information systems and services. Many leaders in the international scientific community became concerned that the prevailing unharmonious trends in handling information were in fact jeopardizing the traditions of international exchange of scientific information. The ICSU and Unesco joined in a three-year (1968–1970) feasibility study, the results of which were submitted to an intergovernmental conference convened in October 1971 and later known as UNISIST I. The recommendations and priorities expressed by member states at the conference gave shape to the UNISIST Intergovernmental Programme of Unesco, designed to stimulate and guide voluntary cooperative action and to facilitate access to and exchange of STI.

Despite early use of the terminology "World Science Information System," UNISIST was from the beginning conceived as a long-term programme. It had as its broad principles:

The unimpeded exchange of published or publishable scientific information and data among scientists of all countries.

Hospitality to the diversity of disciplines and fields of science and technology as well as to the diversity of languages used for the international exchange of scientific information.

Promotion of the interchange of published or publishable information and data among the systems, whether manual or machine, which process and provide information for the use of scientists and engineers.

The co-operative development and maintenance of technical standards in order to facilitate the interchange of scientific information and data among systems.

Promotion of compatibility between and among information processing systems developed in different countries and in different areas of the sciences.

Promotion of co-operative agreements between and among systems in different countries and in diffferent areas of the sciences for the purpose of sharing work-loads and of providing needed services and products.

Assistance to countries, both developing and developed, which seek access to contemporary and future information services in the sciences.

The development of trained manpower and of resources of published information and data in all countries as necessary foundations for the utilization of machine systems.

The increased participation of scientists in the development and use of information systems, with particular attention to the involvement of scientists in the evaluation and synthesis of scientific information and data.

The involvement of the coming generation of scientists in the planning of scientific information systems of the future.

The reduction of administrative and legal barriers to the flow of scientific information between and among countries. [27]

These principles, considered basic for the improvement of the international flow of scientific information, later proved applicable not only to science and technology but also to all fields of human knowledge. The whole UNISIST programme and international movement derived therefrom was based on the firm belief that:

Scientific information embodies the heritage of man's scientific knowledge. It constitutes an essential resource for the work of scientists. It is a cumulative resource; knowledge builds on knowledge as new findings are reported. It is an international resource, built painstakingly by scientists of all countries without regard to race, language, colour, religion or political persuasion. As it is built internationally, so it is used internationally. Scientists who are its builders and users ask only that each other's contributions be verifiable; it is, therefore, not only a source; it is a means through which the world's scientists maintain their discipline. It is a medium for the education of future scientists, and a principle reservoir of concepts and data to be drawn on for application to economic and technological development programs. Unisist is concerned with the cultivation of this resource, with increasing international co-operation to improve its accessibility and use, to the end that, as an international resource, it contribute optimally to the scientific, educational, social, cultural and economic development of all countries. [27]

The twenty-five-year-old UNISIST "credo" is still valid today. It is in fact basic to all present and future efforts of international cooperation to expand access to information on science and technology. While the strategy remained the gradual establishment of a flexible and loosely connected world science information network, based on voluntary cooperation of existing and future information services, UNISIST remained a promotional and catalytic programme organized along the following five programme objectives:
(1) Improving tools of systems interconnection
(2) Strengthening the institutional components of the information transfer chain
(3) Developing specialized information manpower, especially in the developing countries
(4) Developing scientific information policies and national networks

(5) Assisting member states, especially the developing countries, in creating and developing their scientific and technical information infrastructures

During its implementation, increasing attention was given to "technology," in addition to "science," and to the needs of the developing countries. In fact, the feasibility study had come under criticism at the UNISIST I conference for its lack of adequate attention to the specific situations in the developing countries.

The Intergovernmental Conference on Scientific and Technological Information for Development (UNISIST II) convened in 1979 [28, 29] evaluated the work achieved so far under the UNISIST programme. The original recommendations of the 1971 conference, the strategy adopted, and the programme activities carried out were thought to have been sound. Much had been achieved, but a great deal more needed to be done. Many countries had yet to develop coherent national information policies, to set up and coordinate the necessary information infrastructures, and to establish systematic programmes for education of information workers and users, who now ranged from economic planners to "grass-roots" workers in local communities.

The emphasis on "science" in the original UNISIST programme had been thought by the developing countries to indicate a primary concern with the "élite" and thus to bypass many of the basic information requirements and needs of the most deprived international partners. The developing countries had not fully appreciated, in the early 1970s, the emphasis given to information technology, the systems approach, and the accent on standardization with a view to interconnecting systems. At the time, real information concerns in the third world were much closer to the preoccupations of the librarians and archivists facing everyday problems of poor collections, low budgets, lack of adequate space, lack of trained manpower, need for simple equipment, etc. UNISIST was perceived as being too sophisticated for the developing countries. Pure science was felt to be the realm of the industrialized countries, whereas the developing ones needed applied sciences, technology, know-how, and relatively simple solutions to social and economic problems.

In fact Unesco had at the time – in the early 1970s – another programme that addressed these library, documentation, and archives issues. The overlap between these programmes was such that in order to avoid risks of duplication, competition, and conflicting advice and opposing approaches to problems, the General Conference of Unesco combined them and created in 1976 the General Information Programme (PGI) [19].

3.2 The PGI and UNISIST II

The inclusion of libraries and archives, together with a programme conceived for scientific and technical information, under the General Information Programme was accomplished within the basic structure that had been designed for UNISIST:

- promotion of the formulation of information policies and plans
- promotion and dissemination of methods, norms, and standards for information handling
- contribution to the development of information infrastructures
- contribution to the development of specialized information systems
- promotion of the training and education of specialists in and users of information

The PGI was formed concurrently with the launching of Unesco's first Medium-Term Plan (1977–1982). The integration of issues related to library, documentation, and archives services with those related to the transfer of scientific and technical information proved smoother and easier than expected. However, the international scientific community, represented through the ICSU, felt that it had lost its specific programme in this new marriage and was never reconciled with the way the new programme evolved.

By 1979, when UNISIST II was convened, it was obvious that a majority of member states were concerned with the role science and technology played in the development process. It was generally felt that humanity was confronted with a set of problems that needed all the wisdom, intelligence, and generosity it could muster to solve them. To the difficulties caused by the energy crisis were added those created by threats against peace, the deterioration of the environment, the disorder of international commerce, unemployment, political and social tensions, hunger, and the dramatic gap existing between the standard of living in the richer and that in the poorer countries. There was cause for concern, but not alarm. Man can use knowledge to solve these problems. The wise use of knowledge presupposes the efficient management of information [23].

During the 1960s development and progress had been regarded to a large extent as synonymous; and for many developing countries "development" meant striving to reach in two or three decades the stage that had then been reached by the industrialized countries. By the end of the 1970s perceptions of the development process had changed. Developing countries were seeking a type of development that was endogenous, that is, more closely related to their own cultures and traditions; they were concerned with the social and economic consequences of the applications of imported technology. Developing countries wanted information relevant to national needs and objectives. Without relevant information, decision makers cannot choose the best courses of action. If information systems and services were to play an effective role in the solution of development problems, they had to be designed accordingly. It was widely accepted that access to information somehow contributed to development, although as has often been pointed out [14, 21], there has been little research, collection of hard data, or verification of the assumption that there is a well-established correlation between information and development. However, it was known that highly developed countries used 2–3 per cent of the R&D expenditures for STI activities, while for the developing countries this figure fell to a few per mills. Even if the connection between development and information had not been established, it was intuitively accepted as a fact.

It was in this frame of mind that the UNISIST II conference met in May 1979. There was wide agreement at the conference [29] that building up national ability to generate, handle, disseminate, and retrieve information was a paramount task of an international cooperative programme such as Unesco's, since without this ability, such goals as improving access to and the flow and use of information would be difficult to reach. But in this area, as in others, Unesco could act only as a catalyst. The success of its action depended on member states freely accepting their share of responsibility for sustaining effective action.

Political, economic, social, and cultural conditions varied so much from country to country that advice on how to develop information policies and infrastructures could only be of an indicative kind. The experience of developed countries was not necessarily relevant to the developing ones, and a great deal of imaginative adaptation was necessary.

The conference strongly felt that information users deserved greater attention. It discerned a wide variety of users engaged in the development process and advocated the design and supply of tailor-made services to meet the various needs. This implied the selection and evaluation of published information and its presentation in forms suitable for defined audiences. Repackaged information was needed both at the levels of policy makers and planners and at the grass-roots of development in rural areas and small enterprises. Since all sorts of information in a variety of subjects and in different forms and on a variety of supports were thought to be useful for development, the accent was placed for the first time at the intergovernmental level on the social "function" of information. This outlook has since been further developed and expanded to form the notion of "professional information" [14].

Concerning the application of new information technologies, the developing countries needed clear and unbiased explanations of what the new technologies could, and could not, do for them and to have as support further demonstration of services based on new technologies. Developing countries also needed cheap access to on-line services, since long-distance telephone costs were prohibitive for most users. New forms of training were in great demand.

A primary role for Unesco could be summarized as mobilizing "seed" money or "pump-priming" money for the creation of information policies and structures, systems and services, and training programmes in countries of different stages of development, so as to lead them to the point where their progress could become self-generating. This role implied a broad range of activities but, given limited resources, it also implied a strict identification of priorities for action. The priorities were in the areas of education and training and infrastructure building. A criticism often addressed to Unesco/ PGI was that the funds available for the programme were not commensurate with the wide variety of tasks to be performed, which resulted sometimes in spreading the budget very thinly and the risk of minimized impact.

The General Information Programme was gradually modified to approach as nearly as possible the new orientation recommended by UNISIST II. It became an interdisciplinary and intersectoral programme applied to the natural and to the social and human sciences. The Second Medium-Term Plan of Unesco, which defined the conceptual framework, goals, and action strategies from 1984 to 1989, defined the role of the PGI as: "to facilitate general access to information, to promote its free flow and to expand Member States' capacity to exchange, store and use information needed for development." The centre of gravity of the programme remained scientific and technical information, but changes from the previous plan included an insistence on information as a prerequisite for economic and social development; a strong emphasis on user-oriented systems and services and the problem of information underutilization; a marked concern with questions related to new technologies, the creation of databases in the developing countries, and the provision of software packages; an increased emphasis on developing tertiary information sources; an insistence on an adequate balance in the activities between information, libraries, and archives and the importance of regional approaches and collaboration.

The tasks carried out by the PGI were grouped under the following programmes and subprogrammes:
1. Improvement of access to information: modern technologies, standardization and interconnection of information systems
   (a) Development of tools for the processing and transfer of information
   (b) Development and use of databases through the application of modern technologies and normative tools
   (c) Exchange and flow of information: regional and international cooperation among member states and with the organizations of the United Nations system
2. Infrastructures, policies, and training required for the processing and dissemination of specialized information
   (a) National information policies and infrastructures
   (b) Training of information personnel and information users

The impact and achievements of the programme are recorded in several documents, such as the biennial *Report of the Director-General on the Activities of the Organization*, the so-called C/3 series. Another set, known as the C/11 series, constitutes a statement and evaluation of *Major Impacts, Achievements, Difficulties and Shortfalls* for each programme activity of Unesco. Two articles, by Parker [17] and Roberts [19] respectively, provide useful overall reviews, rich with detailed examples of the wealth of guidelines, studies, and publications produced under the programme over the years.

One aspect of this programme deserves particular attention with respect to the theme of the Kyoto symposium: regional cooperation. Under this subprogramme, regional cooperative schemes requested by member states were encouraged and supported. Countries often find it easier to collaborate within the same region or subregion, as we saw earlier in the case of EURONET, for a number of reasons, including language, geographical vicinity, similarity of social and economic conditions, political and legal ties, etc. The projects undertaken have been fairly successful and have received strong encouragement and active participation. The weakness in these regional ventures, sponsored by Unesco in the developing countries, has been the paucity of funds. The seed money made available by Unesco in addition to the national contributions has often not been sufficient to reach the sums required for these schemes to progress as fast as they deserved and meet the regional demands. This is obviously not the case when substantial additional funds are made available from extra-budgetary sources, such as the United Nations Development Programme (UNDP), as was the case for the setting up of the Arab League Documentation Centre (ALDOC).

We may cite as an illustration of successful regional projects CARSTIN, the Caribbean Regional Scientific and Technical Information Network, intended to build up scientific and technical information infrastructures, create a framework for information exchange, and enhance national capacity for handling and using STI. Other interesting examples are the Regional Programme for Strengthening Cooperation among Information Networks and Systems for Development in Latin America and the Caribbean (INFOLAC); the Asia and Pacific Information Network on Medicinal and Aromatic Plants (APINMAP); and the Regional Network for the Exchange of Information and Experience in Science and Technology in Asia and the Pacific (ASTINFO) [16, 20].

The objectives of ASTINFO are (a) to strengthen bibliographic control of each member country's own scientific and technological output, establish databases in subjects of interest to the region, supported by clearing-houses and document-delivery services; (b) to stimulate and promote the creation of non-bibliographic databases in science, technology, and certain socio-economic fields of importance to development in the region; (c) to develop the basis for cross-border exchange of data and information; (d) to improve national information infrastructures; (e) to create in each country a national node; (f) to introduce new and innovative information services; (g) to train information specialists; and (h) to promote and market existing information services.

According to the ASTINFO Independent Evaluation Report [35], ASTINFO's major impact has been in the area of training scientific and technical information personnel, particularly in the use of computerized systems; the distribution and utilization of the CDS/ISIS software package; the creation of a considerable pool of expertise within the region for the utilization of the software package; and the demonstration of the use of information technology, on-line access, the use of CD-ROM, and similar tools.

Regional and subregional information networks are also emerging in specialized fields such as – in the case of Unesco – marine sciences, microbiology, renewable energy, and the chemistry of natural products.

As we pass from the Second to the Third Medium-Term Plan of Unesco (1990–1995), we can witness further significant changes in emphasis and environment.

First of all, the PGI was relocated to a newly created sector on Communication, Information, and Informatics (CII). Each of the three programmes constituting the sector has so far maintained its identity and specificity, but links will be strengthened resulting, on the one hand, from the convergence of technologies and their impact on society and, on the other, from the benefits to be derived from cooperative implementation of projects and activities, whenever feasible. It seems that a whole momentum has been initiated that will lead the way to a movement of cooperation and harmonization between the programmes of Communication, Information, and Informatics. The future will reveal how far this cooperation will go. At the Twenty-sixth Session of the Unesco General Conference, delegates were opposed to integration of the programmes under CII, but they endorsed coordination.

One of the significant changes in the present biennial set of activities for the PGI (1992–1993) is the disappearance of STI as a visible entity [26]. This fact, which was deplored by a large number of delegates at the General Conference, deserves to be analysed.

We have seen how UNISIST, a programme essentially created for the promotion of STI, was conceived in the late 1960s and launched in the early 1970s. We saw how it expanded by the early 1980s to be concerned with information for development, broadly defined, and to achieve a satisfactory balance between activities in information, libraries, and archives.

During the first biennium (1990–1991) of the Third Medium-Term Plan (1990–1995), the activities of the PGI were grouped under the following headings:
– Conceptual and methodological framework
– Information services and networks in science and technology
– Libraries
– Archives
– Intergovernmental Council, subventions to NGOs, PGI Documentation Centre

In the second biennium (1992–1993), the PGI's activities are grouped under the following headings:
– Methodological framework, regional strategies and training
– Libraries and documentation units
– Archives
– Coordination of the PGI

The disappearance of the heading "Information Services and Networks in Science and Technology" triggered at the General Conference a series of interventions deploring the disappearance of STI as an obvious major component of the PGI even though activities dealing with STI remained scattered under various headings of the programme. As we have seen, no agency of the United Nations, other than Unesco, had developed a programme for the promotion and coordination of STI systems and services, at the national, regional, and international levels. Unesco has had the leadership in this area within the United Nations for the last quarter of a century.

At this point, a reference to the ICSU is necessary.

With the encouragement of the ICSU Excecutive Board and the active collaboration of the ICSTI, CODATA conducted a survey in 1991 on current perceptions of problems in accessing STI [2] to see what new role the ICSU could play, in addition to the many activities related to information and data transfer carried out by the unions in their respective disciplines and by CODATA, the ICSU Panel on World Data Centers, and the International Geosphere-Biosphere Programme (IGBP). The survey considered the issue *inter alia* under the following headings:

(a) Restrictions on transmission across national boundaries. No political barriers to the transfer of STI across national boundaries seem to preoccupy scientists today. "Any scientific information publicly available in one country appears to be accessible also to scientists in other countries. In fact, the major information services do operate on an international basis; the location of the computer containing the database which a user is accessing is usually transparent to him" [2]. The great concern with "trans-border data flow" of the early 1980s seems to shade off, although some continue to fear that governments may be tempted to restrict data outflow.

(b) Pricing policies and economic issues. There is concern about the high cost of electronic access to STI. There exists a wide variety of pricing policies and cost-recovery practices. The cost of creating and operating electronic databases obviously must be borne by someone, but wide variations in current practices exist from country to country, between different government agencies within the same country, and from one discipline to another. "The replies [to the survey] suggested confusion and concern but did not pinpoint a clearly-defined problem amenable to solution by ICSU."

(c) Information needs of the academic community. Discounts to academic users are sometimes offered. Most respondents to the survey favoured some form of price reduction for educational institutions. The fact that many university libraries cannot afford to purchase needed journals and books is not a simple problem of access to STI "but an integral part of the much broader problem of inadequate support to basic scientific research."

(d) Barriers that result from efforts of database owners to protect their intellectual property from unauthorized redistribution or other illegal practices. Most vendors seem to be moving toward a philosophy of charging on the basis of the amount of information delivered and permitting recipients to use that information as they wish. There is, nevertheless, a certain tension between the providers of STI who wish to protect the "added value" to the raw data and the scientific users "who reason that the scientific community created the information and therefore should have unhindered use of it." A balance between these points of view "will have to be established, but this will take time, and it is not clear how ICSU can influence the process."

In conclusion, the survey suggested several areas where CODATA and the ICSTI might expand the types of activities they have traditionally carried out (e.g. education and training, preparation of directories, standardization of formats and classification systems). In regard to the ICSU itself, the study does not indicate a need for a new ICSU activity on STI other than "to issue a statement of principle regarding the importance of effective information flow to the health of science" [2].

It should be stressed at this point that two of the above-mentioned issues remain at the heart of the debates in the industrialized countries and are reflected in the current published literature. First is the issue related to the consideration of information as a commodity *versus* a (subsidized) public good made available on a non-fee basis. And second is the problem of restriction to access. For example, in the United States, the 1991 White House Conference on Library and Information Services (WHCLIS) recommended that neither the Congress nor the Executive Branch shall abridge or restrict the right to public information through inappropriate classification, untimely declassification, or privatization of public information, nor should decisions be made to eliminate information collection and dissemination programmes for solely budgetary reasons [6].

A set of questions come to mind: Now that Unesco/PGI has modified its traditional balance in favour of libraries and that STI has lost its visibility, has not a gap been created? Should this gap be filled? Is there a need for an international focal point for STI? Will Unesco recapture this function in the future, or will information be amalgamated more and more with informatics and communication in Unesco international programmes of cooperation? Should not other governmental and non-governmental organizations strengthen their contributions in the area of STI [26]? Is STI an appropriate concept for international cooperation, or should one rather focus on the social use of information [13]?

3.3 UNCSTD

The analysis of Unesco's programmes in STI has sketched the information needs and requirements of the international community and the evolution and changes in strategies of an international cooperative programme. The analysis of the United Nations Conference on Science and Technology for Development (UNCSTD), also known under the name "Vienna Conference," of 1979, will describe another international effort to improve access to STI, using a different approach.

We have seen that the UNISIST II Intergovernmental Conference that met in Paris in May 1979 – three months before the convening of UNCSTD in Vienna – had emphasized the need to strengthen national capabilities to handle and use STI as a prerequisite for the developing countries to effectively participate in international efforts and achieve better access to the international reservoir of scientific and technical information, data, and know-how.

The UNISIST II conference addressed a resolution to UNCSTD in this respect emphasizing the importance of the "national level" and recalling that

. . . (f) given that national and international information services and systems develop in a compatible fashion, it will be technically feasible to establish, gradually and stepwise, flexible, co-operative international net-works of information systems and services for the exchange of scientific and technical information; (g) the establishment of these networks will necessitate substantial resources and will need to be sustained by a con-tinuous effort of goodwill and collaboration among nations and interna-tional systems; (h) the creation, maintenance, and development of national information infrastructures in developing countries necessitate large financial assistance without which it would be impossible to achieve these objectives in a satisfactory manner; . . .

The Unesco resolution further invited UNCSTD, when elaborating guide-lines for future action, to take full advantage of the considerable experience accumulated by Unesco, through UNISIST, and by other United Nations agencies, and invited the conference to "avoid the creation of new pro-grammes and structures within the United Nations system which could duplicate the work of existing agencies; . . . " [29]

This message was not heard in Vienna.

At its closing plenary meeting, UNCSTD adopted a Programme of Ac-tion. This programme dealt, in paragraphs 30 to 33, with scientific and tech-nological information systems. The topic had been the subject of con-troversy. In fact, the conference report contains in Annex 1 "issues of the draft Programme of Action on which agreement was not reached at the Con-ference." As a follow-up, the Intergovernmental Committee on Science and Technology for Development and a secretariat, the United Nations Centre for Science and Technology for Development (also referred to as UNCSTD), were developed.

Regarding STI, the Programme of Action foresaw the setting up of a new mechanism under UN auspices called GIN – the Global Information Net-work [31]. Rather than building GIN gradually from the foundations up as recommended by UNISIST II, the UNCSTD conference advocated what could be called a "top-down" approach. Each country would have a national node, while a global central node would be created under UN auspices. The network would operate as a channelling mechanism facilitating contact be-tween users and suppliers of information. Each national node would have the information-on-information for its country; the global central node would have it for the world at large. In cases of difficulty of obtaining a re-sponse from any other national node, the global node would take measures to ensure that the required information is provided. The accent was not only on published STI, but more particularly on know-how; on "foreign sources of technology supply, its terms, conditions and costs of all major factors and components contributing to the use and application of technology, to enable comparative evaluations to be made"; conditions of licensing, identification of suitable experts, engineers and consulting services, and the like [31].

The implementation of GIN met with many obstacles. Its establishment was an enormous task that could have been possible in the form of an evolu-tionary process taking place over several years during which considerable efforts and large financial resources would have had to be made by national administrations, regional intergovernmental organizations and the United Nations system, as well as the international scientific community at large.

This did not take place. Funds were not made available, probably because potential donors did not really believe in the proposed scheme. The UNCSTD conference had been essentially a political forum, where the voices of the scientists and the technical experts in the information sciences, telecommunication, and informatics were not heard. Also at stake were sensitive areas, such as know-how; terms, conditions, and costs of foreign sources of technologies; licensing conditions; and commercial interests. There was a widespread belief that GIN was an over-ambitious dream for which no serious systems design or cost evaluations had been done. One of the major obstacles, as was pointed out at UNISIST II and that is experienced in every international information effort, is the situation in many developing countries that were invited to become partners in the network. Lack of human and material resources, absence of information flow between decision makers and the productive sector, inadequacy of telecommunications and computer facilities, lack of trained manpower, difficulties in accessing primary documents, and the prevailing difficult overall social and economic conditions prevented many countries from seriously considering the proposed scheme, unless substantial investments could somehow have been made available to the developing countries. Most UN agencies had not made any specific budget provisions for their participation in GIN, considering instead some of their ongoing activities as contributions to the overall effort [24].

In 1989, 10 years after the Vienna Conference, the Intergovernmental Committee on Science and Technology for Development devoted its tenth session to the end-of-decade review of the implementation of the Programme of Action [32]. It was concluded that the accomplishments during the 1980s had fallen far short of the objectives sought by the Programme of Action, except in a limited number of areas. The Global Information Network was not among these areas. As a matter of fact, the concept seems to have been altogether abandoned.

On the other hand, particular attention was given in the report to new technologies associated with development. Two distinct categories were identified: (1) technologies with relatively affordable R&D intensity, such as biotechnologies and energy technologies in which practically every country can hope to participate, and (2) technologies with high R&D intensities, such as information technologies, micro-electronics, microcomputers, and telecommunication and space technologies, whose core aspects of invention and development are presently centred in a handful of countries. Most developing countries and many developed countries could at best only participate in the innovative adaptation and use of these technologies [32].

A lesson that can be drawn from this experience is that an important impetus for successful international cooperation in the transfer of information is *political*. In Vienna, the developing countries insisted on having the

Global Information Network, while the industrialized countries were opposed to it. The concept was retained in the Programme of Action but could not be implemented. The case of INIS, the International Nuclear Information System, also illustrates the importance of international political consensus. Woolston [36] explains that the member states of the United Nations wanted INIS

> not so much for its intrinsic values as an information system, but because it represented a breakthrough from the Cold War and an early step towards a US-Soviet detente in the nuclear field. The politicians were right and, after the agreement on INIS, along came the Nuclear Non-Proliferation Treaty and the Strategic Arms Limitation Talks.

If the political will is there, funds can always be found and enough pressure can be exercised to reach international agreement on technical matters. In the case of INIS, participating countries agreed in a relatively short time on all aspects of the system. This was also the case, as we have seen, for EURONET DIANE, where technical problems were gradually solved once the political will existed.

Another conclusion of the GIN experience may be that since the new information technologies evolve so fast, it may be practically impossible to design such a rigid system on a global level. Flexibility is essential. And, as was suggested earlier, a more realistic approach would have been to build the system from the foundations up, by trying at the same time to cope with some of the most urgent problems and obstacles in the developing countries.

What should not be lost sight of in the above analysis of the attempts and problems encountered in trying to establish GIN is the fact that it represented yet another cry of the developing world for better access to information relevant to their needs. GIN may be criticized on conceptual and technical grounds, but the fact remains that a visible and still growing imbalance does exist among countries in their ability to access and use STI.

In this respect, before concluding, a few further remarks are called for regarding the developing countries, although, on the one hand, many references to their needs, requirements, and problems have already been made and, on the other, the panel in Session 2B, "Achievements and Limitations in International Cooperation As Seen by the Developing Countries," will treat the subject.

4. Difficulties of the Developing Countries: Partners in International Cooperation

"Developing countries" – a term widely used in the literature – comprise over 80 per cent of the nations of the world and include the most populous countries, such as China, India, and Indonesia, and account, therefore, for most of humanity. It has been found convenient to group under one term all these countries in spite of the fact that they are at different stages of development and demonstrate important differences in cultural, political, and social environments and traditions. We should, therefore, be fully conscious of the limitations of such a generalization [10, 14, 21]. As mentioned by Menou [14], there are much greater differences in the national situation regarding STI between Senegal and Guinea-Bissau, for instance, than between Canada and Switzerland.

The published literature reporting on the situation of STI in developing countries tends to describe lacunae but seldom suggests solutions. It is generally observed that developing countries wishing to cooperate in international programmes and systems encounter many obstacles and present many interlinked problems in need of solutions [4, 10, 13, 14, 15, 16, 20, 21, 37]. These obstacles and problems constitute an impressive list that is well known to the organizations involved in international cooperation. They include *inter alia* language difficulties; the high cost of acquiring new technologies as well as primary literature and of linking to international systems; the emphasis on the supply of information rather than demand; legal and administrative barriers; low salaries, lack of trained personnel, and the brain-drain problem; minor relevance of available information to local problems; frequent personnel changes that occur with every government change; and the lack of adequate government support. According to Saracevic [21], "many reports perceive that if there is any one factor to be isolated as the greatest internal and external obstacle to the beneficial use of STI in development, it is the low level or even lack of recognition of its potential role and value, particularly among decision makers and officials of high ranks in the Less Developed Countries." Most reporters concur, although it is remarked that in recent years the third world has to some extent recognized the importance of information since it has used this as an issue in North-South negotiations [14].

One of the serious problems in many developing countries is a lack of coordination among the information systems and services operating at the national level. Many so-called "focal points" are established to cooperate with diverse international information systems or programmes (e.g. focal points for AGRIS, UNEP, INIS, UNISIST, etc.). The situation is particularly dramatic when human and financial resources are scarce, as is too often the case. Better coordination is needed among them, at the country level, and among the supporting programmes at the international level.

To the above obstacles we should add the cultural environment; it is generally agreed that internationally available information products and sources are insufficiently attuned to local cultures and practices. Information technology may be alien to local perceptions and may cause resistance to change. Difficulties of a psychological or an intellectual nature that relate to the presentation of information cannot be neglected. In many cases there is a lack among potential users of a real information-seeking mentality and tradition, which are not conferred by the education system.

Past experience in working with the developing countries in trying to enhance access to information raises a number of questions [13]:

- Should the developing countries be considered as permanent users of information, the bulk of which is produced and made available in the North, while their endogenous production is neglected? If not, then do we know how to harness endogenous information and make it available internationally through existing systems and services?
- Should the developing countries focus on their connection to existing systems and services, internationally available, or invest in creating their own infrastructures?
- If the option selected is the creation and strengthening of national information infrastructures, and considering the limited resources, should priority be given to building a "national memory" for the long term or to providing effective services to the users in the short term?
- Can the information services in the developing countries be sustained and allowed to survive and grow, if information continues to be locally subsidized and handled as a free public good, rather than a commodity?
- Consequently, how could an "information market" be progressively created in a socio-economic environment marked by low income and low resources?

The challenge of international cooperative programmes and systems has been to deal with this set of complex and intricate obstacles and problems and yet produce some tangible results. Will the new information technologies provide new opportunities and new modalities of international cooperation to relieve the developing countries from this burden and improve access to STI?

REFERENCES

1. Budington, W.S. (1971). "Access to Information." In: M.J. Voigt, ed. *Advances in Librarianship*, Vol 2. New York: Seminar Press, pp. 1–43.
2. CODATA and ICSTI (1991). *Barriers to Access to Scientific Information*. Available from CODATA, 51 blvd. de Montmorency, 75016 Paris, 17 p.
3. Commission of the European Communities (1990). *Information Industries: Impact Programme*. (CD-NA-12589-EN-C). Luxembourg: Office for Official Publications of the European Communities.
4. Eres, K. (1989). "International Information Issues." In: M. Williams, ed. *Annual Review of Information Science and Technology* 24. White Plains, New York: Knowledge Industries, pp. 3–32.
5. Fang, J., and A. Songe (1990). *The World Guide to Library, Archives and Information Science Associations*, 3rd ed., IFLA publication 52/53. Munich: K.G. Sauer.
6. Flagg, G., and L. Kniffel (1991). "Delegates' Skill and Tenacity Carry WHCLIS to a Promising Conclusion." *American Libraries* 22 (8): 786–804.
7. Haas, E., and J. Ruggie (1982). "What Message in the Medium of Information Systems?" *International Studies Quarterly* 26 (2): 190– 219.

8. Homet, R.S. (1979). "Communication Policy Making in Western Europe." *Journal of Communication* 29 (2): 31–38.
9. Hughes, G.C. (1991). "The Information Age." *Information Development* 7 (2): 72–74.
10. Keren, C. (1980). "Information Services Issues in Less Developed Countries." In: M. Williams, ed. *Annual Review of Information Science and Technology* 15. White Plains, New York: Knowledge Industries, pp. 284–323.
11. Kranch, D.A. (1989). "The Development and Impact of a Global Information System." *Information Technology and Libraries* 8 (4): 384–392.
12. Lorenz, J.G. (1969). "International Transfer of Information." In: C. Cuadra, ed. *Annual Review of Information Science and Technology* 4. Chicago: Encyclopedia Britannica, p. 379.
13. Menou, M. (1991). Private communication.
14. Menou, M. (1990). "Pour une formation a l'information adaptée dans le tiers monde." *Documentaliste. Science de l'Information* 27 (4–5): 236–239.
15. Moll, P. (1983). "Should the Third World Have Information Technology?" *IFLA Journal* 9 (4): 296–308.
16. Neelameghan, A., and J. Tocatlian (1985). "International Cooperation in Information Systems and Services." *Journal of the American Society for Information Science* 36 (3): 153–163.
17. Parker, J.S. (1984). "Unesco Documents and Publications in the Field of Information: A Summary Guide." *IFLA Journal* 10 (3): 251–272.
18. Parker, J.S. (1985). *Unesco and Library Development Planning*. London: Library Association.
19. Roberts, K.H. (1988). "Unesco's General Information Programme, 1977–1987: Its Characteristics, Activities and Accomplishments." *Information Development* 4 (4): 208–238.
20. Rose, J. (1989). "The Unesco General Information Programme and Its Role in the Development of Regional Co-operative Networks." *Iatul Quarterly* 3 (4): 231–245.
21. Saracevic, T. (1980). "Perception of the Needs for Scientific and Technical Information in Developing Countries." *Journal of Documentation* 36 (3): 214–267.
22. Tocatlian, J. (1975). "International Information Systems." In: M.J. Voigt, ed. *Advances in Librarianship* 5. New York: Seminar Press, pp. 1–60.
23. Tocatlian, J. (1981). "Information for Development: The Role of Unesco's General Information Programme." *Unesco Journal of Information Science, Librarianship and Archives Administration* 3 (3): 146–158.
24. Tocatlian, J. (1984). "Towards a Global Information Network." In: A. Van der Laan and A.A. Winters, eds. *The Use of Information in a Changing World*. FID publication 631. Amsterdam: North-Holland, pp. 5–18.
25. Tocatlian, J. (1991). "Bibliotheca Alexandrina – Reviving a Legacy of the Past for a Brighter Common Future." *International Library Review* 23: 255–269.
26. Tocatlian, J. (1992). "The Future of Information in Unesco: An Assessment of the 26th Unesco General Conference." *Information Development* 8 (2): 69–75.
27. Unesco and ICSU (1971). *UNISIST: Study Report on the Feasibility of a World Science Information System*. Paris: Unesco.
28. Unesco. Intergovernmental Conference on Science and Technology for Development (UNISIST II) (1979). *Main Working Document*. Paris: Unesco.
29. Unesco. Intergovernmental Conference on Science and Technology for Development (UNISIST II) (1979). *Final Report*. Paris: Unesco.

30. Unesco (1991). *The Unesco Clearing-House – Feasibility Study*. Paris: Unesco.
31. United Nations (1979). *Report of the United Nations Conference on Science and Technology for Development*. New York: UN, pp. 55–56, 87–88.
32. United Nations, General Assembly (1989). *Substantive Theme: End-of-Decade Review of the Implementation of the Vienna Programme of Action*. A/CN.11/89. New York: UN, p. 46.
33. United Nations Advisory Committee for the Coordination of Information Systems (ACCIS) (1990). *Directory of United Nations Databases and Information Services*. Geneva: UN, 490 p.
34. Vagianos, L. (1973). "Library and Information Associations in the International Arena." *American Society for Information Science Proceedings* 9: 79–85.
35. Wijasuriya, D.E.K. (1991). *ASTINFO Evaluation Report*. Bangkok: Unesco/ PGI.
36. Woolston, J.E. (1973). "The Future for International Information Systems." *American Society for Information Science Proceedings* 9: 23–24.
37. Woon, L.W.Y. (1990). "On-line Databases and Developing Countries." *Libri* 40 (4): 318–325.

Session 2B

The Technological Experience: Information Resources and Networks

Chairperson: Carlos Correa

Databases and Data Banks

Nathalie Dusoulier

ABSTRACT

After briefly tracing the growth and coverage of the database and data bank field as well as the tradition of cooperation that characterizes their creation and operation, the steps in database production are analysed and the various means of access to databases are discussed. The application of bibliometric methods to the analysis and evaluation of databases is described. The potentials for Hypertext and Multimedia are outlined. Economic aspects of the field are investigated and problems of ownership and copyright are raised.

1. Introduction

Without information, research and industry would decline. To know how to obtain information in the minimum time is an indisputable factor in competitiveness. Databases and data banks (DB), accessible on-line from a microcomputer or a videotext station (Minitel in France), can provide this decisive factor. Nowadays there are thousands of millions of references stored in thousands of databases; this represents an enormous wealth of information, really a supermarket of information, that is still largely underused.

Although the first databases had appeared at the beginning of the century, on-line databases only saw light of day in the 1960s. These early databases were generally in-house, or otherwise not openly available. They included the large limited-access on-line database systems built under contract by Lockheed (e.g. the Nasa-Recon system) and by SDC (e.g. the National Library of Medicine's AIM-TWX system). It is fair to say that no sizeable public access systems were available until about 1972. Growth thereafter was rapid.

2. Some Figures and Definitions

The terms database and data bank are often used interchangeably and in somewhat different senses on different sides of the Atlantic. I will try in this document to standardize these terms, a task becoming more and more difficult with the development of mixed products, and will use the acronym DB as the general term. In the library sense, an "on-line bibliographic database" is generally understood to mean a collection of records held on-line in rapid-access computer store. This concept has become significantly diversified nowadays and DB differ not only in the content of each entity described, but more and more in the media used.

Bibliographic databases can be differentiated by their different contents: simple bibliographic citations, citations accompanied by index terms, or complete references containing citation, abstract, and index terms. The abstracts can contain or be accompanied by factual or numeric data, evaluated or not. Full-text DB are now rapidly expanding. DB also differ by the media in which they are available (diskettes, magnetic tapes, etc.); we can see also a dramatic development of CD-ROM and a diversification of access methods (ASCII or videotext). Nomenclature and description become so complex that increasingly sophisticated DB catalogues have become best sellers.

As for the number of databases available, the latest figures are 5,200 issued by 2,200 producers and available on nearly 800 hosts. If we compare these figures to those of 1980, the growth has been spectacular. In 1979/1980, the *Cuadra Directory of Online Databases* listed 400 DB, 220 producers, and 59 hosts, while today it lists 6,414 DB. We must also note the recent appearance of a new type of database that mixes textual or numeric data with chemical structure diagrams, photographs, weather maps, trademarks, logos, or illustrations. These graphics DB require special equipment for their use.

The Information Market Observatory (IMO) of the European Community recorded in 1989 1,048 different DB produced in the Community and 2,214 produced in the United States. This figure included only DB accessible in ASCII up to the end of 1989, and thus excludes DB that are only produced in videotext. According to the IMO, the United States produces nearly twice as many ASCII DB as Europe, but the European Community has a higher growth rate: 101 new DB, representing 11 per cent growth, in Europe and 151 (7 per cent) in the United States. In Europe, Great Britain dominates with 34 per cent of the production.

It is estimated that there are some 25,000 videotext services within the EEC. Half of them are located in France, which has the largest installed base of videotext terminals (about 6 million).

The CD-ROM market is also growing very quickly: the number of titles published doubles each year. It is expected that the number of titles (about 750 in 1989) will increase to more than 6,000 worldwide in 1992.

As for the typology of DB, production of ASCII DB is dominated by bibliographic data in Europe (62 per cent of new DB) and by full text and factual DB in the United States (76 per cent of new DB). There is no significant difference between the United States and Europe as regards the target markets: the sector with the most products is that of services (especially banking, finance, and insurance). In the industrial sector, chemistry is one of the most important subjects.

The total turnover of the DB industry in 1989 rose to 35 billion francs (US $6.5 billion), of which 10 per cent was from videotext DB. For Europe, it is close to 14 billion francs (nearly US$3 billion). In 1993, these figures should reach 72 billion francs (± US$12 billion) and 28 billion francs (US$6 billion). The United States dominates the market with 56 per cent. English is confirmed as the pre-eminent language for documentation, three-quarters of the DB being accessible in it. Japan imports more electronic information than it exports; this is one of the few areas where its trade balance is negative.

3. Typology of World Databases and Data Banks

This typological study, which is only indicative, has been carried out by analysing information in the Cuadra Directory, mentioned above. The percentages have been calculated from the 6,414 bases in the Cuadra Directory, except for the percentages for the main fields covered, which were obtained by extrapolation from 877 references selected randomly.

The percentages given can only indicate the distribution of the total. They are in fact false because the same base can be cited several times, under different headings. Thus, from the 877 references selected for fields covered, we obtained a total of 1,296 by adding the number of bases by field. In other words, a base is listed an average of 1.5 times. However, in this case the distribution of bases is expressed as a percentage of the number of bases studied (877) and not as a function of the number of bases obtained (1,296). This is why, although a significant number of fields have not been listed, having few references (less than 1 per cent, e.g. urban planning, biotechnology, contracts and awards, etc.), the total exceeds 100 per cent. During the search, 275 different fields were encountered. The principal fields covered have thus had to be grouped together.

Principal fields covered

Finance, economics	29.0%
Business, industry, trade	25.0%
Law, justice	9.0%
Energy, environment	10.8%

Government, defence	4.8%
Chemistry	3.7%
Medicine, toxicology, nutrition, agriculture	5.6%
Science and technology	3.4%
Social sciences, humanities, administration, human resources management	5.6%
Research, innovation	2.6%
Computer, engineering, communications, telecommunications	7.3%
Miscellaneous	16.7%

Principal producer countries

United States	62.4%
England	7.6%
Canada	6.4%
France	5.8%
Germany	4.7%
Australia	2.3%
Spain	2.0%
Italy	1.9%
Japan	1.4%
Sweden	1.3%
Netherlands	1.2%

Principal distributing countries

United States	66.8%
England	13.4%
Canada	7.7%
Germany	6.0%
France	6.0%
Italy	3.3%
Australia	2.2%
Spain	2.0%
Japan	1.5%
Sweden	1.4%
Netherlands	0.9%

Principal types of databases

Reference databases	
–bibliographic	18.8%
–referral	16.4%
Source databases	
–numeric	17.5%
–textual numeric	6.7%

| –full text | 28.2% |
| –images | 4.2% |

Principal languages used

English	72.0%
French	9.3%
German	5.2%
Spanish	3.5%
Italian	2.3%
Swedish	1.4%
Dutch	1.1%

4. Cooperation among Database Producers

We have to go far back in time if we want to retrace the whole history of cooperation between information services. Indeed, it was in 1896 that the Royal Society organized the first Conference for the Joint Production of the International Catalog of Scientific Literature, a complete abstracting and indexing service that was to last 25 years. In 1948–1949, Unesco organized several conferences on abstracting and indexing in biology and medicine and on scientific document analysis. These conferences led to the creation of the ICSU-AB, which was the forerunner of today's ICSTI. The Conference on Scientific Information organized in 1948 by the Royal Society and the International Conference on Scientific Information held in Washington, D.C., in 1958 are two other examples of this will to cope with the new communication needs that scientific information organizations had to face in the science and technology context of the time.

These issues were formalized for the first time on an international basis when the proposals of a three-year (1968–1970) Unesco-ICSU feasibility study were considered at the UNISIST intergovernmental conference convened by Unesco (1971) to prepare the ground for a worldwide scientific information system. In 1971, the keywords, so to say, were enhancement of scientific information as a basic resource and promotion of international cooperation. It was strongly felt that this resource should become more accessible and easier to use, so that as a global wealth, it could best contribute to the scientific, educational, social, cultural, and economic development of all countries of the world.

During this period, a number of approaches to international cooperation were discussed. Ron Smith, from BIOSIS, presented in one of the Miles Conrad Memorial lectures a review of the various panoramas of the situation.

The first approach grew out of some of the work of the International Council of Scientific Unions Abstracting Board (ICSU-AB) and was largely based on bilateral agreements that followed discussions by all member services in a particular discipline. This type of international cooperation is still pursued. An example of this concerns an interaction that developed in an experimental way among three abstracting journals in the area of physics: the English-language journal *Physics Abstracts (PA)*, the French-language journal *Bulletin Signalétique (BS)*, and the German-language journal *Physikalische Berichte (PB)*. ICSU-AB conducted a survey of the journal literature of physics as covered by these three abstracting services. A comparison identified that there were three levels of productivity in the journals that were scanned by the various services. Quite arbitrarily, distinctions were drawn between journals of so-called high productivity (producing more than 100 titles per year), medium productivity (11–99 titles per year), and low productivity (10 or fewer titles per year). In the high-productivity group, there were approximately 80 periodicals common to the three abstracting journals; they were thought of as no more than a useful list and provided perhaps yet another definition of core journals.

In the area of low productivity, the situation was much more interesting. *PA* and *PB* scanned some 1,600 periodicals, and each of their lists contained more than 350 periodicals that fell into the low-productivity category. The acquisition of and selection from these periodicals may be expensive and difficult to justify in terms of return, but may nevertheless be essential if the coverage is to be satisfactory.

This experiment, although very interesting, was not continued for very long and the cooperation between the services was followed up by the creation of a common classification system in physics.

The second approach to international cooperation was undertaken by the ICSU/Unesco (International Council of Scientific Unions/United Nations Educational, Scientific, and Cultural Organization) Committee (UNISIST), in the framework of assessing the feasibility of the previously mentioned World Science Information System. The study was based on the concept that this World Science Information System would be a flexible network based on the voluntary cooperation of existing and future information services. It was headed by a central committee chaired by Professor Harrison Brown and was supported by a number of working groups and an advisory panel that comprised representatives of some of the large existing operating systems. One working group was concerned with questions of standardization of bibliographic descriptions that could serve for classification, indexing, and abstracting; a second working group was concerned with the identification of research problems that had to be studied to achieve an efficient worldwide system; a third group studied the problems of natural and machine languages, especially from the point of view of transferability and mechanized processing; and a fourth group worked on the problems of developing countries and their contribution and access to a worldwide system.

Certain of these actions were pursued later in the framework of Unesco's UNISIST programme and continue under its General Information Programme, but at the overall level the spirit of cooperation no longer exists.

The third approach to international cooperation, which is not attributable to any one organization but which has been fairly widely discussed in Europe, was that of establishing one information service in each of the disciplines into which the existing services would all feed; it was first assumed that the information services established would be in English and that services in other languages would agree to feed material into the system. It was then proposed that instead of having the same information processed a number of times in different languages, it would be processed only once, in one language, and subsequently made available by translation, if necessary. Some of the thinking of this nature was based on the concept that if cooperation of the kind referred to in the ICSU-AB activities were to develop, we might end up with a number of different services dealing with identical material, each in its own language. There may be much merit in the proposal of a one-language system, but there are obvious difficulties too, and not only the cultural or political ones. The current cooperation between Medlars centres is still based on this philosophy.

The fourth approach – the mission-oriented concept – is well known and a few international systems of this kind are operating. One is INIS, of the International Atomic Energy Agency, another is AGRIS, of the FAO; still others are less well known but probably equally effective.

Nowadays, the problems of international cooperation are viewed in a rather different way, since networks or other technologies that allow efficient exchanges of information have rendered obsolete the exchange methods envisaged 20 years ago. Collections of databases are created above all at the level of the hosts, who provide the necessary interfaces with the users and eliminate duplication where necessary.

Database producers get together at national levels to discuss technical problems, the market, and competition not only among themselves but also with other contenders entering the field that the database producers had for a long time considered their own. The most important organization of this type is the NFAIS (National Federation of Abstracting and Information Services) in the United States, which also contains some foreign members. In France, the GFFIL represents the principal producers. Others exist in several countries.

At the international level, the International Council for Scientific and Technical Information (ICSTI) offers great potential. The ICSTI is an international, not-for-profit organization whose purpose is to increase accessibility to and awareness of scientific and technical information. Established in 1952 as the ICSU-AB (the International Council of Scientific Unions Abstracting Board), it has evolved over the past four decades from an organization that was initially concerned with the development of abstracting and indexing services to a wide-ranging forum devoted to the improvement of scientific and technical information transfer.

Scientific and technical information, and specialized information at large, by adding value to the use of science and technology, provide searchers, administrators, and firms with the information necessary for their professional activities. Technological advances have brought many changes to the storage, processing, delivery, and use of information. At the same time, user needs have also evolved and led to the creation of more sophisticated and specialized information products.

The role of the ICSTI is to enable its members to exchange views and ideas on all the aspects of this evolution, to make progress in their comprehension, and to contribute to the development of appropriate tools better able to meet the information requirements of the world community of scientists and technologists.

In the database field, cooperation, whatever its objective or the level at which it is established, is now irreversible.

5. Database Production

Database production has evolved dramatically in the last 30 years. From the totally manual production in the 1960s of references on paper worksheets, of which sections were cut off and sorted manually (slips the size of a postage stamp for authors and slips a little bigger for index terms), to the quasi-automatic production of data distributed on-line, several generations of systems have come and gone.

I will not dwell on the past and will try in this chapter to look at the stages of modern database production, although the processes used are still very diverse. I will aim at a description of the traditional functions and the way in which they are most often handled.

To produce a database, several preliminary operations must be carefully carried out. First is the definition of the subject or subjects covered as a function of the target market or the users to be served. Once the subject has been defined, the scope and depth of coverage are also fundamental points. Determining the number and types of sources (document types, document languages) will also have an influence on the database content, on the costs of the acquisition of these sources and of the data production, and on the number of staff required to process them.

The format definition for each entry is also an important part of the preliminary work: databases can contain simple bibliographic references, indexed or not, complete references with abstracts and index entries, full text, or any combination of these.

The choice of format has a major impact on the budget and the type of staff required, but also on user satisfaction (costs, processing times, etc.). Any derivative products must also be considered here. Once all these options have been decided, database production can begin.

5.1 Bibliographic Description

Bibliographic description is the first operation in the processing of each document. Very precise rules exist for compiling bibliographic references. National and international standards allow producers to process this information in consistent ways, thus allowing exchange. The elements treated are the title and its variants, author names, their affiliations, the titles of the publications from which the references have been taken, their type, editors' names, publication language, etc.

The role of standards organizations such as the ISO and its Technical Committee 46 on documentation, ANSI in the United States, the BSI in the United Kingdom, and AFNOR in France has been crucial. However, because of the slowness of international procedures, other organizations, including the IFLA and Unesco, have taken the initiative, and quasi-standards such as the ISBDS, ISBDM, the *UNISIST Reference Manual*, the CCF, and the ISDS, approved or not by the ISO, have played and still play a major role in these fields.

5.2 Conceptual Processing of Documents

5.2.1 Document Abstracts

The addition of abstracts to bibliographic references enormously enhances a database. For many years, these abstracts have been written by information scientists according to rules learnt in schools of librarianship or information science. Several types can be distinguished: analytic or indicative, objective, or oriented towards a particular type of user. The relative qualities of one or the other have been the subject of many discussions, even polemics. Nowadays most texts are accompanied by author abstracts, and the use of these is becoming more and more common. However, there are two types of use: author abstracts used unmodified and in full, or author abstracts edited and/or shortened. In the case where the database producer decides to create his own abstracts according to the editorial policy of the DB, this work is often subcontracted to outside workers.

Instead of abstracts, some databases, called full-text databases, enter the complete text of the document, whether textual or numeric. The data-capture methods may be different, the current trend being to digitize the text and then, depending on subsequent processing, to store it either in ASCII or image form.

5.2.2 Indexing

Indexing is essential to enable retrieval of the stored documents. The operation varies greatly from one database to another. The quality of retrieval depends on the specificity of the indexing. Current systems vary from simple assignment of a limited number of unstructured terms to multi-level hierarchical indexing from sophisticated controlled vocabularies. Indexing consistency is an important factor for DB quality. Indexing is an expensive operation because it requires a large number of qualified staff. This is why most organizations are looking towards indexing methods, aided or not by expert systems, that will allow the introduction of completely automatic indexing.

5.2.3 Indexing Systems

Before discussing indexing systems, we must define the terms. Indexing is an operation that consists of describing and characterizing a document with the aid of representations of concepts contained in the document. In other words the concepts are converted into documentary language after extracting them from the document by analysis.

Indexing must allow effective searching for information in document sources. It leads to the recording of concepts contained in documents in an organized and easily accessible form, i.e. to the production of documentary search tools (catalogues, indexes, files, etc.).

Indexing is a documentary concept that is still of current interest. However, today a new qualifier more and more frequently accompanies it: "automatic" indexing, or more exactly "computer-assisted" indexing. In fact, the large and ever-growing volume of documents to process in order to make information available to the user as fast as possible makes it necessary to look for ways to speed up the processing.

Other factors encourage automation:
- the existence, in the indexer's work, of repetitive tasks without intellectual added value;
- the search for indexing quality and homogeneity for better access to the information (database reliability);
- the very significant economic realities: "We are arriving at a situation where the costs of human indexing exceed the costs of computer-assisted indexing."

Although indexing is considered an art that requires many qualities on the part of the indexer, the modelling of the thought process, the production of knowledge-based tools (dictionaries, thesauri, knowledge bases, etc.), and the development of computing techniques combine to give efficient help to the indexer. This help is available at different stages of the indexing process, either at the time of concept identification and selection in the document (analysis) or in their conversion into documentary language (coding). The intellectual processes used during indexing cannot be perfectly and completely automated. This is why we must instead speak of "computer-assisted indexing." Research in this field takes several forms. The simplest form is autoposting or generation of co-occurring terms. But it is now more complete and more advanced. Indexing systems that have been implemented fall into several types:
- statistical model: calculation of appearance frequencies of significant terms in a document;
- probabilistic model: co-occurrences method with creation of semantic networks between associated terms (Leximappe or Passat systems);
- procedural model: with a thesaurus and procedural rules allowing conversion of textual terms into thesaurus descriptors (machine-aided indexing or Medindex System);
- linguistic model: with morphological and syntactic analyses (Aleth, Spirit, or Darwin systems).

It will perhaps be a combination of these approaches and an exploitation of their complementary features that will lead to the automatic representation of document content. It is important to remember that indexing is an operation that is not only relevant at the time of document entry but also at the time of searching.

The possibility of user assistance represents a significant challenge for both producers and distributors of information.

5.2.4 Documentary Languages

Documentary languages are indexing tools that allow the transcription, in a concise and standardized form, of the concepts contained in the documents to be analysed. They provide a bridge between the natural language in the documents forming the documentary resource and that of the users' questions.

Indexing tools fall into two main types:
- languages with a hierarchical structure, called classificatory (classification, etc.)
- languages with a combinatory structure (lexicons, thesauri, etc.)

Classification is historically the first type of indexing tool to have been used in documentary systems. These classificatory languages are based on the prior coordination of ideas to express a concept and on the interlocking of classes of concepts. They go from the general to the particular, each class including the previous one, and the whole can be represented as a hierarchical tree. Classifications in general use employ codes or indices (numeric, alphabetic, or alphanumeric) to represent concepts. Examples are the Dewey Decimal classification, the Universal Decimal classification, or the Library of Congress classification.

Classifications have the advantage of offering a general logical framework and the possibility of enlarging or restricting the subject at will by using the hierarchy of classes and subclasses. However, they are awkward to update, complex ideas can only be expressed with difficulty, and their rigid structure requires a formalized approach towards the concept searched for.

The problems posed by classes interlocking with others have led to the development of new types of indexing tools: combinatory languages. The thesaurus is a typical example. This is an organized authority list of descriptors and non-descriptors obeying appropriate terminological rules and interlinked by semantic relationships (hierarchic, associative, or equivalent). This is a vocabulary that is standardized in form (rules for expressing terms: singular, plural, prepositions) and controlled with respect to the sense of terms (synonymous or quasi-synonymous terms represented by a single term). It is organized because it establishes between the terms the hierarchic and associative relationships that constitute the semantic environment. In this it differs from a simple alphabetic list of terms, which is a lexicon.

A thesaurus is very flexible to use because the indexing is carried out by a simple combination of key words. The hierarchic relationships allow indexing at different levels of specificity or generality and the associative relationships allow navigation from one term towards other ideas, thus also leading to a better understanding of the vocabulary. It must be emphasized that updating is simple and allows the evolution of science to be followed step by step, by adding without difficulty new ideas in the form of candidate descriptors.

Documentary searching with combinatory languages is undertaken with the Boolean operators "AND," "OR," and "NOT" and is remarkably flexible, based on intersection, addition, or exclusion of ideas. It can be made more efficient, however, by the addition of syntax to the language. This can be in the form of role indicators, or links indicating the function of a term in the indexing or its relation with another term.

Indexing languages, by the fact that they are a representation of fields of knowledge, have a privileged place in the current development of information systems. Far from being obsolete, the thesaurus retains all its value in the creation of knowledge bases, by the management of the transmitted message, the choice of terms, the richness of their relationships, and their categorization. In addition, linguistics and mathematics today provide indexing languages with firm bases for their description and formalization.

5.3 Data Formats and Capture

Processing formats of databases were the subject of much work and discussion in the 1960s and 1970s; the format most often used is that recommended by the ISO-2709 standard. This offers a general structure, a framework designed specially for communication between systems. Although it was not designed as an internal format for systems, it has strongly influenced them. It provides a structure for data exchange formats such as Unimarc and its derivatives, the Unesco common communication format (CCF), or other formats less generally accepted but used. Currently there is a move towards the replacement of this type of format by one more suitable for electronic publishing, such as the SGML format (ISO 8879-1986 standard).

Character by character data capture from worksheets has been the approach used by most systems for many years and it continues to predominate. The variety of equipment used for this data entry is nearly unlimited. There is a wide choice among microcomputers (PC type), either alone or networked, minicomputers with workstations, or terminals connected to a mainframe. The choice is governed by the size of the database, the flexibility required, and the level of sophistication of the indexing techniques. Capture on formatted screens brings significant advantages. Data capture systems increasingly include checking programmes, spelling correction, or several other facilities to improve quality and productivity. There are many difficulties, however, in dealing with different alphabets or chemical or mathematical formulas, which require special treatment.

This area of database production seems to be the one in which the most significant changes are appearing and will appear in the future. Firstly, the increasingly widespread use by primary journal publishers of electronic publishing should allow automatic capture of data in machine-readable form. Digitization technologies (scanning) are also increasingly used at least for capture of abstracts, which are then processed by an OCR system. More sophisticated expert systems should allow quasi-automatic capture of other elements. The results of these experiments will depend substantially on the standardization of journals.

6. Use of Databases

6.1 Paper Products

Not very long ago, the principal product produced from databases was in printed form. Bibliographic bulletins, under various names, contained the whole or part of the database. Their manufacture and distribution did not require any special advanced technology, and the revenue from the subscriptions, received in advance, represented a financial guarantee for the producers. The products, still widely distributed, are nevertheless being replaced little by little by other methods of accessing the stored information. Microforms, for which a short-term fashion led to the thought that they might replace paper products, have not gone beyond the status of storage media. If they do not disappear altogether, they will instead become an adjunct to paper products.

6.2 On-line Access

On-line distribution of databases began at the end of the 1960s as a part of time-sharing systems. The 1970s saw the beginning of their growth. The first search systems such as STAIRS, RECON, and ELHILL/ORBIT are still recognized names. The National Library of Medicine and NASA were pioneers in the field. The systems were at the time limited by poor telecommunications and by prohibitive storage costs.

Nevertheless, the market began to emerge, as did the idea of hosts, organizations that take responsibility for the distribution of databases produced by others. These hosts provide search systems, storage media, networks, and user training, in general in collaboration with the producers. The hosts can be classified either as supermarkets or as specialists. Some producers are at the same time hosts.

Although relations between database producers and hosts are generally good, there is beginning to be a tendency towards competition or even sometimes confrontation (a recent example illustrates this). A recent article by Harry F. Boyle of Chemical Abstracts Service, published in the *ICSTI Proceedings*, 1991, on the relations between hosts and guests describes in detail what currently happens.

It is appropriate to note that competition between producers is significant in that more than 5,000 databases are available on 800 hosts. The users are becoming more demanding with respect to the services that they want to see. Some of them are beginning to feel that existing on-line systems, based on the Boolean model, are by nature limited. It is true that several other models have been proposed (vectorial, probabilistic, extended Boolean, fuzzy set) that aim to improve the performance of the Boolean model, but none has yet developed into a large-scale commercial application.

Josephine Maxon-Dadd of Dialog Information Services, in *Trends in Database Design and Customer Services*, published by NFAIS, has described the ideal database:

Great currency

Clean data

An easy link to full text

Graphics

Controlled vocabulary (hierarchical) maintained and updated over the whole file

Uncontrolled vocabulary too, perhaps for trade names, proper names, or synonyms

Title, a reasonable number of authors, a good abstract

Bibliographic data fully identified and searchable

Complete coverage of every journal title included

No internal duplicates

Subject classification scheme (text and code searchable)

Cited references

Numeric indexing

User-friendly scientific notation

Multilingual indexing

I think that this eloquent list should make all database producers pause to think, above all when the same databases are more and more used to produce derivative products or are the subject of more and more sophisticated processing.

6.3 CD-ROM

The CD-ROM (compact disc read-only memory) is a database distribution medium that was introduced some years ago for the storage of texts and graphics. It exhibits much the same advantages as the microfiche stores of earlier decades of texts. The disks are relatively easy to produce and to duplicate; they are also easy to ship from place to place, and can therefore be used for local storage of databases and for local retrieval activities. CD-ROMs also provide high-density storage for both text and graphics. A standard disk will store up to 600 million bytes of information.

When coupled with a personal computer, the potential of the medium is greatly enhanced. However, CD technology is somewhat hampered when the size of the database requires more than a disk. In this case, effecting a complete search of the database currently requires the user to change disks or to use jukeboxes.

6.4 Floppy Disks

The floppy disk information delivery medium is emerging as an option for personal computers. Disks are easy to manufacture and can be produced in-house. They can be produced for a variety of operating systems.

6.5 New Methods of Access to Information

Most attempts to improve access to information contained in databases are aimed at moving from documentation to information. In fact, on-line information is little used by companies despite their needs. According to recent figures, databases provide only 7 per cent of the total information processed by companies.

According to Nicolas Grandjean of Synthélabo, what the user really wants is the answer. Real user-friendliness is the relevance of the reply to the question asked, not to the information request. Databases give raw information, where the answer is hidden in primary documents. In addition, the answer is often complex, in that it requires the correlation between several documents. Grandjean does not think that we can stay with these relatively unsophisticated information systems, above all with the volume of information available today. New techniques now under study are providing answers by conceiving a new dimension to information systems.

The sum of the documents contained in a database possesses properties independent of the documents taken separately. These properties can be exploited, both in themselves and to design tools for aiding indexing and searching.

7. Bibliometry Applied to STI or Scientometry

7.1 Definitions

Bibliometry was defined by A. Pritchard in 1969 as "the application of mathematics and statistical methods to books, articles, and other means of communication." "Scientometry" is specialized bibliometry applied to the STI field. A third term, "infometry," was adopted by the FID in 1987 to designate the group of metric activities relevant to information (thus covering both bibliometry and scientometry).

7.2 The Functions of Scientometry

The functions of scientometry are the analysis, evaluation, and graphic representation of STI by means of statistical methods, mathematics, and data analysis. The analysis has the objective of answering the question, "Who is doing what, and where?" from STI. The evaluation that can be done on STI is of two types, one being "metric evaluation" of information flow (articles and journals), the other being "quality evaluation" of information processed in databases. As for graphic representation, this aims to present STI in the form of maps containing both research fields and the participants (researchers, institutions, countries). The objective is to provide a representation of the structure of information at any given time of its development.

7.3 Documentation and Information

We know that the amount of stored documentation is growing exponentially, but its informational content is only growing linearly. Remember that documents and information are not the same thing. By asking the question of the analysis of information and its representation, what we are trying to do is to map the knowledge structures of this information, and not simply to count the documents. It is necessary, however, to present the information in the framework of a relevant cognitive structure. This is why it is important to be able to represent, with the aid of bibliometric techniques, such a framework based on the knowledge contained in the scientific literature.

7.4 Analysis Techniques

The technique used for this purpose is the method of formation of keyword clusters; this allows the structuring of the information and permits its processing in hypertext form, in that, because each cluster (or group of interlinked keywords) indexes a certain number of bibliographic references, this is a means of organizing the information thematically, and thus represents a knowledge structure. Instead of looking through a body of information in sequential order, a simple list of references, a series of bibliographic citations, we have here a method of following a thematic order that can be constructed from the bibliographic data themselves.

   The advantage of the use of bibliometric techniques is that it does not involve classification codes that have previously been assigned and fixed. The development of research and its organization can be followed as they are presented in the scientific literature.

7.5 The Contribution of Linguistics

Many researchers are working on the construction of a new generation of knowledge bases, with the objective of using them in documentary computing. Others are applying themselves to improving access to full-text databases in order to allow multilingual searching, or to showing that it is possible to apply artificial intelligence techniques to database searching.

8. Hypertext

Ted Nelson proposed in 1967 the term hypertext as non-sequential writing that it would be inconvenient to produce or represent on paper. The structure of a hypertext database does seem to provide the flexibility that is required to create a database that would enable researchers to search for information, to follow associations in ways that reflect their normal information-seeking activities. However, the implementation of such a database would also need adequate guidance facilities and more powerful text retrieval capacities.

In the context of information retrieval, awareness of the potential of hypertext needs to be accompanied by recognition of the problems involved in developing effective hypertext-based solutions to retrieval problems. For example, standard indexing techniques employing either controlled or uncontrolled vocabularies can also be employed to help the user navigate the database or identify appropriate areas for searching. Unfortunately, the same problems are encountered with the use of controlled and uncontrolled vocabularies for indexing and retrieval in hypertext as in any other application. Controlled vocabularies require work in their preparation and updating, and users need to index their entries manually, while uncontrolled vocabularies suffer from proliferation of index entry terms and from problems with synonyms and homographs.

9. Multimedia

Databases nowadays cannot be discussed without mentioning multimedia. This concept is so wide that it is nearly impossible to fix its limits. Multimedia systems, which should in due course allow combined manipulation, from a single workstation, of text, sound, and images, should bring together a number of different technologies. The market for this type of product is still unknown. Documentary searching, which in 1989 was almost the only sector concerned with these technologies, should reach a global turnover of US$4 billion by 1994. Nevertheless, it seems that the tools currently available are not adequate for large applications. We know how to navigate in small graphs held in memory but not in multimedia databases of several gigabytes.

10. Economic Problems

10.1 The Position in the Economy

Activities related to database production and distribution were for a long time considered as scientific activities managed by an exchange system, but the current situation is very different. The explosion of information technologies and their rapid distribution into every aspect of economic life has put back onto the agenda a thought already raised in the 1960s. The emergence of new skills, creating many jobs in the database field, has kindled research and discussions. An American researcher, M.V. Porat, carried out a significant statistical study in 1977 in which he estimated that by 1967, 46 per cent of the American GNP was already related to information activities. Using Porat's method, J. Voge estimated that in 1984, information workers represented between 40 and 47 per cent of the workforce of the major industrialized countries. It is true that information production includes not only activities of design and transmission, but also of identification and integration by the user. It is at the same time process and product. Database production and distribution are integral parts of these analyses. Numerous studies have been carried out on the value of information and its special features since it is considered to be an economic connection.

The difficulty of agreeing on some basic concepts and the amount of work carried out on controlling the use and duplication of information demonstrates very well its special features. The flow of international exchanges in the DB field is also the subject of major economic studies carried out by the IMO and certain countries. It is also interesting that in the United States, American DB are mainly used, with only 10 per cent of foreign DB; in Japan, however, 75 per cent of the databases are foreign; in Europe the proportion is 18 per cent.

10.2 The Costs

Regardless of the discussions and studies on the place of databases and data banks in the economy, their production and distribution represent a significant cost. DB production costs fall into four categories: direct production costs, manufacturing costs, indirect costs, and administrative costs.

Production costs include very large staff costs for the library-related and conceptual processing of documents. Analysis and indexing are operations requiring scarce, highly qualified staff that generate high costs. It is in these areas that producers are trying to achieve savings by using author abstracts and by encouraging studies on assisted or automatic indexing techniques.

A significant element, possibly the largest, of produciton costs is allocated to the purchase of sources. Some producers try to minimize these costs by agreements with publishers or libraries in order to acquire the sources free or in exchange for their own products. The data-processing costs involved in database production vary according to the techniques used and on the size of the base. They also vary with the complexity of the system in the case of various types of cooperation on database creation, which require numerous interfaces.

Manufacturing costs essentially include magnetic tape production and the operations required to put them into the different formats of the hosts, the production of derivative products, such as publications containing all or part of the DB, CD-ROM, and diskettes.

Indirect costs include the development costs of new methods and products, promotional costs, marketing, user training and assistance, staff training, and user documentation.

Overhead, as for other activities, varies according to the facilities offered.

10.3 Pricing

Although most of the participants in the information world agree in considering information as a resource, a product, pricing problems are still treated outside real economic considerations. One of the proofs of this is that most database producers, except for business databases, are not-for-profit organizations, and many of these databases are government-subsidized. Pricing problems are also treated differently depending on the products produced from the databases. In fact, what tends to be called the "market price" still plays a role in determining prices.

On-line pricing is most often determined by the hosts, whose own pricing policies are also developing. For some years the division of revenue between database producers and hosts has been the subject of discussion, sometimes acrimonious, and has led to confrontation between the major participants of the two professions.

It is not the object of this paper to enter into the details of pricing, which are extremely complex. It can simply be noted that one of the critical points for database producers is the establishment of a sales and pricing strategy. Whether the strategies are defined by market sector or by product group, or whether price reductions are foreseen for particular users or to create the market, database pricing needs a clear definition of the target revenues by the producer. This requires, evidently, an exact knowledge of the cost elements for each product, and a full knowledge of the market, the competition, and the cross-relations between different subproducts of the same database.

11. Ownership, Legislation, and Copyright Problems

Because of the late recognition by countries and even by producers of the economic value of databases, they are not at present clearly protected by current legislation. The legislation that does exist is varied and in some instances can even impair the establishment and operation of a real market.

Moreover, new technologies and the increasingly widespread availability of localized equipment allow pirating, which, even if not organized in most cases, is very dangerous for database producers. Some could even disappear for this reason.

The Commission of the European Community presented on 29 January 1992, after it had been adopted by the Council, a draft directive relating to the legal protection of databases. The directive is to come into force on 1 January 1993. By databases, the Commission understands a collection of works or deposited material, stored and accessible by electronic means; i.e. on-line data banks are concerned, but more particularly databases, no matter what media are used (except paper), e.g. CD-ROM, videodisc, CDI, etc. This directive is thus essential, as much for producers and publishers of traditional DB as for those in the multimedia and electronic publishing markets. Coming after the Green Paper on copyright and the challenge of technology, published in 1988, this document expresses the European concern about dealing with the problems of intellectual property.

In spite of these efforts, the problems remain complex at all levels. Relations with scientific journal publishers must be settled in the case of full-text databases and the digitization of author abstracts used without added value. Relations with hosts are of a different type but require contracts and licences appropriate for the special situation of DB distribution. These points are examined in detail in a book published by the NFAIS, *Guide to Database Distribution*. Here again, technology creates problems that are difficult to resolve.

The opening of gateways between hosts creates complications that few contracts have so far taken into account. This must be done.

12. Conclusion

In this paper I have tried to show the role and the importance of databases as a component of the economy and as a tool for research and industry. However, this field is in continual technological evolution and in its spectacular growth cannot live in isolation. Firstly there must be thought on how to organize production and development, in order to be always up to date with the technologies. This means a continual monitoring of the quality and security of the data that are produced. DB operation must then be organized in an environment where the actors frequently change their roles. Database production depends heavily on the availability of sources, which means managing the relationship with publishers. DB distribution depends on hosts and users. Relations with hosts must be closely monitored, although they are generally without particular problems. The user is the person who justifies the existence of the database. He also evolves, from the experienced professional to the end user, who has neither the time nor the inclination to experiment with systems that are too sophisticated. It therefore seems essential that in future attention be directed towards him.

BIBLIOGRAPHY

AFNOR standard 247–102, August 1978. *General Principles for Indexing Documents*.

AFNOR standard 274–100, December 1981. *Rules for Creating Monolingual Thesauri.*

Allen, B. (1982). "Recall cues in known-item retrieval." *Journal of the American Society for Information Science* 40 (4): 246–252.

Allen, K.J. (1988). "Online Information and the Needs of Intermediaries." In: *Proceedings of the 1988 International Online Information Meeting.* Learned Information Europe 1: 161–169.

Barbarino, M. (1989). "Similarity Detection in Online Bibliographic Databases." In: *Proceedings of the 1989 International Online Information Meeting.* Learned Information Europe, 111–117.

Berry, J.N. (1991). "The Politics of and Expectations for the White House Conference on Library and Information Services: Lurching Toward . . . Washington." *Library Journal* 1991 (15): 32–35.

Chaumier, J., and M. Dejean (1992). "Computer-assisted Indexing: Principles and Methods." *Documentaliste* 29 (1): 3–6.

Chaumier, J., and M. Dejean (1990). "L'indexation documentaire: de l'analyse conceptuelle humaine à l'analyse automatique morpho-syntaxique." *Documentaliste* 27 (6) 275–279.

Cotter, G.A. (1988). "Global Scientific and Technical Information Network." In: *Proceedings of the 1988 International Online Information Meeting.* Learned Information Europe 2: 611–618.

Czarnota, B. (1991). "The New Europe." *AGARD Lecture Series* 181, 9 p.

"Database Protection Directive" (1992). *Infotecture Europe* 203.

Detemple, W. (1988). "Future Enhancements for Full-Text-Graphics, Expert Systems and Front-end Software." In: *Proceedings of the 1988 International Online Information Meeting.* Learned Information Europe 1: 271–278.

Dreidemy, P. (1991). "Alerte chez les serveurs." *Videotex et RNIS Magazine* (62): 25–35.

Dreidemy, P. (1990). "L'évolution du métier de serveur: l'heure des choix marketing et technologiques." *Videotex et RNIS Magazine* (50): 41–45.

Efthimiadis, E.N. (1990). "Progress in Documentation. Online Searching Aids: A Review of Front Ends, Gateways and Other Interfaces." *Journal of Documentation* 46 (3): 218–262.

Elias, A.W. (1989). "Copyright, Licensing Agreements and Gateways." *Information Services and Use* (9): 347–361.

Fayen, E.G. (1989). "Loading Local Machine-readable Data Files: Issues, Problems, and Answers." *Inf. Technol. Libr.* 8 (2): 132–137.

Holtham, C. (1989). "Information Technology Management into the 1990's: A Position Paper." *J.I.T.* 4 (4), 17 p.

ICSTI Symposium Proceedings: Information, la quadrature du cercle. May 1991, Nancy, France.

"Information Transfer" (1982). In: *ISO Standards Handbook 1*, 2nd. edition, 521 p.

Kennedy, H.E. (1992). "Global Information Trends in the Year 2000." In: *1992 STICA Annual Conference*, March 1992.

Laubier C. (de), and J. Scolary (1991). "Enquête: les bibliothèques face au NTI." *NTI* (29): 21–23.

"La logique d'interrogation des banques de données remise en question" (1989). *Bases* (7): 6–7.

Luctkens, E. (1991). "SIGIR '90: le point de vue de l'utilisateur de systèmes documentaires." *Cahier de la documentation* (1), 11 p.

Lyon, E. (1991). "Spoilt for Choice? Optical Discs and Online Databases in the Next Decade." *Aslib* 25 (1): 37–50.

Marchetti, P.G., and G. Muehlhauser (1988). "User Behaviour in Simultaneous Multiple File Searching." In: *Proceedings of the 1988 International Online Information Meeting*. Learned Information Europe 1: 41–49.

McCallum, S.H. "Standards and Linked Online Information Systems." *LRTS* 34 (3): 360–366.

McLelland, J. (1990). "Computers, Databases and Thesauri." *Aslib Proceedings* 42 (7–8): 201–205.

Melody, W.H. (1986). "The Context of Change in the Information Professions." *Aslib Proceedings* 38 (8), 8 p.

Montviloff, V., and W. Löhner (1989). "The General Information Programme: The Year, Achievements and Future Prospects." *Int. Forum Inf. and Docum* 14 (4): 3–9.

"Multimédia: la Babel technologique." (1991). *01 Informatique* (1162): 44–54.

O'Neill, E.T. (1988). "Quality Control in Online Databases." In: *Annual Review of Information Science and Technology* 23. White Plains, New York: Knowledge Industries, pp. 125–126.

"Protection des données: la proposition de directive du Conseil Européen" (1990). *Infotecture* (213).

"Rapport de l'IMO: 1048 banques de données en Europe et 2214 aux USA" (1991). *Infotecture* (218).

Rasmussen, A.M., and B.A. Hubbard (1988). "Business Information from Databases: A European Perspective." In: *Proceedings of the 1988 International Online Information Meeting*. Learned Information Europe 1: 149–159.

Rivier, A. 1990. "Construction des langages d'indexation: aspects théoriques." *Documentaliste* 27 (6): 263–274.

Rosen, L. (1990). "CD-Networks and CD-ROM: Distributing Data on Disk." *Online* (July): 102–105.

Ryan, H.W. (1991). "Open Systems: A Perspective on Change." *Journal of Information Systems Management* 1991: 62–66.

Salton, G. (1988). "Thoughts about Modern Retrieval Technologies." *Information Services and Use* 8: 107–113.

Schipper, W. (1990). "Through the NFAIS Looking Glass: The Year 2000." *Int. Forum Inf. and Docum.* 15 (3): 29–31.

Schipper, W., and B. Unruh (1990). *Trends in Database Design and Customer Services*. NFAIS report series.

Schmitt, R. (1991). "L'information, un prodigieux minerai à extraire." *Le Monde Informatique* 1991 (11): 35–39.

Schwuchow, W. (1991). "The Situation of Online Information Services Industry in the European Community (with Special Consideration of the FRG). *Int. Forum Inf. and Docum.* 16 (1): 6–10.

Scott, P. (1990). "The Organisational Impact of the New Media." *Aslib Proceedings* 42 (9): 219–225.

Seigle, D.C. (1989). "System Integration in an Image Intensive Environment." In: *Proceedings of SIGED '89*. Paris: Informatics.

Simmons, P. (1990). "Serial Records, International Exchanges and the Common Communication Format." *IFLA Journal* 16 (2): 198–203.

"So Many Databases, So Little Time." (1991). *Infotecture Europe* 1991 (195).

Stamper, R., K. Liv, M. Kolman, P. Klarenberg, F. Van Slooten, Y. Ades, and C.

Van Slooten (1991). "From Database to Normabase." *International Journal of Information Management* 1991 (11): 67–84.

Stern, B.T. (1991). "Adonis Revisited." *NFAIS Newsletter* 33 (11): 140.

Straub, D.W., and J.C. Wetherbe (1989). "Information Technologies for the 1990's: An Organizational Impact Perspective." *Communications of the ACM* 32 (11): 1328–1339.

Strozik, T. (1990). "Managing Technology for Today's Library Service." *The Bookmark* 48 (3): 188–193.

"L'Utilisation des CD-ROM dans les bibliothèques en Europe: où en est-on?" *Bases* 1989 (45).

Woods, L.B., T. Willis, D. Chandler, B. Manois, and P. Wolfe (1991). "International Relations and the Spread of Information Worldwide." *Int. Libr. Rev.* 1991 (23): 91–101.

Communication Networks

Takahiko Kamae

ABSTRACT

A fully digitized communication network, the Integrated Services Digital Network (ISDN), has been expanding and promoting multimedia communication. Video telephone and video conferencing are expected to grow rapidly.

B-ISDN will be an infrastructure in the twenty-first century. New information-offering services may take advantage of B-ISDN.

1. Introduction

ISDN is a fully digitized communication network that is expected gradually to take over the telephone network. ISDN was standardized in detail by the International Telegraph and Telephone Consultative Committee (CCITT), and thus the world-wide connection of ISDN should be easy.

In Japan, ISDN was made available commercially by Nippon Telegraph and Telephone Corporation (NTT) in 1988. Since then, ISDN has been growing and is now available almost all over the country. Multimedia computing is a popular topic among people involved with personal computers and workstations. Multimedia computers are believed to take full advantage of ISDN. In other words, ISDN will promote multimedia communication.

ISDN is based on synchronous transfer mode (STM) technology. To improve multimedia communication features, novel technology, called the asynchronous transfer mode (ATM), is being developed in many countries. CCITT has been standardizing the broad-band ISDN (B-ISDN) on the basis of ATM. One of the most important parts of B-ISDN is the "fibre-to-the-home" (FTTH) concept.

In FTTH, optical fibre cables are extended to customer premises; specifically, optical fibres will replace the metallic twisted pairs now being used for subscriber loops. B-ISDN is expected to push telecommunications strongly toward multimedia services, covering up to high-definition TV (HDTV).

This paper describes the state of the art in ISDN and future trends relevant to B-ISDN.

2. The Narrow-band ISDN

2.1 User-Network Interface

The present ISDN is generally referred to as "narrow-band ISDN" (N-ISDN) to distinguish it from B-ISDN. N-ISDN has two kinds of user-network interface (UNI): the basic-rate interface (BRI) and the primary-rate interface (PRI). BRI has two 64Kb/s channels, called B, and one D channel, and thus it is frequently called the 2B+D interface. PRI has the bitrate 1.536Mb/s, and can be divided into B (64Kb/s), Ho (384Kb/s), and D (64Kb/s). Typical division is 23B+D, and thus it is frequently called the 23B+D interface. However, when high bitrate is necessary, the whole bitrate 1.536Mb/s can be used as one channel, and it is called Hii channel.

The D channel is mainly used for control between user terminals and the network. All information is "packetized" in D channel. A customer can use D channel as a packet communication channel; in this case, control information and customer information are packet-multiplexed in D channel. The packet switching service is also provided through B channel. In this case, the whole bitrate (64Kb/s) in B channel can be assigned to customer packets.

In the long run, ISDN offers switched 64Kb/s service, including telephone, switched 384Kb/s service, switched 1.536Mb/s service, and switched packet service, through both B and D channels. Furthermore, ISDN has various features suitable to multimedia communication.

2.2 Multimedia Communication Features

One of the biggest differences between ISDN and the telephone network is that ISDN is receptive to various communication media, while the telephone network is strongly biased to the telephone service. ISDN is equipped with universal communication protocols according to the standard OSI reference model. The layer 1 interface of ISDN consists of BRI and PRI. The layer 2 and layer 3 protocols go through D channel. The layer 2 protocol mainly establishes the data link between a terminal and the network. The layer 3 protocol is very important to multimedia communication.

Among various commands in the layer 3 protocol, bearer capability, low-

layer compatibility, and high-layer compatibility commands are closely related to multimedia communication. Using the bearer-capability command,
a terminal requests to the network the kind of channel that is necessary;
namely, the bitrate, B, Ho, or Hii, and speech, audio, or unrestricted 64Kb/s
in the case of B. Low-layer compatibility is used to select end-to-end transfer capability. High-layer compatibility is used for match-making of terminals at both ends; namely, high-layer compatibility specifies the kind of terminals: telephone, facsimile, telex, teletex, MHS, etc.

In many cases, various kinds of terminals, e.g. telephone, facsimile, and
videotex, are connected to a BRI. When a call is originated by a Group 4
facsimile, the transfer bitrate is specified as unrestricted 64Kb/s and the kind
of terminal as a facsimile using the high-layer compatibility. At the receiving
side, terminals different from a facsimile do not respond, and thus only a
facsimile gives a response.

Thus, various kinds of terminals connected to the same BRI can communicate independently with the same kind of terminals at the receiving
end. This feature is very valuable to multimedia communication.

2.3 ISDN Application to Video Telephone and Video Conferencing

Figure 1 shows the standard model of video telephone/video conferencing
system. For standardization of video codec, a common interface format
(CIF) for TV signals was defined to enable the interconnection of the 525/60
(North American, Japanese) and 625/50 (European) TV systems, as shown
in figure 2. The standard codec encodes TV signals with CIF. The interchange between a national TV standard and the CIF can be done freely in

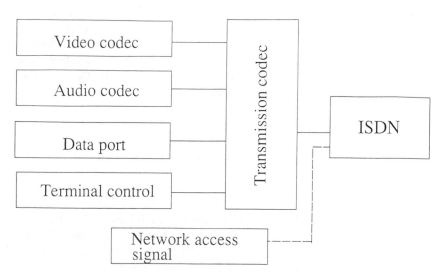

Figure 1 Standard video telephone/video conferencing system

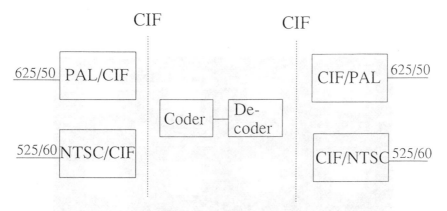

Figure 2 International connection through the Common Interface Format (CIF)

each country. For low-cost video codec, a quarter CIF (QCIF) was also defined.

Table 1 shows typical usages of ISDN channels to video telephone/video conferencing services. The use of B or 2B is likely suitable to video telephone, and Ho or Hii to video conferencing. The combination of QCIF, 48Kb/s video, and 16Kb/s audio may facilitate offering low-cost video telephone service.

Figure 3 shows a one-board video codec developed by NTT. This codec can be used with B and 2B in table 1. Its functions cover NTSC/CIF conversion, video codec, and transmission codec. A 16Kb/s audio codec can be mounted as a child board on this board.

Figure 4 shows a codec for 384Kb/s video. Four newly developed DSPs are used in the former and eight in the latter.

2.4 Colour Picture Transmission and Colour Facsimile

The standard colour picture codec, based on the JPEG standard, was developed as shown in figure 5. This codec has an interface to the VME bus and can encode a picture having maximally 8192 pixels in one direction. By

Table 1 Bitrates for video telephone/video conferencing

Channel	Division	Coding
H_{11}	B(audio) + 23B(video)	audio: SB-ADPCM PCM
H_0	B(audio) + 5B(video)	video: hybrid coding
2B	B(audio) + B(video) 16k(audio) + 112k(video)	audio: LD-CELP
B	16k(audio) + 48k(video)	video: hybrid coding

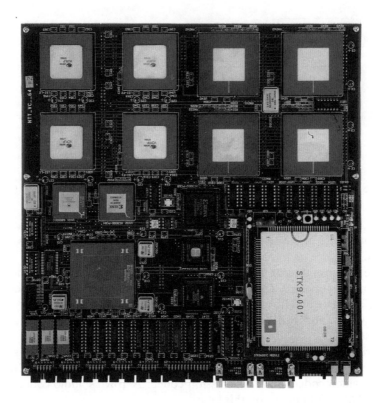

Figure 3 64K/128Kb/s video codec (28 cm × 28 cm)

assigning bitrates suitable to brightness and colour difference signals, the data compression rate can come down to 1/20 without noticeable degradation. An optical disk memory and the JPEG codec board are attached to a UNIX workstation to constitute a colour picture filing system, as shown in figure 6. A colour copier attached to the workstation can be used as an I/O terminal. Colour pictures are transmitted through ISDN. A JPEG codec is useful to save transmission time and storage capacity.

The JPEG standard will stimulate the standardization of colour facsimile. Important points of colour facsimile may be:
– Interworking with existing facsimiles, particularly with Group 4 facsimile;
– the same resolution as Group 4 facsimile in the black and white portion;
– application of JPEG for the full colour portion;

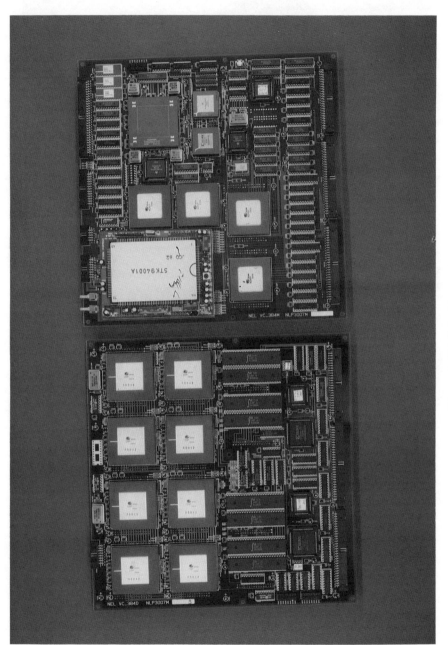

Figure 4 384Kb/s video codec

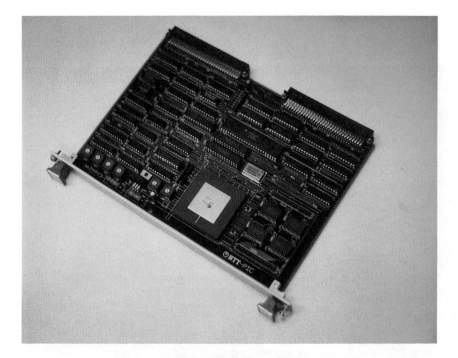

Figure 5 JPEG codec

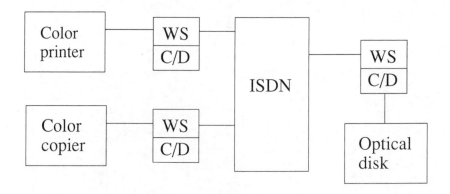

C/D : JPEG codec
WS : Workstation

Figure 6 Still colour picture filing and transmission

– distinction of the black and white and the full colour portions in the scanning process; and
– application of the standard open document architecture to distinguish the
  black and white portion from the full colour portion.
These should be guidelines for standardizing colour facsimile.

3. Broad-band ISDN

3.1 Outline of Broad-band ISDN

The broad-band ISDN (B-ISDN) being standardized at CCITT can be
understood through such keywords as the asynchronous transfer mode
(ATM), the fibre-to-the-home (FTTH), and a multimedia network covering
data up through a high-definition TV (HDTV).

As shown in figure 7, the transmitted data stream is divided into sequences of a small amount of data called a cell. Each cell has a 5 octets
header and a 48 octets information field. At the transmission line, cells are
packed without any gap, as shown in figure 7. Various identifiers in the
header specify a cell and contain sufficient information to deliver a cell.
When a high bitrate is necessary, more cells are captured. In other words,
network customers can define the capacity of a channel at their own will.
This is called a virtual channel. Customers can take as many virtual channels as desired in their subscribers' lines.

B-ISDN is widely believed to be the telecommunication infrastructure for
the twenty-first century. On the basis of B-ISDN, NTT announced the

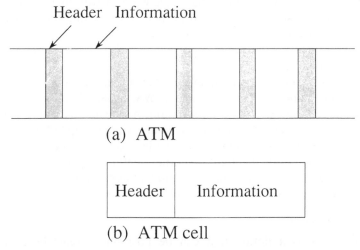

Figure 7 Asynchronous Transfer Mode (ATM)

Table 2 HDTV digital transmission

Usage	Bitrate
Studio grade	624M b/s
Program distribution grade	100–150M b/s
High quality entertainment	40–50M b/s
Application specific	10–20M b/s

twenty-first century telecommunication service vision, by the name VI&P, an acronym for visual, intelligent, and personal communications services.

3.2 Multimedia Service Example

3.2.1 Visual Telephone

Visual telephone is believed to be one of the major terminals for the twenty-first century. Its picture quality will be much better than the CIF-based picture in the N-ISDN environment.

3.2.2. HDTV

Depending on the application of high-definition TV (HDTV), necessary bitrates will become versatile, as shown in table 2. FTTH in B-ISDN will make a wide range of HDTV applications easy.

3.2.3 Personal Multimedia Teleconferencing

A desktop workstation (WS) used in the office will become multimedia. Office workers will be able to do many kinds of work at their WSs. Teleconferencing is an example. In order that office workers with their own WSs will be able to participate in meetings from their desks, multipoint capability is essential. The need for participants in a meeting to see one another makes multi-motion-picture-windowing important. It may be desirable to distinguish who is speaking.

Figure 8 shows a typical screen image of such a multimedia teleconferencing system developed by NTT. From the WS, a participant can send text, scanned image, and telewriting information; his own voice and image; and stored video in colour. The system has audio-window capability. The voice of a person whose picture window is at the left comes from the left, and the voice at the right from the right. Another feature is private conversation capability. Any two persons can talk confidentially during the meeting; this private conversation voice comes from the extreme right.

3.2.4 Network-casting

Network-casting is a notion that can be positioned somewhere between broadcasting and conventional telecommunication. Broadcast media such as

Figure 8 Typical picture in the personal multimedia teleconferencing

radio or TV are called mass media, by which information is distributed to many people, e.g. several thousands of people. Conventional telecommunication media such as telephone and facsimile can distribute information to only a small number of people, e.g. less than 30 people in a certain amount of time, and thus may be called "mini-media." In relation to these, network-casting might be called a midi-medium.

A baseball game in a local stadium may be network-cast to people in a big city through B-ISDN. In an ATM switch, an incoming cell could be duplicated to branch into two or more outgoing cells, as shown in the upper part of figure 8. Then information from the visual source could be distributed to many points. The other method is to branch by using local nodes, as shown in the lower part of figure 9. These two methods could be combined for the economical realization of network-casting.

Different from broadcasting, which is dependent on radio waves, network-casting goes through optical fibre cables, whose capacity is unlimited. TV programmes that are represented by terms such as personalized TV and interactive TV can be distributed through B-ISDN. This is one reason for characterizing network-casting as a midi-medium.

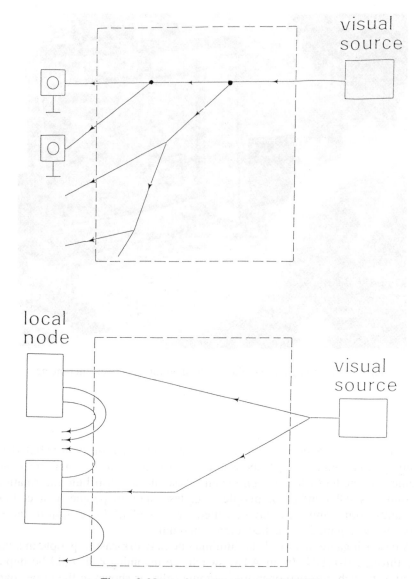

Figure 9 Network-casting through B-ISDN

4. Concluding Remarks

The status of ISDN service and its applications have been described. Multimedia communication, initiated by ISDN, will be promoted by B-ISDN. In addition to this, B-ISDN will facilitate offering a new information service, which might be called network-casting.

The Electronic Library

Masaru Harada

ABSTRACT

The paper reviews the progress already made with the aid of automation and networking towards the creation of "electronic libraries" and then briefly describes various technologies now in the research or development stage. These include intelligent gateways, electronic publications and filing systems, telepresence, and on-line hypermedia systems. Finally, attention is given to suggesting scenarios for the future of the electronic library.

1. Introduction

Prompt and efficient provision of documents or information is the most important function of the library. Librarians have worked for a long time towards performing this function more effectively and satisfactorily in the service of users.

Recent developments in information technology, which could be defined as a sophisticated technology related to the production, transfer, processing, and presentation of information based on a combination of computer and telecommunications technologies, opened a new way for librarianship. Save for some small libraries, it is difficult now to find one not equipped with a computer. The computer has come to be used for almost all tasks carried out in the library: acquisition, cataloguing, searching, circulation – and even for reference services.

Concurrently with the development of computer technology, there have also been the development of communications technology and the construction of extensive communication networks. The library community has worked to exploit this technology as well, and, as a result, library networks, large and small, have been built in many parts of the world. At present, many libraries, having completed the first phase of computerization and networking, are trying to make the systems more intelligent and easier to access and to make use of extensive information resources distributed at many places. The libraries resulting from these efforts are sometimes called electronic libraries.

By the term "electronic library," however, different people mean different things. A glance at the articles published in *The Electronic Library*, a journal dedicated to this subject, will reveal that they deal with a variety of topics, including OPACs, CD-ROMs, library automation, library networks, on-line information services, and multimedia information retrieval.

It could, of course, be supposed that the editor of the journal accepts contributions relevant not only to the electronic library *per se* but also to other technological components that would form part of the libraries of the future. Nevertheless, one might also conclude that at present there is no generally accepted definition of the term "electronic library."

Besides the problem of the definition, another problem arises when discussing the subject; the distinction between the electronic library and certain other concepts is not clear.

One such concept is electronic publishing. Although this concept, too, is not entirely clear, it usually means processes that produce publications in the formats of the new media. Since a library that provides users with this sort of publications may call itself an electronic library, the concept overlaps the electronic library in many respects.

Another related concept is a sophisticated computer-based work environment that enables one to access distributed computers and information resources and to communicate with others. Since provision of required information to the user is a major task that librarians must perform, this has many points in common with the electronic library. At present, intensive efforts are also being made in the area of multimedia information systems.

The ambiguity of the term "electronic library" reflects the fact that so far there have been few systems that many people can agree to call electronic library. In fact, there have been a variety of approaches to the electronic library:
- library automation and networking
- electronic publishing
- computer networks
- hypermedia systems
- more intelligent systems
Actual systems usually adopt some combination of these approaches.

Keeping these points in mind, the author reviews the work carried out so far as well as ongoing projects relating to the electronic library and its technological components. In section 2, contributions made by libraries are discussed, including library automation and networking and Maggie's Place. The author then discusses, in section 3, several other services and projects as examples of the electronic library. Included here are intelligent gateways, collections of electronic publications, electronic filing systems, telepresence, electronic journals, and on-line hypermedia systems. In the final section, after discussing the future of the technology, problems of facilitating access to information in the age of electronic libraries are considered.

Since databases, communication networks, multimedia and hypermedia systems, intelligent front-ends, and human interfaces, which have something to do with the present subject, are to be dealt with in detail by other speakers, the treatment of these points in my paper will be kept to a necessary minimum.

2. Library Automation and the Electronic Library

With the introduction of information technology, there have recently been considerable changes in library operations. Since some of these changes may lead to the realization of the electronic library, the development in this area must be reviewed.

2.1 Library Materials and Electronic Publishing

Libraries collect a variety of materials for preservation and use. They include not only traditional print-on-paper media like books, journals, newspapers, and maps, but also audiovisual materials like gramophone records, audiocassettes, and video cassettes. Now, much of the information and data that has been recorded on the traditional media has also become available in electronic form. Many libraries have begun to collect this type of material, that is, products of a process called electronic publishing.

As with "electronic library," there is no general definition of what constitutes electronic publishing; nevertheless, Feeney [13] has divided the latter into five categories:
(1) Broadcast services (non-interactive):
    teletext, cable services, direct broadcasting by satellite
(2) On-line services (interactive):
    on-line databases, videotex
(3) Stand-alone products:
    videodiscs, compact discs (CD), digital optical discs
(4) Electronic journals:
    integrated publication systems covering writing, editing, refereeing, publishing, and reading
(5) Bulletin boards:
    electronic mail, teleconferencing
As Feeney points out, these categories are not mutually exclusive. For example, videodiscs may be used to supplement an on-line service for graphics backup or CD-ROMs can be made accessible on-line via networks.

Most libraries try to offer these services to users, if their budgets permit them to do so. At the same time, publishers publish their products more and more in the electronic form, and it is forecast that electronic publications will come to form a large percentage of publications, sometimes without their counterpart in printed form. When a large part of the collection of a library becomes available in the electronic form and thus accessible electronically, some might call it an electronic library. Examples of this type are shown in section 3.2.1.

2.2 Library Automation and Networking

Library operations consist of the following major processes carried out with the aim of making the collection and other sources of information available to users: selection and acquisition; cataloguing; classification; indexing; searching; locating and retrieving; and circulation.

During the past 25 years or so, the computerization of part or all of these processes has progressed in most libraries with varying degrees of integration [32].

During the 1970s, a number of so-called total integrated library systems became available on the market, and these have been introduced into many libraries. In addition to providing modules to deal with the processes mentioned above, these systems usually provide access to networks and bibliographic utilities.

Access to outside information resources, e.g., catalogue databases located in other libraries and on-line information services, and communication with other libraries via networks are essential components of today's integrated library systems.

Some people may ask for more; Epstein [12] listed the following features librarians asked to have included in integrated systems: circulation control; public access catalogue; MARC record capability; authority control; acquisitions handling; materials booking; reserve book room control; serials control; local reference files and community information accessible on-line; word-processing; information and referral files accessible on-line; electronic mail; COM catalogue production; remote terminal access; two-way cable television; teletext; payroll and check production; budget control; staff scheduling; personnel records; personnel work statistics; access to outside databases; access to any other automated library system; interface to bibliographic utilities; and high-use indexes.

These features reflect the time when many of the full-text databases and CD-ROM publications currently available were still in development and were less spoken about within the library community. Including these in an integrated library system of today would produce what might be called an electronic library system.

The development of library networks at the regional and national levels, based on computer and data communications technologies, took place concurrently with the automation of individual libraries [23]. At the core of the network were MARC records and union catalogues. The bodies responsible for the administration of the networks also developed hardware and software packages to be installed by member libraries to help automate their housekeeping and networking.

Another development based on modern technology was local area networks (LANs). Through this technology, which enables units within a relatively small area to be interconnected, local area networks have been installed in many libraries, sometimes as part of extensive campus networks, with more efficient resource sharing in view. Examples include public libraries [14, 26], as well as academic and special libraries [1, 2, 33].

Library automation and networking have rarely gone beyond the provision of secondary information like that in bibliographic databases. Today, many libraries are using electronic mail or telefacsimile to transmit requests for interlibrary lending, but those using telefacsimile for interlibrary lending, i.e., document delivery, are fewer, although there are some reports from libraries that have made use of it on experimental or routine basis [15, 31, 37].

2.3 Maggie's Place

If a library acquires electronic publications that can be accessed on-line by any user in the community, provides access to on-line catalogues and commercial bibliographic databases as well as to the community information files prepared by the library, and makes arrangements with other libraries for resource sharing, many librarians would not hesitate to call this library an electronic library.

Probably the best known example of this type would be Maggie's Place, of the Pikes Peak Library District in Colorado, USA [9, 11].

Planned and implemented under the leadership of its Director, Kenneth Dowlin, the Pikes Peak Library drew attention from all over the world. The seven-point role of the library prescribed by Dowlin provides the following services: (1) access to complicated or seldom-used databases; (2) community conferencing and message centre programmes; (3) on-line access to information on library resources; (4) access to community data and community information locations for referrals; (5) access to resources in other libraries via networking; (6) access to high-demand information and materials via computer or videodisc; and (7) access to electronic resources for those who cannot afford home computers or terminals [10].

Almost all these objectives were reached in this library and many librarians saw it as a model for the electronic library. At present, connection is made to two library networks, the CARL system and the MARMOT network, and users can access resources in other libraries via the networks.

The community information files accessible in the library and from home are: (1) calendar file; (2) agency file; (3) club file; (4) local documents; (5)

local authors; (6) courses; (7) senior housing; (8) day care; and (9) facts [22].

2.4 User Satisfaction with the Current Automated Libraries

Library automation and networking have changed libraries drastically, and computers have become essential equipment for librarians. At the same time, electronic publications appear to an increasing extent, and libraries acquire and offer these publications even though most libraries still do this on an experimental basis. Many librarians might agree on the view that these developments are an inevitable step that should in due course lead to the full-fledged electronic library.

There is, however, a different view. With the advent of more sophisticated, user-friendly information systems, most of which are still in the development stage, people may come to prefer these systems to traditional libraries except for reference to old print-on-paper documents. This is the scenario that Lancaster envisaged in his book, *Toward Paperless Information Systems* [19], published in 1978.

This view, I suspect, may have come less from the predicted advancement of technology than from the demarcation that librarians had unconsciously set to their work. The points here are as follows:

(1) Library operations start from the point where publications have appeared. Publications are considered to be data, i.e., things that are given.

(2) The library is concerned primarily with printed documents. Although it has been dealing with non-book materials as well to some extent, this might reinforce the view that the so-called new media constitute simply additional non-book materials.

(3) Major efforts have been made in libraries towards automation for access to secondary information. The primary information has been out of the sphere where the information technology plays a major role except for interlibrary loans sometimes done by telefacsimile.

(4) The automation of a library has been directed to computerizing individual processes carried out in the library. The traditional work flow within libraries has usually been retained without much rethinking about the whole process of library operations, a rethinking that is required in an age when the traditional systems such as those for publication, distribution or dissemination, storage, and retrieval are changing.

(5) The advent of the new technology that is now in the research or development stage will probably change the whole structure of information dissemination. As Alan Blatecky, Vice-President, MCNC, USA, pointed out, "Although communications and computing technologies have been changing rapidly over the last 20 years, the changes we face over the next decade will be radical and will change the way we think and do business; in effect, it will change our culture itself" [3].

3. Other Examples of the Electronic Library

The huge collections of full-text databases that some on-line bibliographic services such as Dialog, BRS, Data-star, and NACSIS provide together with bibliographic databases might be called electronic libraries. In fact, if full texts, including diagrams and photographs, of many more journals were put on the computer and made available on-line with an appropriate tool to assist in obtaining the articles most relevant to the user's inquiry, any information could be retrieved on the desktop within several seconds, which is one of the goals of the electronic library.

The present systems, however, have not yet reached this point due mainly to the restrictions of the technology posed at the time of implementation and that continue. The problems are line speed, access time, the volume of data that can be stored, file structure and software for information retrieval. For these reasons, full-text data are usually limited to the coded character data of texts and therefore at present, diagrams and photographs must be sent to users by mail (snail-mail), telefacsimile, or some other means.

Efforts now being made in this area, however, lead one to expect that this will not continue for long. There are many systems worthy of mention.

3.1 Intelligent Gateways

Although, at present, the databases provided by on-line information retrieval services are mainly those of bibliographic data, recently there has been a significant increase in the number of full-text databases offered by the major on-line service vendors. One possibility of realizing an electronic library is to develop an intelligent gateway, or a front-end, that enables users to search the hundreds of databases provided by a number of vendors located at different places in the world.

EasyNet is one of the commercially available software packages of this type [22, 24]. EasyNet allows the searcher to choose the most appropriate database. There are two options for this; that is, either the software or the searcher may select the database. EasyNet then translates automatically the user's search query into the command language used by the system on which the required database resides; dials the system; logs in; accesses the database; and initiates a search. The limitation of this software is that it usually provides the name of only one system as the most relevant to the user's requirements [27].

Another approach of this type is BioSYNTHESIS, which is under development at Georgetown University, USA, as part of the Integrated Academic Information Management System (IAIMS) project sponsored by the National Library of Medicine (NLM), USA.

BioSYNTHESIS is a two-phase project to develop an intelligent retrieval system. The prototype BioSYNTHESIS II menu includes the following items [4]:

(1) Bibliographic databases
   a. On-line catalog
   b. miniMEDLINE
   c. ALERTS/CURRENT CONTENTS
   d. Full text
(2) Information databases
   e. Drug and poison info
   f. Physician data query (PDQ)
(3) Diagnostic systems
   g. RECONSIDER
   h. DXplain
(4) Communications
   i. Electronic news
   j. Mail Box
   k. External access

The IAIMS project started several years ago with the intention of making the information management systems of academic health sciences centres more intelligent and of connecting them to each other as well as to the National Library of Medicine.

Expected components of the system, which are still in the research stage, are a context selector and an information analyser, which will enhance the system's capability by making it more intelligent and thus allow even the novice user to exploit the system's full capacity with great ease.

3.2 Collections of Electronic Publications: Networked and Not Networked

3.2.1 Optical Disc Installations

CD-ROMs and other optical disc publications have been increasing in number. Although the major part of this category is bibliographic or composed of other reference data, the number of full-text databases is expected to increase rapidly. Since more and more publishers are introducing electronic processing, it is easier to produce publications in electronic form in spite of the fact that the present practice is to have print-on-paper publications.

A factor contributing to the predicted growing share of electronic publications is standardization. Adoption by publishers and printers of standards such as those related to the Standard Generalized Markup Language (SGML), which, while intended to serve for print-on-paper publications as well, will encourage the introduction of electronic processing in the whole publication process and will certainly change the nature of publications. A decision on the medium of publications (i.e. in one or more forms) can be made even at the last stage.

When a variety of electronic publications is available at one place, it becomes a kind of electronic library. From a technical point of view, remote access to a centralized collection of data via a star network is not a big problem, except for matters related to data security.

Thus, several libraries have installed a number of optical disc publications for public use either by asking the user to come to a designated room to operate one of the computers specified for each disc [16], or by allowing the user to access them on-line [25].

3.2.2 Electronic Document Delivery

Ambitious efforts have been made in the publishing community, sometimes in cooperation with other bodies like libraries, with different objectives in mind, to create huge collections of publications in electronic form and deliver the contents to users at remote sites. Examples include Adonis using CD-ROMs, APOLLO by satellite transmission, and HERMES by means of teletex [18].

The Knowledge Warehouse is a project that has been conceived and carried out by Publishers Databases Ltd, United Kingdom, with the support of the British Library and the Department of Trade and Industry. The project aimed to construct a national archive of the electronic form of knowledge works by establishing the methods for preserving the electronic version of works [5, 36].

3.3 Electronic Filing Systems

The electronic filing system refers to a mass storage system that uses optical discs as storage media. There are two types of optical discs: erasable and WORM (write once, read many). At present, many systems are available on the market and users can choose the most appropriate ones according to their requirements for storage capacity or other characteristics. The storage capacity of one model reaches a level of 1,000 Gbytes. The system usually stores, instead of coded characters, digitized images of documents, which require more storage capacity. Many firms make use of the system for storing technical reports, patent information, transactions information, scientific data, and so on.

If the contents of books and journals are stored on a system and the user is able to search any part of a book or any article in a journal, the system becomes a kind of electronic library.

ELNET, the electronic library network, is a full-text database of articles compiled from 37 newspapers and 120 magazines published in Japan. It is considered to be a kind of computerized clipping service. The system allows on-line searching by keywords and free text terms, and if searchers find relevant articles, they can, by using a simple command, order copies of the original documents stored on digital optical discs, which are then transmitted by facsimile in a matter of seconds [17]. As of 1 April 1992, the system contained 2 million articles.

The storage consists of nine jukebox-type optical disc drives, on each of which up to 25 discs can be mounted. The optical discs are of the WORM type and can store, in theory, on average 20,000 articles per disc – in practice, slightly less than this figure. At present, 40 to 50 per cent of the total storage capacity has been filled by 2 million articles. Two thousand articles per day, or 600,000 articles per year, are added to the system.

The database has digitized images of articles and thus includes photographs and diagrams as well as characters. The resolution of the digitized images is 400 dpi, a quality equivalent to the G4 facsimile. For the moment, not all articles are included in the database. Articles written and signed by outside contributors are sometimes omitted because the copyright usually belongs to them.

3.4 Telepresence

Terms such as telepresence, virtual proximity, artificial reality, and virtual reality, though their connotations differ slightly, especially between the first and the second pairs, usually refer to an artificially created environment where people can imagine handling a real object that does not exist in immediate proximity to them, or where people at different locations can sense and talk to each other as if they were participating in a meeting held in one room.

If one could, while sitting at one's desk, visit any library in the world, pass through the entrance to the main hall, walk down the corridors, enter a room, select any book from the shelves, browse in one book after another, or read one from cover to cover as desired, and print any page of which a copy is wanted – all of this simply by mouse-clicking on the icons on the screen and thus have access to huge amounts of data or information regardless of its location, the system would be a realization of telepresence.

Songoku is a prototype system of this type with an emphasis on the browsing process [28]. The name Songoku, or Sunhour in Chinese, is that of a fictitious character in the form of a monkey which, in a novel, followed a famous Chinese priest, Xuanzang, who actually did travel to India in the seventh century to acquire texts of the Buddhist scripture.

In addition to providing almost all the functions listed above, through Songoku the computer may also read texts to the user by means of a speech-generation subsystem. Additional functions that the developers plan to include are:
- to allow the user to browse among the book shelves in search of desired books;
- to create new book shelves filled with books related to the topic that the user commands;
- to see the view of the town where the library is located;
- to allow users to participate in the production of electronic books;
- to develop an electronic management system to protect the copyright; and
- to have a synchronized presentation of texts, music, speech, still images, and moving pictures.

Some of these remain to be realized.

3.5 Electronic Journals

By the term "electronic journals" is meant here a networked computer system that enables people in the research community, whose roles entitle
them to access to materials in process, to write, submit, edit, referee, or
modify the contributions made to a journal, and to read them when "published." From the technical point of view, there are no big differences between conferencing, messaging, and on-line publishing. Two systems are
well known: EIES and BLEND.

EIES (Electronic Information Exchange System) is an electronic journal
system developed at the New Jersey Institute of Technology, United States
[34]. The system had the following "publications": a newsletter called
Chimo; a public conference or unrefereed journal called *Paper Fair*; an
unrefereed publication called *Poetry Corner*; an electronic journal entitled
Mental Workload; and an inquiry-answer system called Legitech (from this
a "brief" is compiled afterwards for each inquiry). The system also included
other functions such as messaging, conferencing, and directory services.

BLEND (Birmingham and Loughborough Electronic Network Development) is another electronic journal system developed by two universities in
the United Kingdom [30]. The system was designed to include the following
publications and messaging facilities: refereed papers; comments and discussion; annotated abstracts or annotated bibliography; LINC News; Bulletin;
cooperative writing of papers; poster papers; enquiry-answer system
between experts; and "Readers Only." At the implementation stage, special
efforts have been made to develop user-friendly interfaces. The project is
well documented.

3.6 On-line Hypermedia Systems

A hypertext system is a computer system that handles textual data and
allows the author to link textual chunks. The system also provides the facilities for the user to manipulate these chunks and to create new links from
one chunk to another chunk, to bibliographic data, or to notes added by the
user. Yankelovich and others point out remarkable features of the hypertext
system as follows:

> using a computer-based hypertext system, students and researchers can
> quickly follow trails of footnotes and related materials without losing
> their original context; thus, they are not obliged to search through library
> stacks to look up referenced books and articles. Explicit connections –
> links – allow readers to travel from one document to another, effectively
> automating the process of following references in an encyclopedia. [35]

Hypermedia is an extension of hypertext that incorporates, in addition to
texts, other forms of presentation such as diagrams, still images, sound, and
moving pictures.

If, as claimed, a user could obtain information regardless of its medium by using this sort of system, then it might be unnecessary to have the books, journals, bibliographic databases, or other information sources that are, at present, provided separately by various agents.

Vannevar Bush's "memex" is considered by many specialists as a prototype of the hypermedia system – although he proposed using dry microphotography for storage of his "mechanized private file and library," which he named "memex" [7]. The essential feature of the memex is "associative indexing," a process of "tying two items together."

Current research efforts are focused on the implementation of Bush's idea on the computer and in a network environment.

There are a number of hypermedia systems on the market, and many more are in the research and development stage. Examples are: Analyst (Xerox Special Information Systems); Andrew (Carnegie-Mellon University); Guide (Owl International); Intermedia (Brown University); Telesophy (Bellcore); Xanadu (Autodesk); and so on. Here I shall refer to only one of these, Intermedia.

Intermedia is a hypermedia system that works in a multi-user, network environment [35]. The system was developed at the Institute for Research in Information and Scholarship, Brown University. It contains five applications: a text editor, a graphics editor, a scanned image viewer, a three-dimensional object viewer, and a time-line editor. A significant feature of the system is a "web," which is a set of links that have been grouped together under a single name. Intermedia is used at the English and biology departments of Brown University and at the Johns Hopkins medical school to create electronic course materials that provide students with new exploratory environments for their fields [6].

4. The Electronic Library of the Future

4.1 The Future of the Technology

4.1.1 Computers and Storage Media

When we deal with images and video of high definition, requirements for their transmission, processing, and storage become far more severe than those met by the current microcomputer systems. The developments that took place in these areas in the past, however, are remarkable, and it is still possible to predict the near future by extrapolating earlier development. It is even forecast, though some might think it too optimistic, that the supercomputer of a few years ago will be available in a laptop type before the turn of the century.

Along with this development, storage capacity of optical media will become much higher. According to a forecast, within a few years, recording density of optical discs will reach a level of 20 Gbits per square inch, more than 20 times that of the present ones.

4.1.2 The Extension of Networks

Communication networks will be set up in many more parts of the world. At the same time, data will be transmitted at a much higher speed from one point to another in the near future. This will enable even moving pictures to be transmitted as digitized data through lines. A project closely related to this is the broad-band ISDN, which will be dealt with by another speaker.

Proliferation of networks adopting different protocols may cause anxiety among people in the library community. But, Internet, "a network of networks," has solved most of the problems that relate to the interconnection of disparate networks.

4.1.3 Hypermedia Systems

The media can be classified by their linearity, i.e., whether only sequential access to information chunks is possible, and by their degrees of integration of different forms of presentation, i.e., characters, sound, still images, or moving pictures. Figure 1 shows where each form of "publications" can be placed.

If, as Vannevar Bush pointed out, a non-sequential multimedia approach is more tenable by human minds, a large part of the media currently provided through libraries might gradually be replaced by hypermedia. For this to happen, however, a far broader band width than that of the current networks and user appliances that enable the user to operate systems with ease would have to be developed.

4.1.4. More Intelligent Systems

The information needs of a user can be met at varying levels of satisfaction. The level of satisfaction is determined by the way the answer to a question is

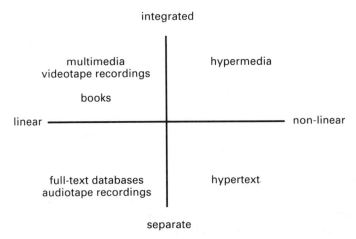

Figure 1 Classification of media by linearity and by degree of integration

deduce answer to question
|
provide answer regardless of its location
|
provide answer from stock
|
refer to document possibly containing answer

Figure 2   Types of answers to a question

provided. There are many types of answers, as shown in figure 2. (See also Lancaster [20].)

Up to the present, libraries have been concerned primarily with the preparation of tools that refer users to documents that may contain the answer to a question. Library catalogues, though many of them have been computerized, are tools to indicate the presence or absence of the documents sought. Referring users to potential sources of information should not be the final goal of the library. In parallel, the library should develop systems that can be used by the user without help. We can do many things to make our systems more intelligent.

The intelligent gateway is one of them. Another is an automated reference/referral system based on knowledge-base system, which is one of the applications intended to realize the deduction of the answer to a question shown at the top of figure 2.

4.1.5 Personalization

The library is not only a place where a collection of publications that is supposed to be of common interest to its users is stored. It should also provide the individual users or groups of users with services they need.

The same applies to the electronic library. If a user could create his own library when and as needed by putting together the necessary parts of the collection(s), he could have a virtual, personal library, or even an ad hoc working library that lasts only for the period it is required. A similar idea was included in the Songoku project, but this has not yet been realized.

4.2 Problems To Be Tackled

The forecast advancement of technologies related to various processes of information transfer might lead us to predict that ubiquitous personal computers connected via networks to any information source regardless of its location will soon make it possible to obtain any required information in a matter of seconds. But, the future is not necessarily programmed to be rosy. It is up to us to make the future a real golden age of knowledge.

Problems to be tackled for the betterment of information transfer are multifarious. The author will deal briefly with some of them: bibliographic control, information gaps, and the right to information.

4.2.1 Bibliographic Control and Universal Access

For the library community as a whole, bibliographic control has been a task of great importance to be carried out as a prerequisite to making required materials available. The preparation of a national bibliography that contains the secondary information on documents published within a country has been an essential part of the efforts. Arrangements for legal deposit have also helped the compilation of national bibliographies.

A number of libraries have started to provide services based on on-line databases and new media like CD-ROMs. In addition, it is predicted that, in the future, many more publications will be available only in electronic form.

Bibliographic control has tended to be limited to visible entities. Because many computer files such as CD-ROMs, floppy disks, and ROM chips are stand-alone products that can be purchased by libraries, they may be treated in similar fashion to conventional audiovisual materials. But, what about on-line databases or videotex? What about the many databases mounted on a computer that is located abroad and accessible on-line from any part of the world? We need to discuss universal access rather than bibliographic control [8].

Efforts are being made by computer specialists to facilitate access to computerized information and human resources located at different sites. With the advent of a network of networks such as Internet, which connects and makes accessible many disparate networks, this sort of tool has become indispensable.

The CCITT X.500 standard is one of the traditional directory services that rely on hierarchical organization, but recent projects aim at enabling more flexible and efficient resource discovery in very large network environments [29].

4.2.2 The Information Gap and the Right to Information

The term "information gap" refers to the fact that certain people or countries are in a better position than others to obtain required information. This may occur between people or regions within a country or between countries. Many people are concerned that the technology gap will worsen the situation because many countries cannot afford to invest in the research and development of advanced computer and communications technologies. They fear that in the future, when most publications become available only electronically, many people will be excluded from access to knowledge.

Another issue related to the information gap is the right to information. The idea that, while this right is not without limits, people must have access to certain categories of information has been well established within the public library community. Recently, however, there has been a gradual increase in the number of public libraries that try to recover the costs of on-line information retrieval services from fees paid by users. If the tendency develops, only those who can afford to pay for it will be able to obtain required information.

Although there are some proposed solutions for these problems, such as the establishment of information funds or the provision of information coupons, the author will leave these problems to the panels, which will consider them in detail.

5. Conclusions

If we trace the history of information transfer, we will find many attempts to collect, organize, and make available the world's knowledge to those who want it. The Royal Society of London, the Institut international de bibliographie (the predecessor of the FID) and H.G. Wells's World Encyclopaedia are but a few examples.

Today, given the advanced technologies such as high-performance microcomputers and workstations, optical discs for mass storage, high-speed interconnected communications networks, multimedia systems, almost within our grasp, we are in a far better position than our predecessors to expect that any person will be able to access information sources at any place in the world from wherever he or she resides. But, this can be said only from a technical point of view. The economic feasibility comes next. Moreover, the technical and economic possibilities do not necessarily mean that any existing new technology immediately comes into our daily lives. There are other factors, such as political, social, and psychological ones. Much work remains to be done to make the golden age of knowledge a reality.

REFERENCES

1. Adams, R.J., and M. Collier (1987). "Local Area Network Development at Leicester Polytechnic Library." *Program* 21 (3): 273–282.
2. Angier, J.J., and S.B. Hoehl (1986). "Local Area Networks (LAN) in the Special Library: Part 1—A Planning Model." *Online* 10 (6): 19–28.
3. Blatecky, A. (1991). "CONCERT: A Communications Network for the Future." *CONCERT Notes* 1 (1): 2–3.
4. Broering, N.C., H.R. Gault, and H. Epstein (1989). "BioSYNTHESIS: Bridging the Information Gap." *Bull. Med. Libr. Assoc.* 77 (1): 19–25.
5. Buckingham, M. (1988). "The Knowledge Warehouse: Technical Issues." *The Electronic Library* 6 (1): 6–9.
6. Bulick, S. (1990). "Future Prospects for Network-Based Multimedia Information Retrieval." *The Electronic Library* 8 (2): 88–99.
7. Bush, V. (1945). "As We May Think." Originally published in: *Atlantic Monthly* 176 (1): 101–108. Reproduced in his *Endless Horizons*. Reprinted ed. New York: Arno Press, 1975.
8. Cochran, P.A. (1990). "Universal Bibliographic Control: Its Role in the Availability of Information and Knowledge." *Library Resources & Technical Services* 34 (4): 423–431.

9. De Lury, N.A. (1987). "Maggie's Place III: The Second Generation of an Advanced Library Computer System." *Library HI TECH News* (22): 1, 7–9.
10. Dowlin, K.E. (1980). "The Electronic Eclectic Library." *Library Journal* 105 (19): 2265–2270.
11. Dowlin, K.E. (1984). *The Electronic Library: The Promise and the Process*. New York: Neal-Schuman.
12. Epstein, S.B. (1984). "Integrated Systems: Dream vs. Reality." *Library Journal* 109 (12): 1302–1303.
13. Feeney, M., ed. (1986). *New Methods and Techniques for Information Management*. London: Taylor-Graham.
14. Hooker, F. (1985). "Local Area Networking: Colorado's IRVING Project." *Wilson Library Bulletin* 60 (Sept.): 38–42.
15. Hulkonen, D.A., and P.J. Hamilton (1989). "The South Dakota Med-Fax Network." *Bull. Med. Libr. Assoc.* 77 (2): 212–215.
16. Jackson, K.M., E.M. King, and J. Kellough (1988). "How to Organize an Extensive Laserdisk Installation: The Texas A&M Experience." *Online* 12 (2): 51–60.
17. King, S.V. (1991). "ELNET—The Electronic Library Database System." *The Electronic Library* 9 (2): 61–72.
18. Lambert, J. (1985). *Scientific and Technical Journals*. London: Clive Bingley.
19. Lancaster, F.W. (1979). *Toward Paperless Information Systems*. New York: Academic Press.
20. Lancaster, F.W. (1982). *Libraries and Librarians in an Age of Electronics*. Information Resources Press.
21. Larsen, G. (1987). "Searching the Intelligent Gateway EasyNet—The End-user's Point of View." *The Electronic Library* 5 (3): 146–151.
22. Larsen, G. (1988). "Maggie's Place III: An Electronic Library Revisited." *The Electronic Library* 6 (6): 404–406.
23. Martin, S.K. (1986). *Library Networks, 1986–1987: Libraries in Partnership*. White Plains, New York: Knowledge Industry Publications.
24. O'Leary, M. (1985) "EasyNet: Doing It All for the End User." *Online* 9 (July): 106–113.
25. "Optical Disk Network Funded by Kellogg Grant" (1986). *Library HI TECH News* (32): 1, 7.
26. Raymond, C., and C. Anderson (1987). "Local Area Networks: Reaping the Benefits." *Wilson Library Bulletin* 62 (3): 21–24.
27. Roussos, A. (1991). "Personal 'Connection Machines'." *The Electronic Library* 9 (4/5): 263–266.
28. Sato, M., and J. Kishimoto (1989). "SON-GO-KU: A Dream of an Automated Library." *Journal of Information Processing and Management* 31 (12): 1023–1034. (In Japanese).
29. Schwartz, M.F. (1991). *Resource Discovery and Related Research at the University of Colorado*. Technical Report CU-CS-508-91. Department of Computer Science, University of Colorado.
30. Shackel, B. (1983). "The BLEND System: Programme for the Study of Some 'Electronic Journals'." *Journal of the American Society for Information Science* 34 (1): 22–30.
31. Stubley, P. (1986). "Experience with Facsimile Transmission in Birmingham Libraries." *Program* 20 (4): 415–419.

32. Tedd, L.A. (1987). "Computer-based Library Systems: A Review of the Last Twenty-one Years." *Journal of Documentation* 43 (2): 145–165.
33. Thayer, C.W., and K.P. Ray (1988). "A Local Network for Sharing Resources and Technical Support: BACS/PHILNET." *Bull. Med. Libr. Assoc.* 76 (4): 343–345.
34. Turoff, M., and S.R. Hiltz (1982). "The Electronic Journal: A Progress Report." *Journal of the American Society for Information Science* 33 (4): 195–202.
35. Yankelovich, N., B.J. Haan, N.K. Meyrowitz, and S.M. Drucker (1988). "Intermedia: The Concept and the Construction of Seamless Information Environment." *Computer* 21 (1): 81–96.
36. Williamson, R. (1988). "The Knowledge Warehouse: Legal and Commercial Issues." *The Electronic Library* 6 (1): 10–16.
37. Wilson, M. (1988). "How to Set Up a Telefacsimile Network—The Pennsylvania Libraries' Experience." *Online* 12 (3): 15–25.

Discussion

The discussion on session 2B began with an intervention by N. Streitz, who wanted to know what the actual usage of databases was and whether they were cost-effective; to his knowledge, "most of the scientific and technical databases are subsidized by governments." N. Dusoulier explained that all databases in scientific and technical fields were in fact underutilized, but that when talking about cost-effectiveness, one should take into account the revenues derived from the various by-products, such as CD-ROMs and publications produced from these very databases. She confirmed that, with a few exceptions, databases in these fields are subsidized by governments or scientific societies.

Next, F. Thompson took the floor to state that "databases can be a costly trap" because once a corporation has invested in the creation of a database, its obsolescence can cause real difficulties; and yet bringing the database up to current technology could be very costly. N. Dusoulier agreed that maintenance of databases is costly, particularly because technologies are moving very fast, updating databases is expensive, and the cost of securing qualified staff is high. This is why "so many databases are created and then disappear."

The next intervention was made by D. Torrijos, who wondered why cooperation in database production was disappearing. She thought that this kind of cooperation was especially useful for developing countries to enable them to actively participate in the production of marketable information products and eventually earn a share of the information market. N. Dusoulier explained that cooperation is not always cost-effective and enumerated some of the current difficulties. It is necessary to agree on the standards to use; one must write interfaces; input is often slow; and consistency has to be checked. Most database producers concentrate on core journals because of the high costs of processing and maintaining databases, as mentioned earlier – a fact that reduces the need for cooperation. As to the share in the market, she added that it is difficult to find a simple way of sharing revenues on real input; the statistics on what is actually used in databases are not well known.

Continuing the subject, M. Dierkes asked two questions: Why is the establishment of databases on a regional basis so weak, and how should decisions on the fields in which to establish databases subsidized by public funds be reached? On the first question, N. Dusoulier thought that perhaps there was no clear incentive given by governments nor were there regional policies on the matter. However, she believed that in Europe, sharing the production of databases will be reached in a few years' time. Responding to the second question, she expressed the belief that the needs and requirements of users should be the driving force for making decisions.

Next, D. Lide focused on the issue of overlap in coverage of different databases. Is the overlap too high, in the sense that it increases total costs, or too low, in the sense that literature on the boundary between two disciplines is not adequately covered? In her reply, N. Dusoulier confirmed that in bibliographic databases the overlap is certainly high on core journals. On the other hand, on fringe fields, "we still have to evaluate the missing information." As to numeric and factual databases, she did not believe that there was a significant overlap.

The discussion then moved on to communication networks. G. Johannsen wondered what will remain of T. Kamae's concepts for the twenty-first century in view of the depression faced by most industrialized countries and assuming that "we need to share with the former East bloc and with the South." He wanted to know which were T. Kamae's priorities. The latter agreed that in the long term, depression was inevitable and consequently the project could be delayed. However, if the project is important to society, the greatest effort should be made to bring it into reality. Without any doubt, he said, regarding the question of priorities, the first choice for ATM (Asynchronous Transfer Mode) implementation will be the business sector.

F. Thompson took the floor to ask when the large reduction in user costs would be realized, recognizing that initial conversion costs to fibres were very high. T. Kamae thought that it depended on what applications would be moved to broad-band ISDN. "If all television is moved to the telephone" he said, "this will happen much sooner." He recognized, however, that this move currently encountered many legal problems.

M. Takahashi then wanted to know what kind of research was being done to clarify the needs of those who would use the new technology in the twenty-first century? The paper's author explained that at present, only engineers were working on the development of ATM-relevant technologies. "We are planning to build a prototype ATM system," he said, "to demonstrate what can be done using available technologies first." On the basis of a common knowledge of the state of the art, specialists from different fields, such as the humanities and the social sciences, will presumably be able to have fruitful discussions on the social impact of new telecommunications technologies.

With reference to the overheads shown by T. Kamae, A. Sage wanted to know what specific interpretation was associated with the term "intelligent networks." T. Kamae explained that the term had roughly two meanings. The first referred to the type developed in the United States. The second, a broader sense, referred to a general network that has some kinds of intelligence, such as a language-translation capability.

Panel Discussion 1: Achievements and Limitations in International Cooperation As Seen by the Developing Countries

Coordinator: Jacques Tocatlian

The panel, chaired by J. Tocatlian, was composed of M. Almada de Ascencio, Centro de Información Científica y Humanística, Universidad Autónoma de México, Mexico; Lian Yachun, Institute of Scientific and Technological Information of China, China; M. Lundu, The Copperbelt University, Zambia; and D. Torrijos, General Information Programme, Unesco, Regional Office, Thailand.

J. Tocatlian introduced the discussion by briefly reviewing the situation in the developing countries and identifying some of the numerous obstacles to be surmounted there in order to provide access to scientific and technical information. He invited the panelists to share experience gained in their respective countries and regions, by drawing attention to the problems, failures, impact, and achievements of international cooperation and assistance. He said that the panel was expected "to clarify the needs, problems, obstacles, and requirements of the developing countries and to try to elucidate why international cooperative programmes work or do not work." In so doing, he invited the panelists to refer to some of the questions raised in his paper, "A Critical Evaluation of Experiences and Strategies," delivered in Session 2A, and to address issues suggested by four main topics:

(1) Outstanding information needs of the developing countries
(2) The role and significance of information technologies
(3) Perception of international cooperation
(4) Recommendations for the future (mostly addressed to panel 2: Towards New Modalities of International Cooperation)

Each of the panelists then made a presentation.

M. Almada de Ascencio used the experience of Mexico and more generally that of Latin America to analyse the needs of the developing countries and illustrate how international cooperative programmes have responded to these needs.

She began by emphasizing the importance for Latin America of having one main language, which has facilitated communication and cooperation throughout the region. Cooperation with international organizations is important because it enhances visibility of the information sector, draws the attention of the higher authorities, and aids in starting projects that require certain information technologies. Also, legislation, which has to be undertaken at the national level, can be triggered by international cooperation, as shown by Unesco's PGI, which promoted the development of national information policies. She enumerated the many information-related problems of the developing countries and said that, "in Mexico, probably our most acute problem has been insufficient professional and technical staff adequately trained in the different information and library skills at the operational, management, and strategic levels."

Reverting to the role and significance of information technologies, she gave a detailed historical perspective of the issue in the region. The more developed countries began using information technologies in the early 1970s. In the 1980s, the use of CDS/ISIS became an important catalyst for database development. "It is difficult to establish how many thousands of small databases have been developed in Latin America." The first computerized, comprehensive Latin American indexes started in 1977. In 1986, the Mexican bibliographic databases were mounted on a Mexican host for national and international access. Since 1989, the indexes have been available on CD-ROM. The larger universities now have their own computer networks for communication between campuses.

"The use of information technologies," she added

> had an important impact on our academic, professional, and technical society, which is now becoming a computer-literate society, as one can find computers even in public libraries in Mexico, and in most schools children are taught to use computers. We now look towards developing an information-conscious society, but this will still take money and, especially, adequate education. Being near computer literacy does not mean that these societies are informed.

In analysing the impact of the introduction of information technologies, M. Almada de Ascencio said that "probably one of the mistakes in the mid-1970s was looking at information technologies as 'development per se' and not as 'ways and means' and as essential infrastructure that should always be integrated into the different development programmes of each sector." It is difficult to maintain the importance of STI and IT if they are not linked to the process of knowledge creation and the sectors where they are to be used. In the 1970s and 1980s, STI was considered a separate sector. Now we understand that it is a component of larger entities. "The problem is how to enhance visibility within each sector and not as IT and STI alone; how to get decision makers, especially in developing countries, to understand that the main component for development, in each sector, is STI and the use of IT."

M. Almada de Ascencio proceeded to analyse the impact of past international programmes in libraries and information in Latin America by giving a detailed account of all types of collaborative actions and underlining the advantages and weaknesses of each.

Turning to the future she made a number of recommendations.

1. International agencies that aid developing countries in starting information services should pay particular attention to the actual demands of users and the dissemination of information as important marketing components to be included from the start in the original design.

2. Donor agencies could establish inter-agency committees to aid in the negotiation and transfer of funds for projects and help in channelling funds from different sources.

3. As nations move into global economies and open markets and governments reduce their participation in different sectors, private investment becomes important. "Thus when we speak of information we should seek both public and private funding."

4. "When we are looking for international and national cooperation, the search has to include the different sectors: industrial, governmental, and academic, because that is the only way to integrate development."

5. Regarding the areas of information, informatics, and communication, there has to be a close relation between them. "I do not think that they have to merge into one area, which would be a mistake, as each one of these fields has its own specific needs and development."

6. International agencies should strengthen their links to be able to focus collectively on high-priority projects and programmes. "Only by aiding the adequate infrastructure and 'infostructure' can there be true high standard development in science and technology, health, education, industry, and business for the well-being of societies. It is by helping developing countries help themselves that we can all look towards a brighter future."

Next, Lian Yachun took as an example the Institute of Scientific and Technological Information of China (ISTIC) to illustrate how an international organization, Unesco, responded to a national need and helped establish the services, giving in his presentation specific examples of difficulties and making recommendations for the future.

The ISTIC began to set up its computerized information service system in 1984 on the basis of recommendations of Unesco consultants. During six years of development, it gradually established a computerized information retrieval network through the public telephone system and the China Public Package Switching Data Network. There are now more than 120 remote terminal users in 50 cities throughout the country that receive information retrieval services, in Chinese and in English, both domestic and international. The network is in fact connected to nine international systems. The ISTIC cooperated with Unesco to create the Chinese version of CDS/ISIS software and with a Swedish company to modify TRIP, a full-text database management system, to enable processing of Chinese characters.

Lian Yachun proceeded to cite examples of areas where international cooperation had helped, such as training, accomplished in cooperation with the UNDP, Unesco, and the British Council; in the creation of a union catalogue of Chinese scientific and technological periodicals, funded by the IDRC; in the area of norms and software, where, he said "the Common Communication Format, CCF, designed by Unesco, offered the best solution for bibliographic data interchange." He affirmed that "the development of the ISTIC's information systems would probably not have progressed to its present state without international cooperation."

Speaking of difficulties encountered, he said that it took more than three years from machine installation to the start of actual service operation – a length of time considered about one-third of the expected machine life. "The delay was due partly to administrative inefficiency and the relocation of the ISTIC's office and partly to the complexity of the IBM system." A major problem is the very low usage of the system. He continued: "If at the beginning we had adopted the strategy of leasing the machine and disk space from an existing computer centre instead of buying our own, priority could have been given to the choice of a software system and to providing effective services." He emphasized the fact that according to his experience, "the selection of software is often more important than the selection of hardware." He also thought that "more emphasis should be placed on library automation, as the retrieval system needs the support of a well-organized library automation system. . . . Since we did not pay enough attention to library management, it is often hard to locate the original documents."

Information services in China have been subsidized by the government and handled as a free public service rather than as a commodity. With budget reductions, most of these services are confronted with the problem of survival. On the other hand, "some new kinds of information services, having charged fees, become prosperous." Having cited several examples, he concluded that: (1) information services in developing countries cannot survive unless information is provided as a commodity; (2) traditional information technologies can still play an important role in developing countries; (3) in China, successful information services are not necessarily the large information centres with advanced computer equipment; and (4) priority in developing countries should be put first on providing effective services to the users in the short term, and then on building the long-term infrastructure.

One of the common problems confronted by developing countries in attempting to access internationally available information is the high cost of international telecommunication. The utilization of CD-ROM represents one way of relieving this difficulty. The other way, rather strategic, is for developing countries to harness domestic information sources and make them available internationally, for example through systems such as the INIS and INPADOC. Lian Yachun concluded his intervention by saying that international cooperation can play three roles in the developing countries: (1) a catalytic role, by bringing in international expertise when planning or designing national information systems and services; (2) technological promotion, through training or experimentation with new technologies; and (3) a promotional role, by drawing attention of higher authorities to the importance of information.

M. Lundu then took the floor and reflected on the origins of the information gap between developed and developing countries, postulating that the information gap is both a myth and a reality, depending upon one's standpoint. At the core of the tradition to characterize third world countries as underdeveloped "is, undoubtedly, the view that these countries have never possessed nor do they possess knowledge, information, ideas, and culture of value that could be transferred to or copied by other countries." Quite a number of theories have been offered to explain this situation. "From one we learn that the explanation of this problem lies in political economy. . . . From the other, emphasis is on racial differentiation. . . . Others stress economic factors, geography, or lack of innovative ideas. . . ." For the past 30 or so years, historian Basil Davidson has endeavoured to explore, from a historical perspective, what tropical Africa's most impressive achievements before European colonial powers invaded the continent. He argues, recalled M. Lundu, that "it was indeed a mistaken notion to believe that African people possessed no institutions, culture, knowledge, and even technology of their own until the white man introduced these on the continent."

Information and knowledge production and transmission from generation to generation have been part of the African way of life from time immemorial. "The unfortunate thing," said M. Lundu "was the fact that this information and knowledge were never organized in permanent records before the coming of foreigners on the continent. This meant that such knowledge and information as existed were scattered. . . ." The colonial powers in Africa decided to change everything, including the method of generating, preserving, communicating, and transferring information and knowledge. Now professionals in the third world depend to a great extent on the information and knowledge generated in the developed countries. The main problem is how to reduce the gap. M. Lundu approached the problem from three different angles: (1) the "two worlds" interdependence, especially in technology and knowledge production, utilization, and transfer; (2) the role of education and training in supporting interdependence; and (3) the need for the third world to identify its own needs and priorities.

On "two worlds interdependence," he said that,

> no doubt, during the last 300 years of contact between Europe and Africa, the North has learnt useful things from the South. Materially, the North gained more from the South during this period. Now, it is the turn of the South to technically and technologically gain from the North using the Japanese-Taiwan model that stresses human resources development as a strategy for the transfer of technology to deprived areas.

On the second angle, he recalled that education and training tend to be, in their negative forms, an effective vehicle for indoctrination. The colonial type of education introduced in the midst of unsuspecting Africans was what Farrell refers to as "static technology" – the kind of knowledge that enables the possessor to carry out certain routine tasks and functions successfully. What is needed are education and training that impart "dynamic technology" – the kind of knowledge that makes the possessor understand the scientific principles governing his work and as such enables the possessor to improve, modify, or change it to suit changing circumstances. With reference to international experts and consultants that try to help Africa, M. Lundu said that, "without proper application of dynamic technology to third world problems, I am afraid the gap will continue to widen." Foreign experts, if indeed they possess dynamic technology, should improve, modify, or change their knowledge to suit the local environment.

On the third aspect, he cited the IDRC as an example of an international organization concerned about third world needs and priorities. Before 1987, the IDRC had approached the issue of developing information systems from the point of view of its objectives as a donor agency. The IDRC Regional Office in Nairobi later recommended increased involvement of African administrators, researchers, and information specialists in the development of a relevant strategy for sub-Saharan Africa, which proved successful. In concluding his intervention, M. Lundu recommended specific objectives for the establishment of "centres for appropriate technologies" in Africa, such as the development of low-cost appropriate technologies and the adaptation of indigenous technologies; the publication and diffusion of appropriate technology material; the provision of technical assistance and education in appropriate technologies and collaboration with local crafts people and farmers as well as with national and international organizations that promote appropriate technologies for improving the quality or life of underprivileged people.

D. Torrijos took the floor to discuss the issue of obstacles impeding the development of and access to STI by developing countries and suggested a number of indicators.

She first recalled that in the information scene, there had been some dramatic changes, especially in countries that had moved up on the world's development scale. Why had these changes been so limited? She suggested it was because "of the inability of the concerned institutions and individuals to 'mainstream information'." She proposed the following as relevant indicators: (1) awareness and appreciation of the value of information as a vital national resource for development; (2) skill in accessing and using information by the general population; (3) political commitment for its sustained development; (4) systematic identification and mobilization of "key players" in STI at various levels; and (5) more active involvement of information providers in the national development process.

Wherever these indicators are substantially developed, the role of information in the development process is visibly significant.

For developing countries, to get to that point is very difficult, because of certain political realities. When resources are severely limited and socio-economic conditions are extremely difficult, short-term, instantly visible projects will get priority attention and not long-range, difficult to measure, and intangible projects, no matter how sound and logical they may be.

It is for this very reason that external assistance is very important, especially when national authorities have only a poor appreciation and understanding of the problem. The practice of tying up assistance strictly with government priorities will automatically eliminate information projects in many cases because they are not on government priority lists. This aspect should be kept in mind.

D. Torrijos further expanded on the role of regional and international cooperation, enumerating many advantages of such cooperation. She recommended that international organizations and donor agencies develop a new form of partnership with developing countries that recognized the increasing stock of competent and able human resources in developing countries. Furthermore, she emphasized the urgent need to improve the skills of information personnel in negotiating and mobilizing support and sources to finance their projects.

She then provided some further information on Unesco's PGI and various international, regional, and national activities undertaken under this programme for improving the flow of information and enhancing the capacities of countries to access and to use information.

Torrijos concluded by commenting on the question raised by J. Tocatlian in his paper, as to whether STI was still an appropriate concept for international cooperation. She first recalled that harnessing science and technology for development had been widely accepted among politicians. Consequently, STI became an excellent guinea-pig for developing various concepts in information management, handling, processing, and dissemination, without touching the more sensitive areas of communication, such as government propaganda and various types of mass media. She asserted that "it is [now] time to integrate STI into the multi-sectoral and multi-disciplinary stream of national life. . . . It is time to integrate the social, human, and economic dimensions of science and technology to give it more practical meaning for all levels and types of information users." She strongly supported the concept of the "social use of information" in J. Tocatlian's paper and spoke against the maintenance of the "purity" of sectoral information. "There is a great need to create the necessary interfaces and interconnections. This is where information technologies will find the most urgent and invaluable applications."

After these presentations by the panelists, discussion began on the issue of information transfer versus technology transfer. Zhou Chaochen recalled that access to science and technology means obtaining scientific and technical information, understanding it, and subsequently using or applying it. Information technology, on the other hand, is mainly concerned with information transfer only. How can one coordinate information transfer with technology transfer?

According to M. Lundu, information transfer and technology transfer are sometimes interchangeable. In his opinion, what is important is the transfer of information or technology that enables the user to modify, adapt, improve, or change it to suit changing circumstances – this being technical know-how.

M. Almada de Ascencio agreed that information and technology transfer should consider not only equipment but also, and above all, know-how. It should consider the adoption or adaptation of imported technologies as well as technological innovations undertaken within the country, taking full account of the local environment. She added that developing countries have to close gaps. They cannot receive information and technology that has become obsolete elsewhere or that is not suited to their own environment.

S. Robertson confirmed that technology is not about equipment but about know-how. Information technologies must to some degree be language-specific, culture-specific, and environment-specific.

C. Correa, commenting on the session papers and discussion in general, remarked that science and technology are too important to permit dealing with the problems of access to both of them under the same principles and rules. Access to science and access to technology deserve to be treated separately. He then reverted to the issue of information technologies, asking how these technologies could affect access to scientific information by developing countries.

D. Torrijos commented that information technology presents a promising possibility for improving the image of library and information services in many respects. For example, it has enabled researchers, scientists, and even policy makers to go to the library and use computer-based services. These same users would never go near traditional libraries but would send assistants instead to make literature searches for them. But the magic of being computer-literate, able to access information using information technology, is very alluring; it raises interest in and ultimately earns support for these services that provide access to science and technology. According to D. Torrijos, this aspect should be exploited to the maximum by information providers.

S. Robertson commented that there are many information technologies – not just computers and telecommunications. Perhaps the earliest information technology is writing. And there are many others that should be kept in mind in such deliberations.

The next intervention was made by W. Rouse, who thought that it was being assumed in the discussion, perhaps only implicitly, that the developed world's information technology worked well and, therefore, was appropriate for transfer to the developing countries. In fact, according to him, these technologies do not work well and the developing countries could do better by focusing on their own needs rather than on the developed countries' technologies.

D. Torrijos reiterated that information technology has an image-building aspect, which can raise awareness and appreciation of the value and use of information, and it must be so exploited to get political support for building information infrastructures and services. The goal in cooperative programmes should be to enhance the capacities of the developing countries to become sufficiently familiar with information technologies to enable them to make intelligent decisions about what to choose and thus avoid the mere mimicking of developed countries' systems. Instead, developing countries will concentrate on what they actually need.

On the general subject of new technologies, S. Robertson made the following comment: new technologies do not in general replace old ones; rather they widen the possibilities of use and applications. Often, competing technologies coexist; new technologies complement old ones. According to Robertson, only very seldom do old technologies die.

The discussion then moved to issues related to the development of information infrastructures and services in the developing countries. Lian Yachun took the floor to state that with the gradual development of science and technology in the developing countries, more and more information will be produced in these countries, for which particular infrastructures will be needed. When the value of this indigenous information is perceived, these services will become international. He continued that in view of the present limited resources, priority should be given in the developing countries to providing effective services for existing users, rather than to long-term targets.

The next intervention was made by C. Cooper, who noted that so far the discussion had been supply oriented. Very little had been said about who makes use of information systems. He said that "until we know more about information needs and requirements in the production and service sectors in the developing countries, we cannot really make suitable judgements about the significance of the evident technology gap."

M. Almada de Ascencio referred to her paper where she had mentioned that the developing countries should now shift to market-oriented, demand-based information systems and that international organizations could help countries by focusing on projects that are user oriented. The importance of the use of front-ends and intelligent computer systems is that these systems are aimed precisely at enhancing the usefulness of information systems. "All information services and products should be planned, designed, and implemented based on current and potential demand, as there is more awareness on the part of the user, even in developing countries, as to the use of information." This does not mean, concluded M. Almada de Ascencio that we have arrived at having information-conscious societies in all sectors.

C. Cooper continued with a question about whether there were information systems for specified sectors, such as industry, and who was using them.

M. Almada de Ascencio explained that in Mexico, for example, it had taken many years for users in industry to feel that they really needed to invest a fair amount of money for information. It has taken time and effort and the evolution of technologies to establish systems that will help intermediaries aid industrialists to solve their problems. "Government protectionist regulations did not induce industry to think they really needed information systems." During the economic crisis and in the opening of markets larger industries realized the importance of investing in the development of specialized information services. However, small industries cannot afford such services.

Next, N. Streitz focused on the information needs of actual researchers. He said that the availability of databases was not enough; actual researchers rarely needed them. Informal communication means, such as e-mail, were necessary. He added that one could find the state of the art in the grey literature and not in the archival publications, which report results that are two to three years old.

Replying to this statement, N. Dusoulier affirmed that scientists needed more than personal contacts and exchanges within the invisible college – which in any case is reserved for the élite. Databases and data banks are the memory of science and give to all scientists in the world the possibility of knowing about research results. Scientists from the developing countries have the same right.

G. Johannsen took the floor to draw attention to the situation regarding telecommunication in the developing world. With reference to new information technologies, he emphasized the crucial importance of good training when introducing these technologies. He said that the costs and quality of telecommunication made even the use of telephone or fax difficult or almost impossible in many countries. Therefore, on-line access to databases or computerized journals will not be practical for a long time for many. He stated that the usage of CD-ROMs will be a better solution. Johannsen pointed out that in East Germany, one very useful achievement after unification was the installation of new telecommunication equipment, which permits normal communication. The situation in many Eastern European countries and developing countries is such that the best possible assistance at this point would be the installation of state-of-the-art telecommunication systems "rather than the newest high-tech, fancy equipment." Speaking of Eastern Europe, he said that science was in a terrible state in that part of the world and that numerous scientists in all kinds of fields may be tempted to leave their countries. International projects have to be organized to promote research in these countries.

Session 3

New Technologies and Media for Information Retrieval and Transfer

Chairperson: Martha Stone

The Potential Offered by "Extended Retrieval"

Michael K. Buckland

ABSTRACT

The traditional form of information retrieval is composed of a single re-source file and a single retrieval mechanism. In the environment created by the new information technology, many resources and many computers are linked by networks. This environment requires an extension of information retrieval techniques to include retrieval from multiple files and the use of multiple retrieval mechanisms. Some benefits and technical consequences of "extended retrieval" are reviewed.

1. Introduction

The traditional form of an information retrieval system is composed of two parts: a resource file and a retrieval mechanism. A bibliographic retrieval system or an on-line library catalogue, for example, is composed of a file of bibliographic records and a retrieval mechanism designed to perform the most commonly desired searches on that file of records, such as a search by author, title, or subject. It is, in effect, a *unitary* system, a single system composed of one resource file and of one retrieval mechanism.

The new information technology is leading to a new computing environment. The cost-effectiveness of computer hardware is increasing, the cost of electronic storage is decreasing, and connectivity through telecommunications is becoming pervasive and less expensive. In the meanwhile, labour costs and building costs continue to rise. These changing conditions are resulting in a new environment in which:

- workstations are becoming widely available;
- very large sets of data can be stored economically;
- many thousands of computers are interconnected over local, national, and international networks; and
- the standards and protocols necessary for effective cooperation are being developed and adopted.

In this situation, we find a rapidly growing number of databases, an increasing use of databases, and a trend for individuals to use a number of heterogeneous databases. The result is increased complexity for the searcher and a greater need for expertise to identify what resources exist and how to use them cost effectively. (For a convenient general introduction, see Lynch and Preston [7].)

This changed information technology has created a new information retrieval environment in which the potential for information retrieval now extends far beyond the traditional form of a unitary retrieval system composed of one file and one retrieval mechanism. I use the term "extended retrieval" to denote this more general form of information retrieval. In this paper, I describe what I mean by extended retrieval and provide examples. Some technical consequences of the extension of information retrieval from a traditional, unitary form to an extended, network environment will be noted.

2. Four Information Retrieval "Architectures"

The generalization of information storage and retrieval beyond the traditional, unitary case of one file and one retrieval system to the more general model of multiple files and multiple retrieval systems can be expressed as four combinations:

1. Traditional, unitary retrieval systems with one file and one retrieval mechanism. An on-line library catalogue would be an example. A search on MELVYL, the on-line catalogue of the nine campuses of the University of California, for example, retrieves 266 records for books using the combination of subject keywords "science" and "Japan."

2. Multi-stage retrieval from a single file. An example of multi-stage retrieval would be when the results of a search by one retrieval system on a file are subjected to additional retrieval operations by a second retrieval mechanism as "post-processing." For example, at Berkeley, an experimental system known as OASIS can be used to refine the results of MELVYL searches [5]. For example, if the results of the previous example, the 266 MELVYL catalogue records for books on "science" and "Japan," are downloaded into OASIS, additional processing can identify numerous subsets defined by date, by language, and by the libraries where copies are held (see table 1).

3. A retrieval system that searches multiple files would be one in which a single retrieval mechanism can search and derive records from two or more files simultaneously. An example is the "Onesearch" feature of the DIALOG retrieval service.

Table 1 Analysis of 266 records by language, date, and campus. Search request: "Science" and "Japan"

Location	At Berkeley			At UCLA			At other campuses			
Language:	In Japanese	In English	Other	In Japanese	In English	Other	In Japanese	In English	Other	Total
1991	3	1	0	0	0	0	1	0	0	5
1987–1990	17	9	1	8	2	0	6	11	0	54
1984–1986	3	8	0	10	4	0	7	8	0	40
1978–1983	3	10	2	7	6	1	10	7	0	46
1972–1977	0	8	1	15	5	0	7	5	0	41
1963–1971	0	7	0	14	7	0	6	4	0	38
1962	1	6	0	19	10	0	3	3	0	42
Total	27	49	4	73	34	1	40	38	0	266

Table 2 Unitary and extended retrieval

Retrieval mechanisms	Number of files	
	Single file	Two or more files
One	a. Unitary retrieval e.g. on-line library catalogue	b. Mutliple file searching, e.g. DIALOG Onesearch
Two or more	b. Single file, multiple processing, e.g. postprocessing	d. Fully extended retrieval

4. Retrieval using multiple files and multiple retrieval mechanisms. The more general case is when multiple files *and* multiple retrieval mechanisms are used. This is the logical consequence of the development of the new information technology environment: It is a networked environment in which *many* different files of resources and *many* different retrieval systems exist and are, in principle, widely accessible over the network.

Note that we are assuming that these systems are heterogeneous. We are not talking about the relatively simple case of distributed database systems designed for compatible, distributed use. We do not and cannot assume that software, hardware, and data structures are standardized. We are concerned with retrieving resources that are related in their *meaning* rather than in their form, so the problems are those of information retrieval rather than data retrieval.

These four cases are summarized in table 2.

3. Illustrations of Extended Retrieval

To illustrate some of the potential of extended retrieval, let us consider two kinds of data: bibliographic data and scientific data.

3.1 Bibliographic Data

A record in a library catalogue will include author, title, location (call number), and a subject heading (e.g. from the *Library of Congress Subject Headings* list). A record in a bibliography representing the same document will likewise include author and title, but may also include an abstract and a subject heading probably from a different list, such as *Medical Subject Headings*. A citation index would include, again, the author and title and the references from the document. These overlapping contents are shown in figure 1. What we have for the same document is three quite different bibliographic descriptions by different publishers, in different formats, and ordinarily searched on different retrieval systems. These three records contain:

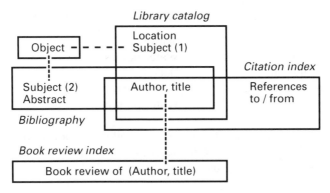

Figure 1 Related bibliographic files

- information that is the same, though possibly expressed differently and not necessarily recognizable as being the same, and
- some information not provided by the others, e.g. the catalogue has the location of a copy; the bibliography has an abstract; and the citation index shows references to and from the document.

The relationship between the records retrieved from the different databases is that they all represent the same document. But this is only one of many possible relationships. Figure 1 also shows two further relationships:

- a book review index may include a record for a different but related document, a book review; and
- outside of the bibliography and library catalogue may be some object that the book is about.

In this way, one's knowledge can be significantly increased by extending one's search to two or more heterogeneous databases. However, although the various bibliographies may refer to a single document, there is no assurance that they will do so in a consistent way. The form and contents of records in bibliographies (like references at the end of papers) vary considerably. This is not normally a problem for human beings, who can recognize what is meant, but it is a serious problem for recognition by a computer.

Differences in subject description can be substantial and significant. Consider, for example, a searcher interested in coastal pollution. A search on "coastal pollution" in the *Library of Congress Subject Headings* in the University of California MELVYL on-line catalogue yielded nothing either as a phrase ("exact subject") or as a pair of subject terms ("subject keyword search"). Nor does either form of search yield anything in the MELVYL file (1988 to date) of the MEDLINE bibliography. Nevertheless, material on coastal pollution does exist in both, and some of it can be found by searching for documents that contain the words "coastal" and "pollution" in their title. Analysis of these records shows that the subject headings actually assigned to these documents include:

LCSH *(MELVYL Catalogue)*	MeSH *(MELVYL MEDLINE)*
Marine pollution	Seawater
Coastal zone management	Water pollution
Water—Pollution	Bacteria
Petroleum industry and trade	Water microbiology
Waste disposal in oceans	Water pollutants

Not only is the plausible phrase "coastal pollution" not used in either set of subject headings, even as a cross-reference, but there is remarkably little overlap in the terms that are used.

3.2 Science Data

Consider the range of different data that could be relevant and available for studying a geographical area such as Kyoto Prefecture or the Sacramento delta:
Topographic: latitude, longitude, altitude
Political map
Satellite image
Land-use map
Gazetteer: place names
Weather: temperature, precipitation, humidity, wind
Textual documents
Census and socio-economic statistics
Photographs, etc.
Handling the retrieval of such diverse kinds of data from quite different sources is a major challenge.

4. Some Technical Issues

We use the phrase "extended retrieval" to refer to the extension of information retrieval to include search and retrieval in multiple files and/or using multiple retrieval mechanisms. In the new environment, a number of interesting problems arise and need to be resolved:

4.1 What Resources Exist?

The fact that many electronic resources exist in many places does not mean it is easy to identify or find them. Files stored on computers are just that: files that are stored. There is, as yet, little or no tradition of cataloguing computer files, so that they can be identified and found, as there is for cataloguing library books and museum objects. The task of developing "directories to the Internet" is not likely to be simple or inexpensive, but it is now receiving increasing attention. The question of identifying which resources one might wish to search is a bibliographic problem, although the describing of electronic resources is still undeveloped. However, there is also a question of which *retrieval system* to choose for any given search if there is a choice. For example, the catalogue records of the library of the Berkeley School of Library and Information Studies can be searched using four different retrieval systems. Different information retrieval systems have different retrieval capabilities. For a specialized search, one may need to select the retrieval mechanism as much as the resource to be searched. This implies a knowledge and understanding of the differing characteristics of different retrieval mechanisms available, with which resources they can be used, and how to use them, singly or in combination. This knowledge is inadequately developed, though Belkin and Croft [1] provide a useful review.

4.2 Search and Retrieve Protocols

It is now possible to access databases at remote sites as well as databases at one's local computer centre. This ordinarily requires establishing a telecommunications connection, a personal account and password, and the use of an unfamiliar command language (as in figure 2a). This is inconvenient and requires expertise. A significant new development is the creation of national (e.g. US NISO Z39:50) and international standards (ISO 10162/10163) for computer-to-computer "search and retrieve" standards. The adoption of these standards will enable one to delegate to one's local retrieval system the extension of a search to some other, different retrieval system. The Search and Retrieve standards translate searches from one retrieval command language to another [3, 4]. This development started among librarians to enable convenient access to each other's catalogues, but it has wider application. The effect is shown in figure 2b.

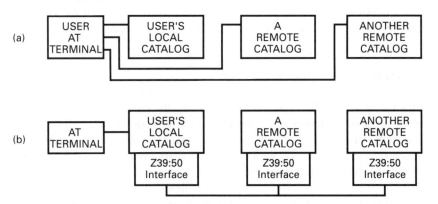

Figure 2 "Search and retrieve" (Z39:50) protocol
(a) A user can connect with various on-line catalogues and must know how to use each.
(b) With the "Search and retrieve" (Z39:50) protocol, the user need only know how to use the local catalogue and to instruct it to extend a search to other systems.

Because different retrieval systems have different capabilities, one could do more than simply extend a search to another database. For example, suppose that the on-line catalogue at library A does not support searching for individual words in titles, but that the on-line catalogue of library B does. A title keyword search desired at A could, instead, be performed on the on-line catalogue at B. The records of any books found at B could then be transferred back to A and, with the benefit of full descriptions, the catalogue at A could be searched to see if they are also held at A. Any such books also found to be held at A would provide the *effect* of a title word search – admittedly probably incomplete – even though the on-line catalogue at A did not support title word searching. The point is that specialized retrieval capabilities available on a remote machine but not available on a local machine could, within limits, be used to enhance local searching.

The idea of a "knowledge robot" or "knowbot" that could be sent off into the networks searching for and retrieving information on any specified topic has aroused interest. The essence of a "knowbot" is the idea of a conditional search command. A searcher at A might send a search command in the following form: Search in Resource B for data with attribute X. If found, retrieve it and transport it to A; if not found, extend the search to Resource C, and so on.

4.3 Questions of Relatedness

Extended retrieval among heterogeneous resources raises difficult questions of relatedness. If a name found in one database is similar to a name in another database, are they variant forms of the name of the same person? If the names look the same, could they still refer to different people? The same problem arises with records in bibliographic databases that may or may not represent the same document [2]. More generally, in extended retrieval in heterogeneous databases, one is concerned with the retrieval of related material but the nature of the relationship may be difficult to define or to determine.

4.4 Anatomy of the Retrieval Process

Retrieval in a unitary retrieval system is easily viewed as an event rather than a process. When considering information retrieval in a networked environment, one might think in terms of local "client" search machines and remote resource "servers," which implies a distinction between a search and retrieve mechanism and a resource file. This may exist when a retrieval system is built to retrieve from two or more files. But information retrieval is, in practice, a complex process including several different components. The question arises concerning how the retrieval process could be optimally divided into different stages on different machines. In fact, when one analyses the individual elements of the retrieval process, considerable complexity and choice emerges. For example, with bibliographic searches:

– different bibliographies represent different, more or less overlapping, populations of documents;
– different bibliographies will have more or less different descriptions, even for the same document;
– the access points ("indexes") that can be searched vary between systems;
– there can be more or less cross-referencing between different index terms ("see," "see also," and other kinds of syndetic relationships); and
– different retrieval systems support different types of searching (matching, comparing), even in the same bibliographic files. Some allow searches for keywords and/or composite Boolean search requests and others do not.

So, correspondingly, one can immediately identify five different classes of reasons for extending searches to two or more on-line bibliographies and/or on-line catalogues. Depending on the circumstances, different options might be chosen:

1. Because different bibliographies represent different populations of documents, one may want to extend a search to another bibliography or catalogue because what was desired was not found in what had already been searched and it would be desirable to extend the search to a new population of documents.

2. Because different bibliographies contain different descriptions for the same document, one may want to extend a search to another bibliography or catalogue for a document that has already been found because differing bibliographic descriptions can be used to accumulate additional information. As noted above, a book might be present in a library catalogue, in a subject bibliography, and in a citation index. All three will have more or less differing bibliographic records for the same document: The catalogue will have a standard catalogue record and a note of the location of each copy; the medical bibliography may contribute a different, detailed subject description and an abstract; and a citation index might contribute a list of other works cited in it and another list of works that cite it. *Combining* these descriptions could improve bibliographic access substantially.

3. For a more complete search, one may want to extend a search to two or more other systems in order to use additional access points. Citations have to be searched in a citation index. The ability to search on other features, such as searching by words within a title or within an abstract, or by the language or date of the document, varies significantly between systems.

4. Because of the complexity and vagueness of language, one may prefer using the system that has the best network of cross-references, the best "vocabulary control," to guide the searcher from the searcher's terms (e.g. "coastal pollution") to the system's terms.

5. It may be worthwhile to extend a search to another system because it has special searching abilities, such as identifying pairs of words that occur close to each other, or to extend the search by downloading results into a personal computer for more detailed analysis ("postprocessing").

There are other possibilities. For example, the extent to which texts are subdivided into fields can affect retrieval performance [6].

We can observe that information retrieval theory remains significantly incomplete, even for unitary retrieval systems, until the effects of changes in any one or more of these variables on retrieval outcomes are properly understood.

5. Conclusion

Automating a catalogue, placing a bibliography on-line, or providing access to any other electronic resource on-line is a substantial technological development. But to think only of an individual on-line catalogue, of an individual on-line bibliography, or of any other *individual* resource – even being aware that there are several different on-line catalogues, numerous individual bibliographies available on-line, and many other resources on-line – is to think in terms of the card catalogues, published paper bibliographies, and the unitary retrieval systems of the past. Instead of thinking of individual retrieval systems, we should now base our thinking on the awareness that there are large and growing populations of electronic resources and of retrieval mechanisms, increasingly connected by telecommunications networks, and containing data sets that can, in principle, be linked, combined, and rearranged. What could happen if, instead of thinking of information retrieval in traditional, unitary terms, as when using *a bibliography* on-line, we were to follow the logic of electronic technology one step further and think instead of *a collectivity of bibliographies* on-line? We need to think in terms of using a whole electronic reference library, even multiple libraries, on-line.

Extended retrieval provides much wider opportunities, but it also increases the difficulties in selecting what is needed from so large and complex a universe.

REFERENCES

1. Belkin, N.J., and W.B. Croft (1987). "Retrieval Techniques." *Annual Review of Information Science and Technology* 22: 109–145.
2. Buckland, M.K., A. Hindle, and P.M. Walker (1975). "Methodological Problems in Assessing the Overlap between Bibliographic Files and Library Holdings." *Information Processing and Management* 11: 89–105.
3. Buckland, M.K., and A. Lynch (1987). "The Linked Systems Protocol and the Future of Bibliographic Networks and Systems." *Information Technology and Libraries* 6: 83–88.
4. Buckland, M.K., and C.A. Lynch (1988). "National and International Implications of the Linked Systems Protocol for Online Bibliographic Systems." *Cataloging and Classification Quarterly* 8: 15–33.
5. Buckland, M.K., B.A. Norgard, and C. Plaunt (1992). "Design for an Adaptive

Library Catalog." In: *Networks, Telecommunications, and the Networked Revolution: Proceedings of the ASIS 1992 Mid-Year Meeting May 27–30, 1992*. Silver Springs, Md.: American Society for Information Science, pp. 165–171.

6. Lynch, A. (1992) "Online Searching on the Internet: The Challenge of Information Semantics for Networked Information." In: *Proceedings of the National Online Meeting, New York, May 5–7, 1992*. Medford, NJ: Learned Information. Forthcoming.

7. Lynch, A., and C.M. Preston (1990). "Internet Access to Information Resources." *Annual Review of Information Science and Technology* 25: 263–312.

8. Stonebraker, M., and J. Dozier (1991). *Large Capacity Object Servers to Support Global Change Research*. SEQUOIA 2000 Report 91/1. Berkeley, Calif.: University of California, Electronic Research Laboratory.

Information Retrieval:
Theory, Experiment, and Operational Systems

Stephen E. Robertson

ABSTRACT

The paper examines the process of scientific communication resulting from users' expressed needs for information and in particular the formal mechanisms for the storage and retrieval of information in response to queries or requests. Formal indexing or coding schemes, Boolean systems, facet analysis, and associative methods, as well as probabilistic models, are reviewed and information-seeking behaviour is discussed.

1. Scientific Communication and Information Retrieval

It is a commonplace that science depends on communication. Science is a social activity; scientists' ideas, models, and results have to be scrutinized by their peers, analysed and tested with the possibility of validation or refutation as well as the construction of further science.

In order to investigate the role of information retrieval and the effect of developing technologies on scientific communication, we may start from a simplified view of the process of scientific communication (see chart p. 145).

Such a diagram is deceptive both in its simplicity and in its circularity. As a publishing scientist, I am clearly not communicating with myself! My potential audience will be not only other scientists (who may indeed feed new publications into the process), but also other users of scientific information, e.g. those who apply the knowledge. The diagram also suggests a system with just one communication channel, and furthermore seems to imply a system that always works! Neither is in general the case.

144

```
     (reading)                      (authorship)
  +-->-------Scientist ------>--+
  I                               I
  Retrieved                      Manuscript
     item                             I
     I   Database            Journal  I
  +-<- record----<----paper -<-+
  (searching)       (indexing)      (publication)
```

The process of scientific communication

We in the information world tend to work with systems (fragments or sub-systems of this larger process) that are supposed to contribute to the whole by providing certain linking mechanisms. By and large we work with rel-atively formal mechanisms (publication, libraries, databases, etc.); we like to think that they are vital to the whole. However, we also know that scien-tists rely extensively on less formal mechanisms (personal contacts, meet-ings, etc.). Furthermore, from the scientist's-eye view, there are many sources/channels of information that may be selected or rejected at different times, for all sorts of reasons. One of our concerns in developing a science of information must be the scientist's perception of the information environ-ment, and the selection and use made of sources, channels, and modes of both obtaining information and communicating ideas.

The concern in the present paper is mainly with formal mechanisms for the storage and retrieval of information, in response to queries or requests. But the wider aspects of the communication process will be kept in view. I start with a perspective on the situation that gives rise to the request; at the end, I will return to some features of human information-seeking behaviour.

2. Anomalous States of Knowledge

We must first of all ask the question, Why does a user (scientist) approach an information retrieval system? The simple answer must be because of a need or wish or imagined need for information. However, the user's percep-tion of this information need deserves some exploration.

In Taylor's classic paper, "Question-Negotiation and Information Seeking in Libraries" [14], four stages are identified:

(1) the visceral need (i.e, the user's gut feeling of a need for information);
(2) the verbalized need (the user's first attempt to put the information re-quirement into words);
(3) the formalized need (the user's expression of the requirement in terms acceptable to the system);
(4) the compromised need (the user's revised expression of the requirement after negotiation with the system).

The last two stages relate to the user's interaction with the system, which is discussed later.

Belkin has further analysed the origins of the visceral need. A user has a state of knowledge of the world, an internal knowledge structure of great complexity. The perception of an information need arises from a perceived problem with some part of this knowledge structure (which may not be a simple gap but some internal inconsistency, conflict with evidence, or whatever). Belkin has called this the "anomalous state of knowledge," or ASK [3].

The ASK hypothesis potentially has strong consequences for the design of information retrieval systems. Most systems in effect demand that the users specify the piece of information that they require, and aim to provide the items that fit the specification. The ASK hypothesis suggests instead a problem-solving approach, where the system cooperates with the user in an attempt to solve the perceived problem (or resolve the anomaly).

Although some of the ideas discussed below predate the ASK hypothesis and involve a rather more traditional approach to IR system design, the ASK idea will inform my discussion throughout. Something like the problem-solving approach will recur in later sections.

3. Relevance

One of the most important concepts for an understanding of our present ideas of information retrieval is that of relevance. Relevance is an extremely difficult idea to pin down and has been the subject of much work over a large number of years [12]. It originates from well before the ASK hypothesis, and one way to think of it might be in terms of correctness. An item might be regarded as relevant to a request if it is in some sense a "correct" response to that request. There are indeed IR theorists who take a modern version of that view: an item is relevant to a request if the request can be inferred from the item, rather in the fashion of theorem proving, but with an appropriate logic [15].

However, probably the dominant view of relevance (and one that is rather more in sympathy with the ASK hypothesis) is a much more subjective one. An extreme version of the subjective approach would be to say simply that an item is relevant to a user's information need or ASK if the user says it is, or in other words if the user would like the system to retrieve this item. More commonly, in our experiments we rely where we can on end-users making relevance judgements in relation to their perceived needs, according to some descriptive scale that we devise; we also do some laboratory experiments with expert judges judging relevance to stated requests.

Even without agreement as to what exactly relevance is, the idea of relevance is of central importance to theory and experiment in IR, and it is becoming important to IR practice as well. From the experimental point of view, where the concept originated, we need it in order to evaluate different approaches and methods in IR. From there it has fed into theory; many recent theories in IR depend upon it. Finally, some methods of IR ask the

user to make relevance judgements on-line and use that information internally as one kind of clue to help formulate a new search.

The major assumption is that users *can* make relevance judgements or
recognize relevant items, even if not necessarily with absolute confidence. It
is hard to imagine an information-seeking activity where the user was in
principle unable to assess whether an item is appropriate or not, though of
course users may suspend judgement or change their minds in particular
cases, perhaps until or after they have more information about the item or
have read some other item.

4. Early Experiments in IR

The idea of the experimental evaluation of IR systems is central to both
theory and operational system development. Perhaps surprisingly, this idea
is only about 35 years old. (Admittedly, 35 years is a long time in the history
of *computer-based* IR systems; but some kinds of IR system, such as library
classification schemes, predate computers by at least two-and-a-half mil-
lennia!)

We are not so much concerned here with whether the system works in a
technical or physical sense as with something that might now be described as
its cognitive functioning. In other words, the question as to whether the system will succeed in locating items with specific characteristics (words, codes)
is not generally at issue. The question is, Do those items with the specific
characteristics actually serve the information need or resolve the ASK? This
will depend, in general, on the ways the system offers of specifying characteristics. In the earliest experiments, the question took the form: Does the
system retrieve the "correct" answer in response to a query? Setting aside
the problem of ASKs and relevance, the implied model of the IR system
was what might be described as an input-output model – feed in the query,
get out the answer. It was a model that fit well with the early computer-based
systems. (Actually, they would very likely be human-assisted; the searcher
would send the query off to a library or information centre, where an expert
would formulate it in system terms, run the search, and return the results.)
In retrospect, however, it seems like a temporary aberration. Both older
systems (card catalogues, printed indexes, etc.) and newer ones (highly interactive on-line systems) exhibit characteristics that do not fit too well with
the input-output model, particularly if the searcher is the end-user, that is
the person needing the information.

The early experiments told us a little about the design of IR systems, but
they also focused attention in certain areas, and it may be argued that their
lasting influence lies in this focusing process. One area is the one already
mentioned, that of relevance: the necessity to devise an operational definition of a "correct" answer was a major stimulus to the reconsideration of the
notion of relevance. A second kind of focus was on the particular aspects of
IR system design that seemed important. A number of such aspects that had

been endlessly debated in the 1940s and 1950s now seemed to be of relatively minor importance; by contrast, some aspects that had received little consideration now became central. One of the later outcomes of this process, as we shall see, has been the concern with highly interactive systems.

5. Language

Received wisdom in the 1950s was that IR required some kind of formal indexing or coding scheme. (Exactly *what* kind of scheme was one of the topics of endless debate.) Thus, items required indexing/coding in terms of the scheme, which probably in turn required a human indexer (though a machine might be taught to do it). A similar process was required at the search stage, in respect of the query or need, though an end-user might possibly learn enough about the formal scheme to conduct a satisfactory search.

The results of the early experiments, together with the developing technology and changing perceptions of how it might be used, caused a backlash against this received wisdom. It became feasible to throw into the computer larger and larger quantities of text, and retrieve on the basis of words in text rather than assigned keywords or codes. At first sight, the necessity for any kind of indexing scheme, at either end of the process, seems to disappear: the user can use "natural" language to search a "natural" language database, without any interference from librarians.

We have since come to a much more balanced view of language, though the debate continues to generate new questions as we develop our highly interactive systems. It is clear that natural language searching is a powerful device that can often produce good results economically. However, it places a large burden on the searcher, and besides, certain kinds of queries are not well served. In recognition of these points, many modern databases include both formal indexing and searchable natural language text.

Formal artificial languages (in which category I include library classification schemes) represent particular views of the structure and organization of knowledge. One idea that emerged from the analysis of such languages, and that is central to modern indexing languages as well as to the practice of searching, is that of the facet. Once it is recognized that topics and problem areas are potentially highly complex, it becomes essential to approach the problem of describing them via different aspects, or facets, and combining the resulting descriptions in a building-block fashion [8]. (The idea of a faceted classification scheme, while originally due to Ranganathan in the 1930s, was put in its most concise form by Vickery; B.C. Vickery, *Faceted Classification*, London: Aslib, 1960.)

Many modern indexing languages, while not necessarily following the rules of faceted classification, reflect an essentially facet-based approach to the organization of knowledge. But the approach also has value at the searching stage, whether or not the database being searched is indexed by such a language. This theme is taken up again below.

6. Boolean Logic, Search Strategy, and Intermediaries

Long before either natural language searching or faceted classification, and certainly long before the modern computer, it became apparent that certain kinds of information retrieval would benefit from the ability to search on combinations of characteristics, in a way that we would now normally represent by the Boolean operator AND. Indeed, much effort went into devising mechanisms to allow such searching. Prime examples were Hollerith's mechanically sorted punched cards of the 1890s and a scheme of optical coincidence cards first invented in 1915 [13].

Most currently available computer-based systems allow searching using Boolean logical search statements, together with a few extensions to Boolean logic appropriate to searching text; for example, an operator to indicate that two words should not only occur in the same record, but also that they should be next to each other. In this respect, they appear to differ little from the punched-card systems of the 1930s. However, we may point to two major differences. Firstly, as discussed above, we have the possibility of searching natural language text. Secondly, the systems are designed specifically to allow and encourage certain kinds of feedback during the search. In other words, it is not expected that a searcher will be able to specify, precisely and a priori, the characteristics of the desired item(s). Rather, the search is expected to proceed in an iterative fashion, with the results of one (partial) search statement serving to inform the next search.

Feedback in Boolean systems is of a very limited kind (this point will be taken up again below). Furthermore, the use of Boolean logic seems to suggest an analogy with traditional database management systems, where feedback is not normally an issue. However, the use of even a crude form of feedback is a recognition of the cognitive problems discussed earlier and a departure from the simple input-output model of information retrieval.

The problem of formulating a search and developing a search strategy for a Boolean system has received a great deal of attention [2]. This work has been informed by theoretical developments (the ASK hypothesis and facet analysis) and by the results of experiments. But one would not describe such work as theoretical or experimental so much as a codification of good practice. Searching such systems is best described as a skilled task. For this reason, it has not been the norm for scientists themselves to conduct their own searches. This does not mean, of course, that it does not happen. The companies offering search services on large text databases have been predicting for many years the dominance of end-user searching and the demise of the search intermediary. However, the intermediary has signally failed to disappear! (An intermediary is normally a specialist in the art of searching large text databases, perhaps but not necessarily with some subject knowledge.)

The impact of experimental work on the study of search strategy has not been very great. However, there was one experimental result from the 1960s that encapsulated, in a surprising way, the problems subsequently addressed by the ASK hypothesis and that was seminal in our understanding of the search strategy problem. It is therefore worth describing this result.

The experiment involved the Medlars Demand Search Service and the National Library of Medicine (NLM) in the United States [6]. At that time, in order to conduct a search on a medical topic, the user would have to communicate with the NLM, either directly or via a local expert. Local experts existed at various places around the world to help users make the best use of the service.

The requests that were collected for the experiment could be divided into those where the user talked face-to-face with an expert and those where the user wrote a letter requesting a search. The expectation was that the face-to-face communication would be beneficial to the search by enabling the development of a better search formulation. However, in the event, the reverse was the case: on average, the letter-based requests performed slightly better than the face-to-face ones.

The experimenters' explanation of this result, after studying the data, was to suggest that users often came to face-to-face meetings without clearly verbalized requests, and that intermediaries tended to suggest formulations that were easy in system terms. The letter writers, on the other hand, were forced to articulate their needs more systematically before encountering the constraints of the system.

This result made a clear link between the Taylor/Belkin view of information retrieval and the practical problems of searching and had a major impact on the training of search intermediaries.

The relation between Boolean searching and facet analysis is a simple one: an analysis of the search topic from a facet viewpoint fits naturally with a canonical form of search statement, as follows:

$$(A_1 \text{ or } A_2 \text{ or } \ldots) \text{ and } (B_1 \text{ or } B_2 \text{ or } \ldots) \text{ and } \ldots$$

Here the separate facets (or component concepts of the topic) are A, B, . . . , and each search term A_i is one of the ways of representing concept A. A number of "intelligent" front-end systems for helping users to construct searches assume this canonical form. It is, however, an extremely limited form if interpreted literally: for example, it fails to allow that one of the A_is might itself be a phrase or a combination of concepts.

7. Associative Methods

Experimenters and theorists in IR have been working for many years with alternatives to Boolean search statements. In particular, they have been using what might be described as "associative" methods, where the retrieved documents may not match exactly the search statement but may be allowed to match approximately. For example, the search statement may consist of a list of desirable characteristics, but the system may present as possibly useful items that lack some of the characteristics. Then the output of the system may be a ranked list, where the items at the top of the list are those that match best in some sense, but the list would include items that match less well.

Until relatively recently, this work had had little impact on operational systems. Most such systems continued to use Boolean (or extended Boolean) search logic. However, some recent systems have adopted some associative retrieval ideas. For reasons that will become apparent, I believe that associative retrieval offers far better possibilities for systems that genuinely help end-users to resolve information problems or ASKs. Therefore I welcome this development and indeed see it as long overdue.

There is a wide range of possible approaches to the problem of providing associative retrieval. This section will give a very brief overview of some of the approaches, and the following section will look at one particular model in somewhat more detail.

The associative approach with the most substantial history of development is the vector space model of Salton and others [11]. In this model, the documents and queries are seen as points in a vector space: retrieval involves finding the nearest document points to the query point. This model leads naturally to various associative-retrieval ideas, such as ranking, document clustering, relevance feedback, etc. It has been the basis of a number of experimental systems from the early 1960s on, and many different ideas have been incorporated at different times and subjected to experimental test. Mostly these tests have been along the lines described in section 4 above, with the system treated in input-output fashion.

A second approach is suggested by Zadeh's Fuzzy Set theory. There have been a few attempts to apply fuzzy set theory to information retrieval, though it has not received nearly as much attention as the vector-space model [19]. (This reference describes the original fuzzy set theory. Much related work has occurred since in, for example, fuzzy logic or fuzzy decision-making. An example of an application in IR is: W.M. Sachs, "An Approach to Associative Retrieval through the Theory of Fuzzy Sets," *Journal of the American Society for Information Science*, 27: 85–87.) The main attraction for the IR application is that it seems to present the possibility of combining associative ideas with Boolean logic, although there are actually some serious theoretical problems in that combination [4]. There is a conspicuous lack of any attempt to evaluate fuzzy set theory-based systems.

A third approach is that based on statistical (probabilistic) models. Although statistical ideas have been around in IR for a very long time, most such work nowadays is based on a specific probabilistic approach, which attempts to assess the probability that a given item will be found relevant by the user. In this sense it belongs firmly with the evaluation tradition discussed in section 4 and with the ideas of relevance that emerged from that tradition, although it turns out to fit very naturally with more recent ideas of highly interactive systems. The probabilistic approach is discussed in more detail in the next section.

It is not strictly necessary to regard these three approaches as incompatible. It is possible to devise methods that make use of ideas based on more than one approach. However, they do suggest very different conceptions of the notion of degree of match between documents and queries.

8. Probabilistic Models

Once it is assumed that the function of an IR system is to retrieve items that the user would judge relevant to his/her information need or ASK, then it becomes apparent that this is essentially a prediction process. These judgements of relevance have not yet happened. Or rather, if any items *have* been seen and judged for relevance, then those items are no longer of interest from the retrieval point of view because the user already knows about them. The system must in some fashion predict the likely outcome of the process in respect of any particular item should it present that item to the user. On the assumption that relevance is a binary property (the user would like to be informed of the existence of this item, or not), the prediction becomes a process of estimating the probability of relevance of each item and of ranking the items in order of this probability [9].

Translating this idea into a practical system depends on making assumptions about the kinds of information that the system may have on which to estimate the probability and how this information is structured. A very simple search-term weighting scheme, collection frequency weighting, seems to derive its power from being an approximation to a probabilistic function [5]. But more complex techniques may depend on the system learning from known judgements of relevance, either by the current user in respect of the current query or by other users in the past. The latter prossibility has not yet, to my knowledge, been put into effect in any operational context, but the former is the basis for more than one operational system.

This is the idea of relevance feedback: after an initial search, the user is asked to provide relevance judgements on some or all of the items retrieved, and the system uses this information for a subsequent iteration of the search. Once again, the idea of relevance feedback is not exclusive to the probabilistic framework but fits very naturally within it. Indeed, the idea was first demonstrated in the context of the vector-space model.

Relevance feedback information can be used by the system partly to re-estimate the weights of the search terms originally used, but mainly to suggest to the system new terms that might usefully form part of the query. These new terms can again be weighted automatically, and might then be used automatically or presented to the user for evaluation. Thus, on iteration the search statement may not only be imprecise, it may also be actually invisible to the user. The system can locate items that the user might want to see on the basis of criteria of which the user is not aware.

Although relevance feedback seems at first glance to be not too far removed from the input-output model (being an explicit form of feedback within the same framework), and also seems to embody a relatively mechanical notion of relevance, its implications are actually revolutionary. We begin to perceive the user not as feeding in a question and getting out an answer, but as exploring a country that is only partially known and where any clue as to location in relation to where the user wants to be should be seized upon. This concept of retrieval is explored further in the next section.

An example of a system that incorporates relevance feedback is OKAPI [17]. Although an experimental system, it functions in an operational environment, with a real database of realistic size and real users, in order to allow a variety of evaluation methods to be applied. Some results of a recent experiment using OKAPI will help to inform the next section.

9. Information-seeking Behaviour

A recent experiment investigated various aspects of searching or information-seeking behaviour, including the behaviour of repeated users of the system [18]. The system was accessible over the network in an academic environment and was available to many users through terminals on their desks or very close by. There was no direct cost to using the system, and since it is very easy to use (being designed so that someone walking in off the street could be expected to be able to use it), there was no barrier of any kind to its repeated and frequent use. Individual users were logged.

What was found was that a number of users made repeated use of the system, quite often (surprisingly) starting with a query that was very similar to or even absolutely identical to their previous query. It is clear that they were not simply asking the same question again, but rather using the entry point that they already knew about as a way into this somewhat unfamiliar country, a familiar starting point for a new exploration. Relevance feedback (which is just one of the mechanisms of which they might make use) is not a matter of saying "this is correct," but rather of saying, "supposing we try this direction, where will it take us?"

Thus it seems that starting with a theoretical approach based on a traditional, input-output model of IR has led us to methods and techniques that fit very well with the ASK hypothesis and a problem-solving or exploratory view of IR. We have arrived at the right answer, but for the wrong reasons!

There are, of course, researchers in the field who are entitled to say "I told you so!" Examples include Oddy's THOMAS system and Swanson's view of retrieval as a trial-and-error process [7]. However, we do now have evidence that we are capable of providing information retrieval systems that can have a genuine impact on information-seeking behaviour in a broad sense. One task that faces us is to develop our methods and ideas of evaluation to take into account this broader view. We, researchers in information retrieval, need to know much more about how users (including scientists and technologists) approach their information-seeking or problem-solving tasks, preferably over a period of time rather than simply in response to a suddenly perceived information need [10].

Indeed, I have found it instructive now to revisit some work that was (when it was undertaken) right outside the field of information retrieval: T.J. Allen's work on communication in science and engineering [1]. What is critical here is the user's perception of his or her information environment and the sources and channels of communication that are open. One of Allen's conclusions concerned the relative importance of informal as against formal channels. The more we can design systems that appear to the user to be less formal, perhaps the better we shall be able to serve him or her. An information retrieval system should be as accessible and as easy to communicate with as a colleague in the next office; only then will the real breakthrough occur.

10. Intelligence

It may be noted that I have not yet mentioned any of the work in the artificial intelligence (AI), expert system or knowledge-based system (KBS) areas. There have indeed been many attempts to apply such ideas to information retrieval, though there is in my view less evidence for their effect or effectiveness in the context of operational systems.

The possible role(s) for knowledge bases in IR is the subject of much debate. One approach is to treat the expert intermediary as the source of knowledge, in other words to try to encapsulate the intermediary's skill in a system [16]. However, a major component of the intermediary's expertise, at least as represented in such systems, seems to be the manipulation of Boolean search statements. If we can get by without such statements, then much of the point of these systems seems to be lost.

The other kind of knowledge that, in principle, should be of use would be that embodied in a thesaurus, classification scheme, or other formalized indexing language. But such knowledge does not seem to fit very easily with established KBS ideas.

My own opinion, for what it's worth, is that the way forward may be to incorporate selective and small-scale "intelligent" (or moderately clever) methods into the associative retrieval framework, without attempting to go all the way to an intelligent system. Cleverness need not take the form expected in the current KBS tradition: a relevance feedback system based on the probabilistic model already seems quite clever to the user. Perhaps the central point is that we are attempting to provide tools to help the user solve his or her own problems; we are not attempting to solve their problems for them. Relatively simple tools may be best suited to that purpose.

REFERENCES

1. Allen, T.J. (1968). "Organizational Aspects of Information Flow in Technology." *Aslib Proceedings* 20: 433–454.
2. Bates, M. (1987). "How to Use Information Search Tactics Online." *Online* 11: 47–54.
3. .Belkin, N.J. (1980). "Anomalous States of Knowledge as the Basis for Information Retrieval." *Canadian Journal of Information Science* 5: 133–143.
4. Buell, D.A. (1985) "A Problem in Information Retrieval with Fuzzy Sets." *Journal of the American Society for Information Science* 36: 398–401.
5. Croft, W.B., and D.J. Harper (1979). "Using Probabilistic Models of Document Retrieval without Relevance Information." *Journal of Documentation* 35: 285–295.
6. Lancaster, F.W. (1968). "Evaluation of the Medlars Demand Search Service." Bethesda, Md.: National Library of Medicine.
7. Oddy, R.N. (1977). "Information Retrieval through Man-Machine Dialogue." *Journal of Documentation* 33: 1–14; D. Swanson, "Information Retrieval as a Trial-and-Error Process." *Library Quarterly* 47: 128–148.
8. Ranganathan, S.R. (1937). *Prolegomena to Library Classification*. Madras: Library Association. 2nd ed. London: Library Association, 1975.
9. Robertson, S.E. (1977). "The Probability Ranking Principle in IR." *Journal of Documentation* 33: 294–304.
10. Robertson, S.E., and M.M. Hancock-Beaulieu (1992). "On the Evaluation of IR Systems." *Information Processing and Management*. Forthcoming.
11. Salton, G. (1971). *The SMART Retrieval System: Experiments in Automatic Document Processing*. Englewood Cliffs, N.J.: Prentice Hall.
12. Schamber, L., M.B. Eisenberg, and M.S. Nilan (1990). "A Re-examination of Relevance: Toward a Dynamic, Situational Relevance." *Information Processing and Management* 26: 755–776.
13. Taylor, H. (1915). "Selective Devices." U.S. Patent no. 1165465.
14. Taylor, R.S. (1968). "Question-Negotiation and Information-seeking in Libraries." *College and Research Libraries* 29: 178–194.
15. Rijsbergen, C.J. (1989). "Towards an Information Logic." *Proceedings of the Twelfth ACM SIGIR Conference on Research and Development in Information Retrieval*, 77–86.
16. Vickery, A., H.M. Brooks, B. Robinson, and B.C. Vickery (1987). "A Reference and Referral System Using Expert System Techniques." *Journal of Documentation* 43: 1–23.
17. Walker, S., and R. DeVere (1990). *Improving Subject Retrieval in Online Catalogues, 2: Relevance Feedback and Query Expansion*. London: British Library.
18. Walker, S., and M. Hancock-Beaulieu (1991). *OKAPI at City: An Evaluation Facility for Interactive IR*. BL Report no. 6056. London: British Library.
19. Zadeh, L.A. (1965). "Fuzzy Sets." *Information and Control* 8: 338–353.

Computerized Front-ends in Retrieval Systems

Linda C. Smith

ABSTRACT

The paper explores the role of expert systems and gateway software in retrieval systems as aids to individual researchers. Covered are the definition of front-ends, taxonomy of front-ends, examples of and an evaluation of existing front-ends, directions for future research and development, and the implications for developing countries.

1. Introduction: The Information Environment

Information technology – the set of computer and telecommunications technologies that makes possible computation, communication, and the storage and retrieval of information – has changed the conduct of scientific, engineering, and clinical research [24].

While information technology offers "the prospect of new ways of finding, understanding, storing, and communicating information"[24], there are a number of barriers to realizing that potential. The number, diversity, and scatter of information sources and systems now available in electronic form challenge the ability of individual researchers to make effective use of these resources. This paper explores the role of computerized front-ends in retrieval systems as aids to individual researchers seeking to locate information.

The complexity of the electronic information environment is well known.

156

Databases now may contain bibliographic records, full texts, numeric data, images, or combinations of these types. The retrieval systems on which they are mounted are diverse, with differing interfaces and means of query formulation. The interconnectivity of telecommunications networks internationally provides the potential for interactive access to retrieval systems throughout the world. While most of the early front-ends were developed as aids to accessing a small number of commercial systems and the databases mounted on such systems, these resources now represent only a fraction of the resources of potential value to the researcher. Many universities and government agencies now provide public access to on-line catalogues, campus-wide information systems, and other electronic data repositories that supplement the resources provided by commercial vendors. While access to commercial systems is billed and restricted to those who are recognized users, many of the other resources on the network do not currently have such access restrictions.

Within the scope of this paper it is not possible to review in detail the many front-ends already developed. The reader is referred to two recent reviews by Drenth, Morris, and Tseng [10] and by Efthimiadis [12] for extensive lists of references to prior work. This paper provides a framework for understanding the potential roles of front-ends in retrieval systems by exploring: definition of front-ends, taxonomy of front-ends, examples of front-ends, evaluation of front-ends, directions for research and development, and implications for developing countries.

2. Definition of Front-ends in Retrieval Systems

Turning first to dictionaries for aid in understanding what is meant by "front-ends in retrieval systems," one finds the following definitions in the *Macmillan Dictionary of Information Technology* [21]:

> Front-end system. Synonymous with Intermediary system. (p. 227)
> Intermediary system. In online information retrieval, a kind of expert system used to assist end users searching online databases. Such systems offer assistance with query definition, database selection, search strategy formulation and search revision. *Compare* Gateway software. Synonymous with Front-end system, Intelligent intermediary system.
> Gateway software. In online information retrieval, dedicated communications software that acts as an interface between end users and online databases. The software typically offers automatic dialing and logon, offline search formulation, downloading and may also allow text or data processing. *Compare* Intermediary system.

While these definitions attempt to clearly distinguish between software providing assistance with communications (gateway software) and software providing expert assistance in query processing (front-end systems, intermediary systems, intelligent intermediary systems), in practice the terms have not been used consistently, and a single system may perform multiple functions. For the purposes of this paper, the term "front-end" will be used to encompass all forms of assistance "which in some way, and to some degree, make the differentness or difficulty of database use *transparent* to the user" [40]. This assistance can range from the purely clerical, such as automatic dial-up, to fully intellectual, such as selecting the best way to modify an unsuccessful search query. The front-end controls what the user can request and how, as well as what the user is given from the external system and the way it is presented.

3. Taxonomy of Front-ends

In order to characterize what has been accomplished in the development of front-ends to date as well as to suggest new directions for research and development, it will be helpful to provide a taxonomy by which front-ends can be classified. A number of dimensions are useful in distinguishing among the efforts to provide computer-based assistance to users of retrieval systems.

3.1 Resources Accessible via the Front-end

The front-end is the user's "window" on the world of information available on-line. The front-end may be highly specialized, intended to help the user in answering a particular type of question. In this case, it may link to a single database on a single system and focus on only a part of its contents (e.g., locating literature on particular cancer therapies in a database like Medline, which covers many other aspects of clinical medicine as well). A slightly more general front-end would support full use of a single database on a single system. Increased accessibility would be provided by front-ends linking to multiple databases on a single system, and finally to multiple databases on multiple systems. In this context, it is also important to note the types of resources supported: bibliographic, full text, numeric, and/or image. Recognizing the value of networks to support communication between individuals as one important type of information resource, front-ends may also assist users in navigating directories of other people accessible on the networks.

3.2 Location of the Front-end

"Front-end" implies that the software is located somewhere between the user and the system being accessed, but there are multiple possible locations. Front-ends may reside on the user's workstation, on another computer on the network, or on the host system itself. Meadow [22] has provided a detailed analysis of the trade-offs in the location of a front-end. Front-ends on the user's workstation have the advantages of reducing costs by allowing local editing of queries, supporting graphic displays [25], and performing functions such as data analysis that may not be available in the host system. However, it may be difficult to update such front-ends to reflect changes in the resources to which they are providing access. While front-ends on the host system can be maintained by the managers of the host, they may be a more costly mode of access for the user than using that same host with its "native" command language. They also limit the user to databases available on that particular host. Front-ends resident on the network computers can simplify access to multiple host systems. Use of such a front-end may add to the cost of performing a search, and the user is dependent on the developer of the front-end to accommodate any changes in the host systems being accessed.

3.3 Types of Assistance Provided by the Front-end

The definition of "intermediary system" given above indicates that front-ends could offer assistance in query definition (concept identification), data-base selection (identification of databases to be searched and the systems where they reside), search strategy formulation (selection of subject terms, selection of other types of terms, selection of logical operators), and search revision (broadening, narrowing, or other changes). In practice, not all of these forms of assistance may be provided by a single front-end, which may instead focus on a particular step in the process, such as database selection. Going beyond information retrieval, a front-end may also support post-processing functions, ranging from formatting records for easier reading to performing statistical analyses on the content of records. Sormunen, Nur-minen, Hamalainen, and Hiirsalmi [31] have developed a quite complete re-quirements specification for a front-end. They analyse the types of informa-tion and decisions as well as the information sources associated with each of seven on-line search tasks: setting terminal configurations; database selec-tion; search profile formulation; log-on and log-off; search execution; eval-uating and editing; and post-processing of results. This is followed by a statement of functional requirements for the front-end together with iden-tification of problems and issues for further development. The list includes both clerical and intellectual tasks. Several operational front-ends are lim-ited to assisting with clerical tasks because they are easier to automate.

There have been efforts to characterize the knowledge that needs to be in-corporated in front-ends to provide assistance with the various search tasks. Pollitt [26] suggests four categories: (1) system (the command language and facilities available in the search system); (2) searching (strategy and tactics to be employed in searching); (3) subject; and (4) user (knowledge about each individual user). Often there is a trade-off, with some front-ends having general knowledge to support searches of multiple databases on multiple systems while others have in-depth knowledge to support specific categories of questions and/or databases.

3.4 The Nature of Assistance Provided by the Front-end

When examining the user assistance provided by front-ends, it is helpful to distinguish between two possible roles for the computer, i.e., computer-assisted vs. computer-delegated [7]. In computer-assisted mode, the system provides advice to the user in making decisions. In computer-delegated mode, the system makes decisions automatically, given some initial input from the user. Thus database selection could be handled by suggesting some alternatives to the user or by automatically making the selection of which database(s) to search, given some initial information about the query to be answered. Buckland and Florian [7] suggest that a computer-assisted approach is likely to be more effective because the intelligence of the system and the intelligence of the user ought to augment each other. Bates [3] has also identified the need to provide optimal combinations of searcher control and system retrieval power, arguing that many users would not want to delegate the entire search process to the front-end.

3.5 User Modelling Capabilities

Front-ends are designed to simplify access to electronic information resources by masking some of the complexity. A further level of assistance could be provided by user modelling capabilities, such that responses would be tailored to particular users. As Allen [1] notes, while the term "user model" emphasizes the information about the person, situational, task, or environmental information may also be encoded in the model. In front-ends, user models could be employed to adapt explanations to the user's level of expertise as well as to adapt to user preferences. They may affect database selection, query formulation, and the natural language interaction with the user. Brajnik, Guida, and Tasso [6] point out that information about the user may be obtained through two different mechanisms: (1) external acquisition, where information is obtained in response to questions posed by the front-ends, or (2) internal derivation, where information about the user is obtained through inference from the search session. Borgman and Plute [5] call these forms of models stated vs. inferred. They also distinguish user models as being either static (an unchanging model that is embedded in the system) vs. dynamic (changes throughout the search session and over a period of time to incorporate new information about the user). They caution that user models make assumptions about users' goals and intents and make decisions for them. Therefore, "while accurate models indeed are helpful and reduce the burden on the searcher, inaccurate user models may do more harm than good" (p. 189).

4. Examples of Front-ends

Several front-ends and related knowledge bases are briefly described to illustrate the state of the art, with particular attention to support for accessing scientific, technical, and medical information.

4.1 Medicine: Grateful Med and Loansome Doc

To allow physicians and other health care professionals to search a variety of medical databases, such as Medline available on the National Library of Medicine's (NLM) MEDLARS system, staff at the National Library of Medicine have developed Grateful Med for the PC [30], with Version 6.0 scheduled for release in June 1992 [23]. It assists with menu-driven off-line entry of strategies. Once on-line, it automatically reformats the terms entered into MEDLARS commands, executes the search, saves the results, logs off the system, reformats and displays the citations. The Grateful Med software generates suggested controlled vocabulary terms (Medical Subject Headings) based on retrieved Medline citations. When a strategy results in zero retrieval, a help screen is available that offers suggestions for modifying search strategies. COACH, an expert searcher program to help Grateful Med users improve their retrieval, is currently under development.

Since Grateful Med accesses bibliographic databases, users also need assistance in locating the actual documents. Loansome Doc, introduced in 1991, allows the individual user to place an on-line order for a copy of the full article for any reference retrieved from Medline. If the user's library can fill the document request directly or if it is filled through interlibrary loan, the user receives a photocopy by the preferred delivery method (e.g., mail or fax).

4.2 Medicine: Unified Medical Language System

The goal of the National Library of Medicine's Unified Medical Language System (UMLS) project is to give easy access to machine-readable information from diverse sources, including the scientific literature, patient records, factual data banks, and knowledge-based expert systems [17]. The barriers to integrated access to information in these sources include: the variety of ways the same concepts are expressed in the different machine-readable sources (and by users themselves), and the difficulty of identifying which of many existing databases have information relevant to particular questions. The UMLS approach to overcoming these barriers is to develop "knowledge sources" that can be used by a wide variety of application programs to compensate for differences in the way concepts are expressed, to identify the information sources most relevant to a user query, and to carry out the telecommunications and search procedures necessary to retrieve information from these information sources.

The three UMLS knowledge sources are: (1) a metathesaurus of concepts and terms from several biomedical vocabularies and classifications; (2) a semantic network of the relationships among semantic types or categories to which all concepts in the metathesaurus are assigned; and (3) an information sources map that describes the content and access conditions for the available biomedical databases in both human-readable and machine-readable form. Objectives for the next three years of the UMLS project are to develop and implement important applications that rely on the UMLS knowledge sources, to establish production systems for ongoing expansion and maintenance of the knowledge sources, and to expand the content of the knowledge sources to support the applications being developed. The NLM plans to develop these capabilities within its existing user interface, Grateful Med, and in COACH. For example, COACH uses the metathesaurus to augment user search terms and to help find new terms.

4.3 Environment: ANSWER

ANSWER is a stand-alone microcomputer-based workstation designed for use by health professionals and related personnel in US state and federal agencies responding to hazardous chemical situations. It was developed by the Toxicology Information Program of the National Library of Medicine for the Agency for Toxic Substances and Disease Registry. ANSWER illustrates the possibilities of using local data access and front-end capabilities in support of problem solving in emergency situations. ANSWER includes: a CD-ROM database with information on the medical and hazard management of exposure to over 1,000 hazardous substances; a database of information on previous chemical emergencies; a gateway to the National Weather Service's on-line information (automatic dial-up, log-on, and data capture for state, regional, and local weather information); an air dispersion modelling package for determining plume path and dispersion; a front-end for access to chemical, toxicological, and hazardous waste files located in various governmental and private sector on-line systems; and a report generation capability for editing, sorting, merging, and transforming retrieved data files.

4.4 Environment: Eco-Link

Eco-Link was developed as an electronic research system to take advantage of electronic sources of information on the environment and to coordinate their acquisition, storage, and presentation [39]. It integrates a wide variety of data from electronic sources relating to the environment. The heart of the system consists of download-filter-manage software routines that automate access to electronic databases and process the acquired information so as to merge data from a broad range of different sources in a common set of locally constructed databases. Eco-Link standardizes output from on-line catalogues reachable through the Internet, bibliographic citations from locally mounted databases for newspapers and journal articles, and information from commercial vendors of full text, directories, news sources, and statistical data.

4.5 Chemistry: Graphics Front-ends

Chemical structure searching presents a need for customized front-ends, allowing the scientist to use two-dimensional chemical structure diagrams. Graphics front-ends support the off-line building of chemical structure graphics and subsequent uploading to a host computer, as well as the capture (downloading) of retrieved records. Warr and Wilkins [37] have reviewed the key features of a number of these graphics front-ends, such as STN Express, the front-end software that provides access to STN international databases. STN Express enables one to prepare off-line the strategy formulation (including structural query formulation) and then upload the search strategy line by line after logging on. Off-line chemical structure building is menu-driven. In addition to the ability to create search strategies off-line, the program provides predefined search strategies for general subjects, such as toxicity, that take advantage of individual databases provided by STN. The MOLKICK software package allows the user to enter chemical structures and then translates them into the proper format for searching in three different host systems (STN, Télésystèmes Questel, Dialog) [2].

4.6 Engineering: Ei Reference Desk

Engineering Information Inc. (Ei) has been developing an integrated software package that is designed to bring together both the searching and retrieval of documents [4, 28]. Users will have a choice of browsing through electronic tables of contents for engineering journals, searching COMPENDEX PLUS on CD-ROM, accessing other databases through a telecommunications link, and marking documents for automatic ordering and delivery from Ei's document delivery service. Each function of the Reference Desk has been implemented as a separate application but integrated within the Windows graphical interface. A planned enhancement is an electronic mail function.

4.7 The Livermore Intelligent Gateway

The Livermore Intelligent Gateway creates a framework that links distributed, heterogeneous computer resources and provides a single user interface such that a "virtual information system" can be tailored to any user's needs [8]. In addition to extensive data access capabilities, the Gateway system provides powerful analysis and processing tools to complete the creation of an integrated information environment. Once connected to the selected host, the user may interact in the system's native mode, use a Gateway overlaid common command language, or execute a fully automated search and retrieval procedure for routine tasks. Having simplified access to and retrieval of information, be it bibliographic, numeric, or graphic, the Gateway provides a tool kit to further analyse and repackage the information. Post-processing tools fall into two major categories: analysis of numeric data through statistical, mathematical, and graphics software, and analysis and restructuring of text through translation and analysis routines. In addition to the analytical tool kit, the Gateway provides sophisticated electronic mail capability as well as a wide variety of Unix utilities such as text editors and document preparation subsystems. The menus that a given user or group of users sees on the Gateway can be tailored to create a customized environment.

4.8 TOME SEARCHER and IMIS

TOME SEARCHER is microcomputer software that seeks to provide the inexperienced on-line user with a series of facilities [34]: choice of database(s) in relation to the subject of a search; guidance in formulating the scope of the search; natural language input of the search topic; guidance in clarifying and/or amplifying the topic; automatic conversion of the topic into a Boolean search statement; automatic inclusion of synonyms and spelling variants in the search statement; estimate of likely yield of a search statement; and guidance in narrowing or broadening the statement if the estimated yield does not match the output specified by the user. All this takes place off-line. The system continues by providing automatic dial-up, automatic transmission of search statements to the host using the appropriate command language, display of dialogue with the host, automatic downloading of search output, and the ability to browse through the downloaded records. Much implementation of TOME SEARCHER is customized to a particular subject area, such as electrical and electronics engineering. TOME SEARCHER is one component of the more ambitious IMIS project to develop an intelligent multilingual interface to databases, mounted on an IBM PC and accessing a number of European hosts [36]. IMIS will be designed to support interaction in English, French, German, and Spanish.

4.9 EasyNet

Perhaps the best-known front-end is EasyNet, which offers access to multiple databases on 13 hosts, including many science and technology databases [32]. It gives searchers the option of selecting a database themselves or allowing EasyNet to do so based on answers to a series of questions related to the subject and type of material required. Searching can be accomplished using menus to assist in constructing a search strategy or with commands based on the Common Command Language. Users are responsible for selecting their search terms and also for selecting Boolean logical operators to relate these terms. EasyNet translates the strategy into the command language of the host selected and logs on. After the search is completed and the data downloaded to EasyNet's computer, the user is logged off from the host. On-line help from professional reference staff is available by typing SOS. A customized version of EasyNet is marketed by BIOSIS as the Life Science Network, providing access to more than 80 databases [29]. Dyckman and O'Connor [11] report the results of a study analysing user problems handled by the SOS help service. Their analysis revealed that users seeking human help found the front-end's assistance inadequate in wording their search statements, using features of a specific database, or deciding which database to use.

4.10 Wide Area Information Servers

The Wide Area Information Server (WAIS) project seeks to determine whether current technologies can be used to create end-user full-text information systems [19]. The WAIS system is composed of three separate parts: clients, servers, and the protocol (Z39.50) that connects them. The client is the user interface, the server does the indexing and retrieval of documents, and the protocol is used to transmit the queries and responses. Questions are formulated as English-language queries, which are then translated into the WAIS protocol and transmitted to a server that translates the encoded query into its own query language and then searches for documents satisfying the query. The list of relevant documents is then encoded in the protocol and transmitted back to the client, where they are decoded and the results displayed. The user may modify the query or mark some of the retrieved documents as being relevant. The system can then attempt to find other documents that are similar to those judged relevant. A single interface provides access to many different information sources. With WAIS, the user may select multiple sources to query for information. The system automatically asks all the servers for the required information with no further interaction necessary by the user. The documents retrieved are sorted and consolidated in a single place, to be easily manipulated by the user. To support selection of databases, an on-line Directory of Servers is maintained. It can be queried to identify potential sources on a topic.

5. Evaluation of Front-ends

Front-ends are designed as tools for users. To assess their performance and to identify areas in need of improvement, it is necessary to evaluate them. As noted above, front-ends may function in computer-assisted or computer-delegated mode. For those decisions that are computer-assisted, one must determine if the advice is helpful. For those decisions that are computer-delegated, one must determine if the computer's decisions are appropriate. Where assistance is not offered, one must determine if the targeted user group has the necessary expertise to function unassisted. Because the front-end controls what the user can request, one must determine whether it is "habitable" [38], where a habitable language is one in which its users can express themselves without straying over the language's boundaries into un-allowed questions. Furthermore, one must consider whether the output op-tions meet the users' needs. In addition, there is a need to analyse what effort is required to use the front-end. Van Brakel [33] suggests using the activities of a human intermediary as a framework for evaluating the capabil-ities of a front-end.

Because information retrieval is a complex task, it is difficult to develop front-ends to achieve human levels of performance. Buckland and Florian [7] caution that "delegation, with computers as with people, invites the possibil-ities of undesirable decisions by the person or machine to whom the deci-sion has been delegated." Two examples can illustrate limitations of front-ends. The first, drawn from early versions of Grateful Med, illustrates the difficulty of anticipating all possible variations in input that must be handled. Grateful Med allows the user to type in author names as initials followed by the surname. It then is programmed to translate this into the form required to search the Medline database, i.e., surname followed by a space and then the initials. Initially Grateful Med did not properly handle some input names. Entering "D.A.B. Lindberg" resulted in the translated string "B. Lindberg DA" because the front-end did not expect more than two initials prior to the surname. Clearly this is an error, but the user did not have a way to override this error. The second example illustrates limitations in the database selection capability of EasyNet. A study by Hu [16] analysed this capability of INFOMASTER, a version of EasyNet. Database selection is accomplished by narrowing down the subject selections for a particular query using menu choices made by the human searcher. Then INFOMAS-TER selects a database seemingly at random from among a group of data-bases in a particular subject field. Hu discovered that in some cases, the same menu selections by different searchers for the same query led to incon-sistent databases selected by INFOMASTER. This meant that searches done at different times would be conducted on different databases, with varying levels of completeness and no indication to the user that there might be additional databases with better yields. Given the possibility of such errors or poor advice in front-ends, systematic evaluation is needed to char-acterize their strengths and weaknesses and to pinpoint areas in need of improvement.

6. Directions for Research and Development

While the examples given in this paper indicate that there are already a number of front-ends available to scientists and others who wish to do their own searching, additional research and development are required to create more useful front-ends. In addition to completing evaluations as described in the previous section, a number of other issues need to be investigated as outlined below. It should also be noted that development of front-ends may be aided by new computing tools such as user-interface management systems (UIMS) [15].

6.1 From Directories to Resource Selection Aids

There is currently a great deal of activity in developing databases of databases or directories. Examples include the Directory of Biotechnology Information Resources maintained by the National Library of Medicine and the Listing of Molecular Biology Databases maintained by Los Alamos National Laboratory. Other examples are guides to library catalogues and the Internet Resource Guide. Work with WAIS and UMLS is exploring ways to use directories for resource selection. This is an important area for further research, since sophisticated assistance with search strategy formulation is of little value if an inappropriate resource has been selected.

6.2 Coping with Null Sets and Large Sets

Studies of end-user searching frequently reveal that such users often formulate strategies that retrieve nothing or that retrieve very large sets. Prabha [27] reviews a number of strategies for reducing large sets. Front-ends must be designed to help users modify strategies to avoid both situations.

6.3 Front-ends for Non-bibliographic Databases

Much of the effort to date has focused on creating front-ends to handle query formulation for multiple bibliographic databases. As Järvelin [18] points out, there is also a need for assisting with access to multiple numeric and other types of non-bibliographic databases. While some of the front-end functions would correspond to forms of assistance needed for accessing bibliographic databases, others, such as data conversion between varying data representations, must be dealt with in accessing numeric databases.

6.4. Multilingual Facilities

As Vickery and Vickery [35] remark, a needed further enhancement of front-ends would be provision of multilingual facilities. They suggest that such a front-end could have the following features: (1) screen displays would be available in all the languages covered by the system; (2) the interface would accept input of a user query in each of these languages and would refine the query by interacting with the user in the language of input; (3) the terms of the refined query would be translated into the language of the selected database(s); and (4) retrieved records would be translated into the language of the user. As one example of such multilingual support, Halpern and Sargeant [13] describe a front-end for bilingual searching of Medline that has been developed by INSERM (Institut National de la Santé et de la

Recherche Médicale) and the host Télésystèmes-Questel. It supports bilingual access at both the command and query level. At the command level, menus and help screens are presented in French only. At the query level, subject searching can be performed in French or English, taking advantage of a French translation of the controlled vocabulary Medical Subject Headings.

6.5 Knowledge Acquisition

One of the problems involved in the design of more sophisticated front-ends is the lack of extensive knowledge gathered from experts about how they search. As human intermediary expertise is better understood, the problem of knowledge acquisition by the front-end remains. If performance of front-ends is to improve over time, then some provision for modification or learning must be implemented.

6.6 User-Friendly Systems

Vickery [36] has remarked that a front-end "accesses existing online systems, with all their constraints and deficiencies, so it can only be as successful as the online search system allows it to be. An interface does not address the problem of restructuring the database or the search system to make retrieval more intelligent." Harman [14] argues that attempts to develop more user-friendly front-ends are inherently limited by the design of the underlying retrieval systems. WAIS is a current example of efforts to create both a more user-friendly front-end and a system that is easier to use in searching, based on statistical methods, allowing natural language input, and returning lists of records in order of likely relevance. Research is needed to evaluate and further develop these alternative approaches to retrieval so that future front-ends are not so constrained. Knowbots [9] – knowledge robots transporting the user's request out into the universe of digitized information where outlying knowbots will search for answers – may well be part of that future, but there still must be provision for human oversight of the information retrieval process.

7. Conclusion: Implications for Developing Countries

Keren and Harmon [20] suggest that nearly all publications that deal with information work in the developing countries cite one or more of the following problem areas: "lack of appreciation by national decision makers for the role of STI in development; the absence of an adequate infrastructure for information storage and processing; the absence of an adequate infrastructure for information use and absorption by users; and economic, administrative, technological, cultural, educational, and structural barriers to an adequate information flow." Given this context, it is necessary to ask what role computerized front-ends might play in overcoming barriers to information access and use in developing countries.

A key lesson is that there is the possibility with front-ends of accommodating some differences (e.g., language spoken) among users of information systems and of providing some guidance in the use of information systems tailored to the needs of particular user groups. As the telecommunications infrastructure gradually develops, making possible access to remote information resources, it will be necessary to investigate how best to design front-ends to meet the needs of particular user groups in specific countries or regions. In addition, as indigenous information sources are developed, front-ends have a role to play in integrating access to indigenous and external information sources where both are relevant to the users' needs.

REFERENCES

1. Allen, R.B. (1990). "User Models: Theory, Method and Practice." *International Journal of Man-Machine Studies* 32 (5): 511–543.
2. Badger, R., C. Jochum, and S. Lesch (1988). "MOLKICK: A Universal Graphics Query Program for Searching Databases with Chemical Structures." *Proceedings of the National Online Meeting* 9: 7–8.
3. Bates, M.J. (1990). "Where Should the Person Stop and the Information Search Interface Start?" *Information Processing & Management* 26 (5): 575–591.
4. Berger, M.C. (1989). "Engineering Information Workstation." *Proceedings of the National Online Meeting* 10: 33–35.
5. Borgman, C.L., and Y.I. Plute (1992). "User Models for Information Systems: Prospects and Problems." In F.W. Lancaster and L.C. Smith, eds. *Artificial Intelligence and Expert Systems: Will They Change the Library?* Urbana, Ill.: Graduate School of Library and Information Science, University of Illinois, pp. 178–193.
6. Brajnik, G., G. Guida, and C. Tasso (1990). "User Modeling in Expert Man-Machine Interfaces: A Case Study in Intelligent Information Retrieval." *IEEE Transactions on Systems, Man, and Cybernetics* 20 (1): 166–185.
7. Buckland, M.K., and D. Florian (1991). "Expertise, Task Complexity, and Artificial Intelligence: A Conceptual Framework." *Journal of the American Society for Information Science* 42 (9): 635–643.
8. Burton, H.D. (1989). "The Livermore Intelligent Gateway: An Integrated Information Processing Environment." *Information Processing & Management* 25 (5): 509–514.
9. Daviss, Bennett (1991). "Knowbots." *Discover* 12 (4): 21–23.
10. Drenth, H., A. Morris, and G. Tseng (1991). "Expert Systems as Information Intermediaries." *Annual Review of Information Science and Technology* 26: 113–154.
11. Dyckman, L.M., and B.T. O'Connor (1989). "Profiling the End-User: A Study of the Reference Needs of End-Users on Telebase System, Inc.'s Easynet." *Proceedings of the National Online Meeting* 10: 143–152.
12. Efthimiadis, E.N. (1990). "Online Searching Aids: A Review of Front Ends, Gateways and Other Interfaces." *Journal of Documentation* 46 (3): 218–262.

13. Halpern, J., and H.A. Sargeant (1988). "A New End-User Interface for Bilingual Searching of MEDLINE." *Proceedings of the International Online Information Meeting* 12: 427–443.
14. Harman, D. (1992). "User-friendly Systems Instead of User-friendly Front-Ends." *Journal of the American Society for Information Science* 43 (2): 164–174.
15. Hix, D. (1990). "Generations of User-interface Management Systems." *IEEE Software* 7 (5): 77–87.
16. Hu, C. (1988). "An Evaluation of a Gateway System for Automated Online Database Selection." *Proceedings of the National Online Meeting* 9: 107–114.
17. Humphreys, B. (1991). "Unified Medical Language System: Progress Report." *National Library of Medicine New* 46 (11–12): 7–8.
18. Järvelin, K. (1989). "A Blueprint of an Intermediary System for Numeric Source Databases." In S. Koskiala and R. Launo, eds. *Information*Knowledge* Evolution*. Amsterdam: North-Holland, pp. 311–320.
19. Kahle, B., and A. Medlar (1991). "An Information System for Corporate Users: Wide Area Information Servers." *Online* 15 (5): 56–60.
20. Keren, C., and L. Harmon (1980). "Information Services Issues in Less Developed Countries." *Annual Review of Information Science and Technology* 15: 289–324.
21. Longley, D., and M. Shain (1989). *Macmillan Dictionary of Information Technology*. 3rd ed. New York: Van Nostrand Reinhold, pp. 277, 231.
22. Meadow, C.T. (1992). *Text Information Retrieval Systems*. San Diego, Calif.: Academic Press.
23. "NLM Plans June Release of Grateful Med Update" (1992). *National Library of Medicine News* 47 (3–4): 1–3.
24. "Panel on Information Technology and the Conduct of Research" (1989). *Information Technology and the Conduct of Research: The User's View*. Washington, D.C.: National Academy Press, pp. 1, 3.
25. Percival, J.M. (1990). "Graphic Interfaces and Online Information." *Online Review* 14 (1): 15–20.
26. Pollitt, A.S. (1990). "Intelligent Interfaces to Online Databases." *Expert Systems for Information Management* 3 (1): 49–69.
27. Prabha, C. (1991). "The Large Retrieval Phenomenon." *Advances in Library Automation and Networking* 4: 55–92.
28. Regazzi, J.J. (1990). "Designing the Ei Reference Desk." *Proceedings of the National Online Meeting* 11: 345–347.
29. Seiken, J. (1992). "Menu-driven Interfaces Simplify Online Database Searching." *The Scientist* 6 (8): 18–19.
30. Snow, B., A.L. Corbett, and F.A. Brahmi (1986). "Grateful Med: NLM's Front End Software." *Database* 9 (6): 94–99.
31. Sormunen, E., R. Nurminen, M. Hamalainen, and M. Hiirsalmi (1987). *Knowledge-based Intermediary System for Information Retrieval: Requirements Specification*. Research notes no. 794. Espoo, Finland: Technical Research Centre of Finland.
32. Still, J. (1991). "Using EasyNet in Libraries." *Online* 15 (5): 34–37.
33. van Brakel, P.A. (1988). "Evaluating an Intelligent Gateway: A Methodology." *South African Journal of Library and Information Science* 56 (4): 277–290.
34. Vickery, A. (1989). "Intelligent Interfaces for Online Searching." *Aslib Information* 17 (11/12): 271–274.

35. Vickery, B., and A. Vickery (1990). "Intelligence and Information Systems." *Journal of Information Science* 16: 65–70.
36. Vickery, B.C. (1992). "Intelligent Interfaces to Online Databases." In F.W. Lancaster and L.C. Smith, eds. *Artificial Intelligence and Expert Systems: Will They Change the Library?* Urbana, Ill.: Graduate School of Library and Information Science, University of Illinois.
37. Warr, W.A., and M.P. Wilkins (1990). "Graphics Front Ends for Chemical Searching and a Look at ChemTalk Plus." *Online* 14 (3): 50–54.
38. Watt, W.C. (1968). "Habitability." *American Documentation* 19: 338–351.
39. Weiskel, T.C. (1991). "Environmental Information Resources and Electronic Research Systems (ERSs): Eco-Link as an Example of Future Tools." *Library HiTech* 9 (2): 7–19.
40. Williams, M.E. (1986). "Transparent Information Systems Through Gateways, Front Ends, Intermediaries, and Interfaces." *Journal of the American Society for Information Science* 37 (4): 204–214.

Multimedia Technology: A Design Challenge

James L. Alty

ABSTRACT

The term multimedia used in this paper refers to different presentation media such as sound, graphics, text, etc. The possible benefits of a multimedia approach are discussed and some examples are given of how different media affect knowledge comprehension in human beings. An indication is given of what multimedia facilities are currently available. The PROMISE multimedia interface project is described, and in particular a possible approach to the formulation of a multimedia design methodology is proposed. Some initial guidelines on multimedia interface design are given. Although some of the ideas expressed here come from work in process control, they are likely to be generalizable to wider domains.

1. Introduction

The term "multimedia" has two possible meanings. Firstly, the "media" can refer to storage media such as WORMs, CD-ROMs, and disks. Secondly, it can refer to the presentation of information using different media such as sound, graphics, text, etc. [10]. In this paper the second meaning applies throughout.

The idea of using multiple media to improve communication between humans and computers is not new. A paper of 1945, "How We May Think" [7], suggested a multiple media approach that was later reassessed [8]. Some well-known early experiments with multimedia were carried out at the MIT "Media Lab" in 1977 [6], and Maekawa and Sakamura [14] also described an

172

early multimedia machine that was being implemented at the University of Tokyo. The system had an optical disc, 100 Mbyte disc, a high-speed resolution graphics display, a TV camera, and sound input/output. These collections of multimedia devices were very expensive and it is only recently that a host of multimedia tool kits have entered the market-place at a relatively low cost.

But why should we be interested in multimedia interfaces? Well, even a cursory study of human beings communicating information between each other will show the importance of the use of multiple media in the communications process. Human beings often use at least two sensory channels (visual and auditory) but frequently use a third – touch – as well, and within these communication channels, a rich variety of media are employed. When one artificially reduces the richness of the set of communication media being used (for example by taking a tape recording of a meeting), the reduction in communication power is obvious. Since human beings do seem to be able to communicate more effectively between each other than with computers, the idea of employing additional media in human-computer interaction seems a sensible one, whose inclusion is likely to improve the communication process. The early stumbling blocks that prevented the implementation of multimedia facilities on computers were lack of power and storage. Now these problems have largely been overcome and multimedia tool sets are thus becoming available on both personal computers and high-performance workstations.

The technical problems of providing a variety of media with acceptable performance and at a reasonable cost are not the only stumbling blocks that may prevent the exploitation of multimedia facilities. Another key problem is that of devising a methodology to aid multimedia interface design. Being able to provide multimedia interfaces is not enough. One must also be able to know when to use which media and in what combination to solve a particular interface problem.

I will first discuss multimedia facilities generally, indicating what media are and why there may be a problem associated with the design of multimedia interfaces. I will then examine the PROMISE project (a large collaborative project), which is particularly concerned with the development of a methodology of multimedia design. Although many of the ideas in PROMISE come from work in process control, they are likely to be generalizable to wider domains.

2. What Are Communication Media and How Do They Differ?

Although we often think initially about the physical aspects of communication media (screens, colour, sounds, etc.), the main attribute of a medium is that it provides a language for communication. This language will involve a syntax, semantics, and pragmatics, with the physical aspects of the medium providing constraints on the syntactic possibilities. Traditional computer media have usually been restricted to one sensory channel (e.g. text [visual], graphics [visual], sound [auditory]), but even within one sensory channel, a rich variety of media can be supported. The visual channel, for example, can support graphics, tables, diagrams, pictures, maps, graphical animation, and 3-D graphics, all of which communicate information in different ways. The auditory channel can support speech, verbal gestures, realistic sounds, artificial sounds, and music. The sensory channel can involve vibration and sensory input that might result from, for example, a data glove.

Some information already exists to guide us on the relative claims for different media for effective information transfer. Auditory media, such as radio, make dialogue salient. For example, children given a story in audio only or with a visual + audio combination (same soundtrack) recall dialogue better when it is given as audio only. Television presentation seems to be better for action information, which has improved recall when presented via television [5]. Whilst audio information seems to stimulate imagination, spatial visualization is better handled visually, as might be expected. Other studies have indicated that diagrams are better at conveying ideas, whereas text is better for detail [22]. However, the visual channel does not always dominate. Walker and Scott [24] found that human beings judge light as being of a shorter duration than an identical tone, and when these are presented together, the auditory channel dominates. Pezdek [21] carried out experiments to determine if the visual channel dominated over the auditory channel in the comprehension of information on television. Whilst there was evidence of visual domination, the presence of the auditory channel actually improved comprehension and vice versa.

Text is often better for communicating complex information to experts, whilst pictures are better for exploratory learning. In particular, visual representations are excellent for synthesis. These ideas are shown in figure 1, where the usability of different channels (visual and auditory) is contrasted with the previous knowledge and experience in the knowledge area. For some tasks, text can be a very effective medium of communication. Each different medium of communication, therefore, has properties that will enhance or restrict its capability for transmitting particular types of knowledge.

At a higher level, we can form new media by combining existing ones. When two media are combined, new syntactic and semantic units become possible. These higher level media often use more than one sensory channel. Examples would include movie films and animated diagrams with verbal talk-over. It might be thought that a more complex (or rich) medium would always be preferable to a simpler one, but this is not always so. For example, experiments with televised weather forecasts and radio weather forecasts have shown that the auditory medium is superior in many cases. One can also remember radio plays that are more "vivid" than television plays. As Kosslyn [13] states,

Multimedia technology can deliver information like a fire hose delivers water. Just as drinking from a fire hose is not an efficient way to quench

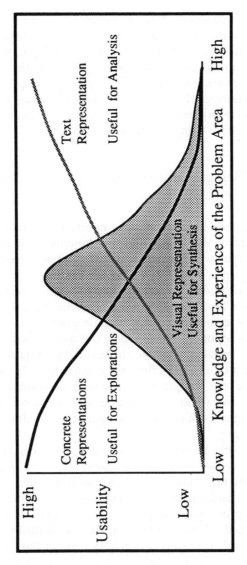

Figure 1

one's thirst, high powered multimedia presentations can overload the senses and fail to communicate information effectively. Multimedia relies on the essential truth in the Chinese proverb that tells us a picture "is worth more than a thousand words." Unfortunately some pictures do not help to control the flow of information, and actually make it worse.

Work on the effectiveness of different media for communicating information has been carried out over many years, but interpretation of the results is not straightforward. Washburne [26] found, for example, that graphs were easier to interpret than tables, whilst Vernon [23] found the opposite. These apparent divergencies are not necessarily surprising. Later work [11, 9] has shown that the usefulness of the different display formats is highly dependent upon the tasks being performed. Effectiveness is dependent upon the nature of the information being sought by the reader.

3. Are Human Beings Aware of the Capabilities of Different Media?

Human beings in general, and computer interface designers in particular, are not really aware of the capabilities and limitations of different media. Early computer output was restricted to text. This is a medium that most people understand, since it has been used for hundreds of years. Therefore we have an intuitive understanding of how to communicate with text, at least at a fairly basic level. When graphics became possible on computers, the situation did not materially change because graphics had also been used by human beings to communicate (often in conjunction with text) for many years. When colour became available, however, the situation deteriorated. Colour was used in a totally indiscriminate manner in early interface design. It was not surprising that the quality of interfaces deteriorated. Most people cannot choose their own wallpaper or clothes with a real colour sense. It was much later that designers realized that the spare use of colour was the key to good interface design.

It is quite interesting to reflect on why human beings (in the Western world in particular) do not have good colour sense. Mary White [25] makes the interesting point that "imagery" (which in her definition encompasses paintings, sculpture, and television as well as computer graphics and video) has been used as a primary learning tool for centuries. The invention of the printing press changed all that, and the primary mechanism for learning became the printed word. Ironically, towards the end of the twentieth century we have moved back to a world where imagery is important again, and it has replaced the word as a major political communication medium. There is, however, little research on imagery and its effects.

We are now about to enter the era of multimedia interface design, and I expect the experience with colour to be repeated, writ large. The quality of home videos testifies to the level of most peoples' fitness for designing interfaces using moving pictures and voice! Since human beings are only vaguely aware of the differences between media, we need a methodology that helps to answer the question, "What medium when?" to achieve a particular interface goal. I will return to the issue of the methodology later.

4. What Can the Technology Do Now?

The technology has moved rapidly in the past few years. The enabling technologies required to support a multimedia workstation go far beyond provision of the media themselves. A typical list [15] would include:
- Bit-mapped displays
- Audio and video support
- Device independent graphic rendering
- Control and synchronization
- Data compression
- Networking support
- High CPU/disk performance
- Authoring tools

To illustrate how far current workstations have come in being able to deliver such a collection of enabling technologies, Malleo-Roach cites the SUN SPARCstation as an example and provides the figures given in the table.

To give an idea of the costs associated with multimedia support, he suggests three possible configurations: "minimal," "enhanced," and "operational," with costs at about US$10,000 and US$35,000 for the first two (the third not yet being available). He expects an "operational" version to be available in 1995 for about US$10,000.

Bit-mapped display:	1152 × 900 bit-mapped display, most models providing 8-bit pseudo-colour
Device independent Graphics:	All SPARCstations support OpenWindows 2.0 with X11/NeWs imaging
Audiovisual:	Every SPARCworkstation has built-in audio capabilities (8 kHz single channel. Videopix (a video frame grabber) is also available)
Data compression:	JPEG compression support for VideoPix. Uniflix-video compression/decompression tool kit for full-motion video
Control synchronization:	No complete and comprehensive solution, but a number of mechanisms are offered such as shared memory, interprocess communication, times, semaphores, and threads
Networking:	Client-server support. Multimedia as natural extension of the client-server model
Authoring:	A number of tools are available including HyperNeWS, Cats Meow, and Mediawrite.
High CPU/disk: Performance	Sufficient

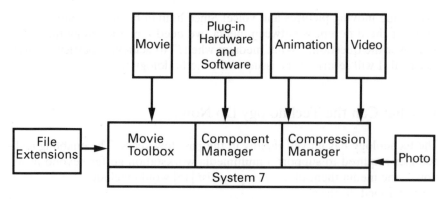

Figure 2

Multimedia capabilities have also already arrived on most personal computers. There are multimedia extensions to Windows, and Apple now offers "Quick Time" on the Macintosh. Quick Time offers three new components: the Movie Toolbox, the Image Compression Manager, and the Component Manager (figure 2).

The Movie Toolbox provides facilities for creating, editing, and playing back moving images. The Component Manager enables new devices (such as VCRs, camcorders, etc.) to be attached as well as third vendor software modules. The Image Compression Manager uses one of three compression techniques for animation, video, and photos. The JPEG (Joint Photographic Experts Group) algorithm is used to compress photographs. It removes data that is redundant to the human eye. Compression ratios of between 500:1 and 5:1 can be achieved with a full screen colour image taking about 10 seconds to compress. The video compressor can achieve ratios of between 25:1 and 5:1 giving about 15 seconds of compressed video per megabyte. Boards are available for capturing video directly from cameras.

Compression techniques are essential since a standard television frame contains nearly 750,000 bytes of information. In uncompressed form, even a compact disc can only store 30–40 seconds of video. Furthermore, since a compact disc delivers data at an error free rate of 150 Kbytes per second, and 10 per cent of this needs to be reserved for audio, each picture needs to be reduced to about 5,000 bytes to maintain motion. This corresponds to a compression rate of 160. This can be done by DVI (Digital Video Interactive) algorithms.

5. User Centred or Design Centred?

Articles on multimedia tend to stress the technological side rather than utility aspects. There are many who appear to view multimedia techniques as a technology looking for an application. This technological-push view of multimedia is summed up in the introduction to an extended set of articles on multimedia in a recent edition of *OpenLook* [19]:

> For while multimedia has the potential to transform computing, many of the components that create it are already available – it's just that users haven't brought them together yet. . . . Admittedly, the one thing that multimedia's been waiting for is applications. But they . . . are now coming on stream.

In other words, we have the technology, why don't you users want it? Because of the availability of the CD-I ROM and the richness of the media, educational technology has been a major target area for multimedia development. Many people extol the virtues of multimedia education, however one does tend to get the impression that the designers start with a multimedia tool kit and then try to find an application. As a result, a number of people are rather sceptical about current developments and the progress really made.

> Tools for the construction and manipulation of collections of multimedia material are becoming widely available on successively cheaper hardware platforms. In terms of the educational use of these systems, however, it is unclear that any progress has been made. The effort in developing the technology has not been matched by a similar concern with the pedagogy. . . . At present it is an article of good faith that multimedia is a good thing for education and training. There is no evidence that multimedia enhances learning, or makes it more cost effective. [12]

Thus, the importance of multimedia applications lies not in what it can do but in what user goals it can solve more effectively. In Process Control, for example, the goals of the operators could be interpretation accuracy; problem comprehension; task performance; decision quality; speed of comprehension; decision speed; recognition and recall; or viewer preference [11]. Some of these goals will be more important than others – e.g. recall is usually less important than task performance. It is vital that interface designers always keep these user goals (rather than technological goals) uppermost in their minds. If this approach is not adopted, we could find that things get worse rather than better. The challenge of multimedia technology is to understand how to minimize information overload by the appropriate use of different media.

6. The PROMISE Multimedia Interface Project

The overall objectives of the PROMISE project (PRocess Operators Multimedia Intelligent Support Environment) are twofold:
(1) the design, construction, and evaluation of a multimedia tool set for im-

proving user interfaces to advisory and diagnostic expert systems used by single operators in process control environments;
(2) the development of a methodology to support designers using the tool set.

6.1 The Media Supported

The PROMISE system is multimedia, as it supports many ways of communicating information, and multimodal, as many styles of interaction are supported. The system supports a number of multimedia options [2], including:
– text, graphics, and sound output
– full colour
– live and still video output
– text and Mouse input
– two-dimensional animation
– unmediated video
– natural language output (speech)

The media "unmediated video" [1] perhaps require some explanation. The term subsumes many others, including algorithm animation, data fusion, visual realization, program visualization, and abstract representation. It is best illustrated by an unmediated audio example. In the 1960s, ICL engineers linked the data bus of a 1900 series computer to a loudspeaker, no doubt for hardware debugging purposes. The loudspeaker was left in the operator console of the computer and could be heard by increasing the volume control. It is almost certain that this information was never intended for operator use, but it soon became an extremely useful operator aid. Although the connection between the sounds emitted and the machine activity was never explained, operators were rapidly able to use the information to tell if the machine was idle, which job was running, and if a job was in a loop. At night, this use of the auditory channel also enabled such monitoring to take place in the operator tearoom adjacent to the computer room, probably the first recorded example of a multimedia interface advantage. There are other, more visual examples of this approach that have been used in process control.

6.2 The Environments

The multimedia approach in PROMISE is being applied in two distinct types of environment that will help us in evaluation – a real-time application and a simulated one. The real environment puts serious constraints on the evaluation environment, since the clock cannot be stopped and unanticipated events can occur. Such an environment also poses other problems from an evaluation standpoint. Errors cannot be deliberately introduced or reproduced, and controlled experiments are difficult. Therefore a simulated environment where we can control the nature of the upsets and monitor operator performance will also be used. The two application areas chosen are control of a chemical plant (DOW Benelux) and the use of power station simulators (Scottish Power).

6.3 The Unique Features of Multimedia Usage in Process Control

Most multimedia ideas grew out of work in computer education. The educational multimedia approach involves the use of videocomputers, interactive television, and electronic books to provide a richer educational environment. It is envisaged that a user will be able "to browse through vast libraries of text, audio and visual information" [4]. These libraries will be highly interlinked using a technique known as "Hypermedia," a concept developed from Hypertext that was first defined in the 1960s and led to the Xanadu system [18]. Most of the current literature about multimedia approaches derives from the educational sector. These educational applications tend to be activated by the learner alone where the user of the multimedia system is in complete control.

In contrast, the process control environment places extremely demanding requirements on the architecture of a multimedia presentation system.
- The operator may be engaged in many distinct tasks simultaneously, for example monitoring, tracking alarms, etc.
- New tasks and hence interactions may be instigated at any time by either system or operator.
- The process state may be extremely dynamic, so information may need to be presented rapidly and in a form that is readily understood by the operator.

The designer of a user interface cannot know in advance what combinations of multimedia resources might be required. The problem may therefore be thought of as a resource management problem, managing the competing requirements of different media for the limited resources of the interface. The existence of additional media provides problems and opportunities. The problems are those of "media-clash" – for example, several alarms wish to use the audio channel at the same time. The opportunities come from the ability to switch media to overcome the problem.

6.4 The Architecture

The overall architecture is shown in figure 3.

The actual realization of an object is derived from:
- designer options: the designers specify alternative renderings of an object and provide measures of their preferences for such rendering;
- resource limits: the availability of the physical rendering resources;
- multimedia options: given a choice, what would be the preferred rendering?
- operator preferences: are there special operator requirements (i.e. colour blindness)?

At the far right are the dynamic system to be controlled and the "super-

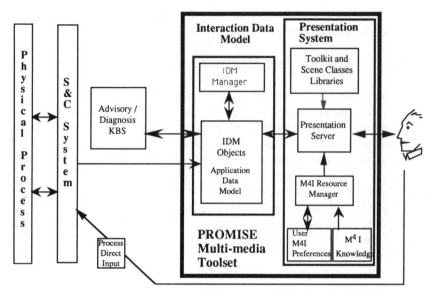

Figure 3

visory and control system" that controls it. The PROMISE system is an advisory system so that the operator can completely ignore it if required and can control the process separately. Indeed, no direct control of the system is possible through the PROMISE terminal. An "advisory/diagnostic knowledge based system" is shown in the figure, though such a system is not actually part of PROMISE. At the heart of the tool set is the "interactive data model." This module models the relevant portions of the plant and all objects rendered to the operator. It is therefore a key player in consistency maintenance. The "presentation server" actually renders the images, sounds, etc., under control of the "resource manager," which decides what media will be employed taking into account the designers' preferences, the current resource usage, the user preferences, and general rules about rendering (the M^4I Knowledge in the diagram).

Currently, this tool set is being installed in the chemical plant and in a nuclear power station simulator so that experiments can be carried out on the effectiveness of multimedia interfaces. In parallel, a series of experiments are being carried out in the behaviour laboratory at Loughborough University.

7. How Does One Design a Multimedia Interface?

Although the project is ongoing, we can already provide some guidelines as to how to design a multimedia interface. Our approach is one of experimentation combined with pragmatism (figure 4).

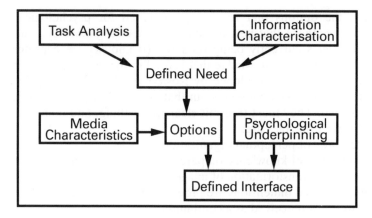

Figure 4

A task analysis [20] will be carried out on the operator task to establish the information needs. These needs then must be characterized in terms of a defined information processing need using a knowledge characterization scheme. The options that the various media provide are then matched to the needs to obtain a set of multimedia possibilities. Finally, an envelope of defined interfaces can be derived by checking the possible interfaces against our knowledge of the human cognition system.

The knowledge characterization scheme identifies a number of important information communication features that describe the knowledge to be communicated to the operator, for example. More details are given in Alty and Bergan [3].
- The number of dimensions needed:
- Ordering or sequence:
- Relationships:
- Dynamic:
- Static:
- Persistence:
- Instance:
- Information content:
- Urgency:
- Context change:

There are, of course, many more. We then map these information requirements onto appropriate media using another list of media properties, for example:
- ICON is 0-D, has no ordering, limited relationships, static, is persistent, and has limited information content.
- PICTURE is 2-D, strict ordering, probably strongly related to real world, static, persistent, and has high information content.
- VOICE OUTPUT is 1-D, strict ordering, strongly related to real world, dynamic, non-persistent, and has medium information content.

Finally, the options that result are examined for psychological appropriateness, i.e. those that best exploit the human cognition system. At this point, all objects currently being rendered (if known) are considered. An example of the sort of knowledge that is used in the psychological underpinning is the set of rules developed by Kosslyn et al. [13] – five principles of articulate graphics. Kosslyn points out that graphical displays must be articulate to succeed. They must transmit clear, compelling, and memorable messages. His five principles are:

(1) Gestalt laws of perceptual organization
(2) The influence of knowledge on perceptual organization
(3) Incremental transformation of images
(4) Different visual dimensions are processed by different channels
(5) Colour is not perceived as a continuum

For example, incremental transformation of images is important for creating and maintaining mental images. In the real world, objects do not disappear and reappear elsewhere. So pictures on VDU displays should change incrementally. Other important psychological principles are the 7 ± 2 rule of Miller [16] and issues concerning the coherence or interference between visual and auditory channels. Application of the latter two rules, for example, make audio explanation with video highlighting a superior mechanism for explaining a complex diagram than the use of the visual channel only.

8. Some Initial Guidelines

Multimedia interface design has only one purpose – that of achieving some goal in a "better" manner. It is therefore important to clearly establish the goals to be achieved and what is meant by "better." In the PROMISE application, we decided early on that the key problem in process control was information overload with too little time for cognitive processing. Thus our major goal is to use multimedia in such a way as to buy time for the operator by rendering the information in such a manner as to enable it to be assimilated more quickly. Other goals include error reduction and better job satisfaction. Multimedia design is therefore critically bound up with the tasks being performed and the skills of the operators at the interface. Even so, we can make some general observations:

8.1 Real World Sensing

Multimedia interfaces allow us to reconnect the operator/user with the real sense world. The history of process control has been the gradual separation of the operators from the system they are controlling. In former times, operators were able to see, hear, and feel the plant. We have created a media-poor world that the multimedia approach can recreate. When operators close a valve they should see and hear it close. In PROMISE, we expect to have a microphone icon that can be moved round the process diagram. The operator can then hear information from selected parts of the plant.

8.2 Maintaining Information Source Integrity

Human beings often do not know how they use information. From a set of sensory inputs they "know" what decision to take. It is therefore important to preserve the integrity of information sources, particularly if we are not absolutely sure what is important. In the PROMISE project, for instance, our studies revealed that converting a particular test to a number lost a great deal of important visual information. Now we preserve a video still of the test so that the operators can still make a full judgement.

8.3 Exploiting Human Cognition

There are certain aspects of human cognition that lend themselves to a multimedia approach. Overloading of one sense can often be avoided by using a second sense in parallel. In particular the audio + visual approach can be particularly powerful.

8.4 Changing Context

This aspect does seem to be best dealt with using the audio channel. Auditory warnings are most effective and allow the operator to concentrate on other aspects of the task.

8.5 Simulation

The visual approach is, of course, particularly well suited to simulation. A useful combination is that of real video and overlaid graphics.

8.6 Use of Non-verbal Sounds

Very short sounds are useful for alarms or status identifiers. Continuous sounds are more difficult to use effectively. They are best used when identifiable changes really matter. Their use for absolute measurement should, of course, be avoided. The critical issue is whether the important changes can be recognized by the human observer. Continuous sounds can also be very irritating.

9. Conclusions

I hope that this paper has shown that the key issue facing multimedia technology implementers is not primarily a technical one (although technical performance is important) but a design one. We need to know where and when particular media combinations will enable us to achieve our goals more effectively. Because the advantages (or disadvantages) of multimedia usage depend upon the tasks being performed, it is difficult to set up precise laboratory studies and apply the results directly to real life situations. The answer here is probably more longitudinal studies in real situations. However, we are now reasonably confident that our list of guidelines will eventually extend into a methodology.

There are issues that have not been addressed in this paper. For example, dynamic allocation of multimedia resources raises major issues about interface consistency. We believe that, provided the representations change only occasionally and only in critical resource situations, this need not be a serious problem. However, this is a matter for debate.

10. Acknowledgements

The work reported here is part of the PROMISE project funded by the European Commission as project 2397 in the ESPRIT II programme. Their funding support is gratefully acknowledged. Thanks are also due to the partners in the project, namely Scottish Power (Scotland – Prime Contractor), Tecsiel (Italy), Dow Chemicals (Netherlands), University of Leuven (Belgium), IDS (Spain), Work Research Centre (Eire), University College Dublin (Eire), and Algotech (Italy).

REFERENCES

1. Alty, J.L. (1989). The concept of "Unmediated Video and Audio" was first discussed in the original PROMISE Technical Annex (1989) (ESPRIT II Project 2597).
2. Alty, J.L., M. De Winter, P. Del'omo, and M. Zallocco (1989). "D2 – Interface Definition for First M4I Toolset (PRO/6)." Deliverable no. 2, Esprit Project 2397.
3. Alty, J.L., and M. Bergan (1992). "The Value of Multimedia Interfaces in Process Control." To appear in the Proc. of MMS '92, 5th IFAC, IFIP, IFORS, IEA Symposium on Analysis, Design and Evaluation of Man-Machine Systems, The Hague, Netherlands, June 1992.
4. Ambron, S. (1988). "What is Multi-media?" In: S. Ambron and K. Hooper, eds. *Interactive Multi-media*. Washington, D.C.: Microsoft Press, p. 5.
5. Beagles-Roos, J. (1985). "Specific Impact of Radio and Television on Adult Story Comprehension." Presented at a meeting of the American Psychological Association, Los Angeles, September.
6. Bolt, R.A. (1977). *Spatial Data Management*. Interim Report, Architecture Machine Group, MIT. DARPA Contract MDA903-77-C-0037.
7. Bush, V. (1945). "How We May Think." *Atlantic Monthly* 176 (1): 101–108.
8. Bush, V. (1969). "Memex Revisited." In: *Science Is Not Enough*. New York: Apollo Editions.
9. Carswell, C.M., and C.D. Wickens (1987). "Comparative Graphics: The Historical Development and Experimental Comparison of Alternative Formats." Aviation Research Lab., Savoy, Illinois. Unpublished manuscript.
10. Cawkell, A.E. (1991). "Multimedia and Hypertext." In: *World Information Technology Manual* 2. New York: Elsevier, p. 674.

11. DeSanctis, G. (1984). "Computer Graphics as Decision Aids." *Decision Sciences* 15: 463–487.
12. Elsom-Cook, M. (1991). "Multimedia: The Emperors New Clothes." *DLT News* 6: 1–2.
13. Kosslyn, S.M., C.F. Chabris, and S.E. Hamilton (1990). "Designing for the Mind: Five Psychological Principles of Articulate Graphics." *Multimedia Review* 1 (3): 28–29.
14. Maekawa, M., and K. Sakamura (1983). "Multimedia Machine." In: R.E. Mason, ed. *Proc IFIP 9th World Congress*, Paris, Sep. (Information Processing '83). Amsterdam: North Holland.
15. Malleo-Roach, J.A. (1992). "Bringing Multimedia Technology to the SPARC-station." *SUNEXPERT* 3 (2): 60–65.
16. Miller, G.A. (1956). "The Magic Number Seven, Plus or Minus Two: Some Limits of Our Capacity for Processing Information." *Psychological Review* 63: 81–97.
17. Moray, N., P. Lootstein, and J. Pajak (1986). "Acquisition of Process Control Skills." *IEEE Trans on Systems, Man and Cybernetics* SMC-16 (4): 497–504.
18. Nelson, T.N. (1981). *Literary Machines*. Swarthmore, Penn.: Edition 87.1. Available from the author.
19. OpenLook (1992). "Multimedia: The Key to a Thousand Doors." *OpenLook Editorial* 3 (2): 39.
20. Payne, S., and T. Green (1989). "Task-Action Grammars: The Model and Its Developments." In: D.A. Norman and S.W. Draper, eds. *Task Analysis for Human Computer Interaction*. Hillsdale, N.J.: Lawrence Erlbaum.
21. Pezdek, K. (1987). "Television Comprehension as an Example of Applied Research in Cognitive Psychology." In: D.E. Berger, K. Pezdek, and W.P. Banks, eds. *Applications of Cognitive Psychology: Problem Solving, Education, and Computing*. Hillsdale, N.J.: Lawrence Erlbaum.
22. Rewey, K.L., D.F. Dansereau, L.P. Skaggs, R.H. Hall, and U. Pitre (1989). "Effects of Scripted Cooperation and Knowledge Maps on the Processing of Technical Material." *J. Educ. Psychol.* 81: 604– 609.
23. Vernon, M.D. (1952). "The Use and Value of Graphical Material in Presenting Quantitative Data." *Occupational Psychology* 26: 22–34.
24. Walker, J.T., and K.J. Scott (1981). "Auditory Visual Conflicts in the Perceived Duration of Lights, Tones and Gaps." *J. Exp. Psychology: Human Perception and Performance* (7): 1327–1339.
25. White, M.A. (1990). "Imagery in Multimedia." *Multimedia Review* 1 (3): 5–8.
26. Washburne, J.N. (1927). "An Experimental Study of Various Graphic, Tabular, and Textual Methods of Presenting Quantitative Material." *J. Educ. Psychology* 18 (6): 361–376.

Discussion

The discussion on session 3 began with an intervention by F. Thompson about the paper on "The Potential Offered by 'Extended Retrieval'." He said that the result of retrieval was expected to be a "file" that is to be directly used by the requestor. In the future, however, the computer itself will need to retrieve an item of data to be used in conjunction with many others to "compute" the response to the user. He asked: "How are 'addressing' into files to be developed and used?" M. Buckland explained that data within files may be found through indexes in the files themselves, and that inferences from the data might be made or aided by computer processing.

N. Streitz wondered about research conducted in library science, computer science, and computer linguistics in an attempt to solve the problems users face in looking for information. M. Buckland said that research was in progress in all three fields and that "successful research is likely to be a combination of these and other fields."

W. Rouse then took the floor to ask why the same technology could not be used to improve source documents themselves. In other words, could computer technology not be used to assure that documents are understandable? – otherwise they would not be published! M. Buckland agreed that new files derived from existing files could well be useful. He said that greater standardization of documents could well be helpful, but difficult to achieve.

Commenting on W. Rouse's position that authors should have responsibility for writing their articles so that better retrieval would be possible, J. Alty informed the audience that the Turing Institute had used the SEML Makeup language to do just this. SEML was used to mark up documents to show structure. The marking scheme was defined by the author. The Institute extended this idea to "semantic" marking – that is, words or fields in the document were marked up and could subsequently be retrieved. "This approach," Alty continued, "provides significant advances in information processing." He explained that they were able, for example, to code up the essential "if-then" rules in operator documents and later extract them and check them for consistency and completeness.

The next comment on the same paper came from C. Cooper. He expressed that M. buckland's paper left the impression that a system that supplies more informaton on each item is self-evidently a better system. "This surely cannot be correct: what we need is just that critical amount of information that is required to make sensible choices." In his reply, the author recalled that historically, information retrieval systems had been designed to retrieve everything with some specified attribute – a fact that can be useful in limited circumstances, such as patent searching. However, it would be better to design systems so that they would normally select the references best matching the user's preference – with an option for requesting more, if need be.

In concluding this part of the discussion J. Kendrew found the idea of restricting the language in documents difficult to accept. "Documents – all documents – are literature; they may be bad or good, but they are literature. Is it not an arrogance to suggest rewriting them? Shakespeare? James Joyce?" It is the job of the information scientists, according to J. Kendrew, to cope with the infinite variety of language as it is, rather than to make life easier for themselves by restricting and so destroying language.

Moving to the paper on "Information Retrieval: Theory, Experiment, and Operational Systems," W. Rouse referred to Thomas Allen's work mentioned therein and commented that traditional information retrieval systems cannot meet users' desires if all they do is provide large numbers of pointers to possible answers. While the information retrieval system need not provide the answer, "more value added is needed – a few good pointers. What do you think?" In his reply, S. Robertson agreed, but underlined that pointers, as opposed to answers, are a valuable function for information retrieval systems, though perhaps Allen's work might suggest pointers to people rather than documents.

N. Streitz then took the floor. He said that while S. Robertson reported on new and innovative ways of providing better access to information retrieval systems, commercial, large-scale systems do not essentially change, staying with their "not very user-friendly interfaces, requiring knowledge about Boolean queries." He asked whether there was any reasonable explanation why this was still the case. Why do they not make use of graphical user interfaces or new retrieval techniques? S. Robertson replied that there were several explanations, though not necessarily good ones. One was that the major host systems use very old software, modified but not fundamentally changed over 20 years. A second reason was that operational systems designers were frightened of the possible computational requirements of associative methods – although some associative methods can in fact be implemented in computationally undemanding ways. Nevertheless, said S. Robertson, "there are now signs of change."

Making a general comment, I. Wesley-Tanaskovic referred to the need for a general concept of information technologies enabling improved or changed conduct of science and engineering, such as post-processing in "front-end" systems by enhancing the delivery of messages or communicating knowledge. S. Robertson agreed and remarked that we have a vast amount of stored human knowledge, in various forms, a very small proportion of which is in machine-readable form. "We have to work with our history," he remarked; "we are not in a position to start from scratch."

In concluding the discussion of his paper, S. Robertson made a general comment on databases. He said that full-text databases, such as press cuttings, have many of the characteristics of bibliographic databases, except that they contain the "information" rather than a pointer to it. Reference works also exist as databases. One challenge is to combine text-retrieval facilities with those associated with relational databases.

D. Lide then addressed the third paper of the session on "Computerized Front-ends in Retrieval Systems." He thought some projects that had developed front-ends to lead the user to the most appropriate database for answering his query have encountered resistance from database vendors, who feel that this might put them at a competitive disadvantage. He wanted to know if this was still a problem. L. Smith replied that there was an example in the front-end EasyNet system of a solution that may be acceptable to the database producer but not to the user. In this particular front-end, database selection involves having the user go through a series of menus, specifying the subject areas and type of material of interest. However, very often more than one database may satisfy this specification. In these cases, the front-end was programmed to divide usage among the different databases, choosing each base with equal frequency but without informing the user that there was more than one possibility in the particular search under consideration. Thus, a user might be led to a better or poorer database, depending on the day the system was used – a situation that satisfies the database producers but cannot be adequate from a user's perspective.

M. Takahashi then asked two questions. First, he wanted to know if there were front-ends that include post-processing capabilities to process the documents drawn from databases. L. Smith explained that some front-ends provide capabilities to build local databases with downloaded records. She added that there was experimentation, as described in papers by N. Dusoulier and M. Buckland, to analyse output using certain attributes, such as year or language of publication. She concluded by saying that "we need to learn more about what users would like to do with text or bibliographic records to further develop this capability of front-ends."

The second question of M. Takahashi concerned non-professional people. He wanted to know what kind of front-end systems were good for them. L. Smith said that it was easiest to design front-ends in support of access to particular databases for particular tasks. "This will be easier for a user of such a front-end since they are working in a particular context."

The discussion on front-ends was concluded with a comment from W. Rouse, who believed that the front-end can be more friendly than the back-end if we separate traditional searching and downloading from specialized and tailored searching of local, smaller databases. "While downloaded databases cannot be any better than the source database, their dramatically reduced size can allow much more sophisticated and friendly processing."

Regarding multimedia technology, W. Rouse asked if it had been found that colour, graphics, animation, video, etc., may be more useful for selling a system and training in initial use, but less so once someone became a "power" user. J. Alty in his reply admitted that "fancy tricks" do help to sell an initial system. It is also true that when new techniques or facilities are provided, designers usually tend to "over use" them, as was the case with the initial use of colour. He said there was an urgent need for a multimedia methodology, because the number of combinations of media is huge. Since media can often assist in transmitting "difficult" knowledge, they are particularly useful for new users or those in need of training. It has been found in experiments that more complex media do not always help. Texts may be found useful in many situations whereas in others graphics, colour, and sound may be crucial. The methodology ought to help us identify the situations where multimedia are useful.

N. Streitz next took the floor to make two comments. The first on the figure in J. Alty's paper plotting "usability" against "knowledge and experience of the problem area." According to N. Streitz, another dimension was needed: the task. J. Alty agreed, but said, however, that the diagram was not originally his; it was from Marmollin and could not be changed. The other question was about the "comparison of texts." N. Streitz felt that there were better ways to do this, such as DIFF programs. J. Alty replied that the example he gave was meant to make a point for multimedia – how, by using properties of the human cognition system, we can significantly simplify a task. If the task becomes more complex, such as paragraphs getting out of order, then DIFF may not work well. He said that they were hoping to set up a system where one document is read over whilst the other is visually followed. There will also be visual aids to help with paragraphs out of sequence. At that point, "we will campare it with DIFF," concluded J. Alty.

Session 4

Intelligent Access to Information: Part 1

Chairperson: Hisao Yamada

Simulated Man-Machine Systems As Computer-aided Information Transfer and Self-learning Tools

Gunnar Johannsen

ABSTRACT

Several negative side-effects of socio-technical changes in industrialized as well as in developing countries are recognized. The potential of information technology for bettering these situations is discussed, particularly with respect to improved training for acquainting people with technical systems and their practical use. Human interaction with integrated automation in man-machine systems is surveyed. This involves discussion of control and problem-solving tasks, decision-support systems, man-machine interfaces, and several stages of knowledge transfer. The main contribution of the paper deals with knowledge-based information access by means of simulation and self-learning tools. The author suggests simulated man-machine systems for training supervision and control tasks as well as computerized tools for self-learning of problem-solving tasks and for information transfer of broader man-machine systems engineering issues in a societal context. The cultural aspects of such tools as well as the needs for future research and for future socio-technical development are also emphasized.

1. Introduction

Presently we are living in times of rapid socio-technical changes with certain dangers of possible instabilities. These changes refer, on the one hand, to
- progress in science and technology, particularly in information and communication technologies, and, on the other, to
- inequalities in the standard of living, for example, with respect to food

supply, housing, health care, and mobility, as well as in individual and social freedom,
- environmental and social consequences of thoughtless technological developments,
- the need for conversion of many military and very high-risk technologies,
and
- the need for trust, responsibility, tolerance, and mutual support in individual, social, and international relations.

One may come to the conclusion that during the last two to three decades,
many problems on our globe have worsened. Certainly, parts of the world
have attained higher standards of living, more individual and social freedom, and higher mobility than ever experienced before. However, larger
parts of the world are far from reaching these standards and have even suffered deteriorations in several cases. Thus, the poor and the developing
countries are sometimes lagging further behind, and they see much larger
gaps between themselves and the highly developed countries than a few decades ago.

What about the industrialized countries themselves? There, technological
developments have also brought about a large number of negative long-term
effects. Environmental problems due to the consumption of energy and
materials, traffic, and waste (radioactive, toxic, and non-toxic) have created
many concerns with respect to the future qualities of air, water, and soil.
Other technological developments have led to social consequences such as
unemployment, unreasonable distribution of work, and isolation. Traffic
congestion in cities and on highways as well as fatal accidents with cars, aircraft, power plants, and industrial processes may remind us of our limited
capabilities and understanding. Some experts have recently estimated that
the delayed consequences of the reactor catastrophe of the Chernobyl
nuclear power plant in 1986 will amount to between one and two million casualities over the next years. Additional risks have only started to become visible to the general public, since the end of the Cold War between
the West and East blocs of the industrialized countries makes most of the
industrial-military complex obsolete on both sides. The need for conversion
of institutions, systems, and materials has been recognized as a huge challenge.

It seems that the amount of clearing work to be done in the near future
will be overwhelming. To distrust science and technology because of their
negative side-effects will certainly not be an appropriate solution. Instead,
science and technology are badly needed to perform the clearing work in an
appropriate way and also for improving the standard of living – substantially
in poor and developing countries, but also moderately in developed countries. Often, several degrees of development exist within a single country.
Therefore, suitable strategies are required that try to avoid earlier errors,
when further developing technologies and societies. In order to accomplish
this enormous task for mankind in the near future of the next two to three
decades, we need a spirit of trust, responsibility, tolerance, and mutual sup-

port in individual, social, and international relations. And we all need to share in the extra work to be done as well as in the financial and socio-psychological burdens. I believe that information, communication, and truthfulness together with the adoption of a realistic long-term balance of costs and benefits will contribute to building up this spirit and the willingness to share as a prerequisite for peaceful developments.

Information and communication technologies will therefore play an important role for humane development all over the world. However, the very many possible misuses of these technologies need to be banned internationally as much as does the violation of human rights. Each individual human being must have the possibility of access to any information he or she wishes to receive and to understand. Because of the many individual limits to understanding different kinds of information, special tools have to be available for navigating through the information.

In this paper, I will restrict myself to a special kind of tool for computer-aided information transfer and self-learning. Through a further restriction, the paper mainly concentrates on simulated man-machine systems that are to be used as information tools. They are dynamic and interactive and, thus, can demonstrate and teach how technical systems – the machines – are to be operated and maintained. Combined with regularly renewed licensing of human task responsibilities, such tools may possibly help to avoid at least some of the most fateful accidents. Before discussing these tools in more detail in section 3, the most prominent technical and cognitive aspects of man-machine systems will be introduced in section 2. The paper concludes in section 4 by pointing out the need for further work on realizing the suggested tools that now exist at best only in laboratory versions.

2. Human Interaction with Integrated Automation in Man-Machine Systems

2.1 Control and Problem-solving Tasks

All dynamic technical systems that are operated by one or several human beings can be viewed together with these human operators as man-machine systems. Thus, all kinds of vehicles, continuous and discrete industrial processes, power plants, robots and manipulators, business and public information systems, biomedical support systems, and many more are man-machine systems in this sense. To achieve the prescribed goals of safe and efficient operation of man-machine systems, two main categories of tasks, controlling and problem solving, have to be performed. In principle, both task categories can be performed by the human operator(s) as well as by automatic computerized systems.

The control activities comprise reaching, open-loop and closed-loop control in the narrower sense of control theory, monitoring, and lower supervisory control functions such as intervening in automated processes. On higher

Different levels and phases of human and automatic controlling and problem solving
(after ref. 13)

Behavioural levels Information-processing phases	State-oriented level	Context-oriented level	Structure-oriented level
Categorization	Signal (and alarm) detection, state estimation	Pattern recognition and matching, analysis of sequential observations	Situational and system identification
Planning	Fixed related (between categorization and action)	Script selection	Plan generation and adaption
Action	Stereotyped automatic control	Symptomatic rule application	Topographic rule application

cognitive levels, problem-solving activities have to be performed. These include fault management, particularly with fault diagnosis and correction, goal setting and hypothesis generation, planning, and the higher supervisory control functions such as teaching. The distinction between control and problem-solving tasks is especially advocated by Johannsen [11,12]. The latter is a task-oriented concept, whereas the overlay of the supervisory control paradigm is an interaction-oriented concept. The latter consists of a hierarchical, decentralized structure with task-interactive computers at the lower level and human-interactive computer(s) at the coordinating, higher level [26]. The five generic supervisory control functions are planning, teaching, monitoring, intervening, and learning.

Human and automatic controlling and problem solving show several similarities, but there are also differences between both forms of the two task categories. A further classification of these task categories is illustrated in the table. The table shows human and automatic activities separated into three succeeding information-processing phases, namely categorization, planning, and action (see also Rouse [21] and Sundström [30]). These information-processing phases can be cycled through on three distinct behavioural levels, namely the state-oriented, the context-oriented, and the structure-oriented levels, which correspond in some way to the cognitive levels of behaviour as suggested by Rasmussen [18, 19].

The main domains of automation systems are the complete state-oriented level as well as at least some parts of the categorization phase of the context- and structure-oriented levels. Of course, human operators can also be engaged in these activities. However, they are particularly superior to automation systems in the planning and action phases of the context- and structure-oriented levels. These areas of table 1 correspond more to the problem-solving tasks and, thus, require higher cognitive behaviour. This is even true when prescribed plans, so-called scripts, have to be selected and rules for actions have to be applied on the basis of observed symptoms. The human operator is particularly indispensable when new plans have to be generated and rules have to be applied on a topographic or completely structural relationship because an unforeseen situation has to be dealt with.

The recent aircraft accident near Stockholm is a good example of the latter case, although it also has elements of the context-oriented level. After the engines no longer worked as the aircraft neared the ground, the pilots had only a very short time to plan and execute an emergency landing on a field just behind a forest. All passengers survived because of the excellent human performance. Where to glide down with the aircraft had to be decided on the structure-oriented level, whereas many of the subordinated guidance and control activities were probably done with some kind of script selection. These scripts were built up during previous simulator training. It is a general policy that pilots are well trained by means of simulated critical events in order to build up more automated scripted behaviour that can be reproduced much faster than the generation of new plans but will hopefully never be needed in reality.

The higher cognitive functions as described with table 1 can also, at least partially, be handled with new information technologies. There are knowledge-based decision support systems. Similarly, like human operators, these decision support systems mainly process symbolic information about contexts and structures rather than numerical information about signals and states. Thus, it is necessary to realize some kind of signal-to-symbol transformations and vice versa to be performed by the human operator or the computerized system when freely navigating through the whole plane of table 1.

Appropriate function and task allocations determine which tasks will best be performed by the human operators and which will best be assigned to the computerized systems. With an integration between traditional automation systems and the more recent information technologies of the decision-support systems, human-centred designs and dynamic forms of task allocations are possible (see, e. g., Rouse [22]).

2.2 Integration between Traditional Automation and Decision-Support Systems

Human-centred designs of integrated automation will lead to improved man-machine systems in the very near future. These perspectives occurred because of knowledge-based information technologies. The term integrated automation means that traditional automation systems and knowledge-based decision-support systems cooperate closely in a suitable way. The criteria for what is suitable, however, need to be defined and require the human-centred approach that is also part of the integrated automation concept.

As elaborated in more detail by Johannsen [11], an extended Operator (User) Interface Management System architecture for dynamic technical systems allows the description of the integrated automation concept with a number of separated levels (see also Alty and Johannsen [2]). Figure 1 shows these levels of the UIMS architecture. The features of man-machine interfaces will be reviewed in the next subsection. Here I discuss the different levels of the technical system as shown in figure 1. The information selector 1 processes numerical information, whereas the information selector 2 transforms signals into symbols for the information processing at the decision-support level. This level cooperates with the traditional automation of the supervision and control level.

Decision-support systems can be subdivided into those that are more application oriented and those that are more human operator oriented. In figure 1, they are called application model support and operator model support, respectively. The application model support systems deal with situations in the technical process and the supervision and control system. The information that is processed for these situations is then communicated to the human operator as a decision support. Operators will normally perform a dialogue with such decision-support systems in order to satisfy their information needs. These application model support systems can also be viewed with respect to the behavioural levels as suggested in table 1. Examples for these three levels are the decision support for heuristic control [17], for fault diagnosis (e.g., Borndorff-Eccarius [3]), and for plant management; see Johannsen [11] for more details.

Similarly, the operator model support systems deal more directly with human operators' behaviour and help them to perform as well as possible. Again, three behavioural levels can be distinguished. Examples are human error evaluation (e.g., Hessler [9]), plan or intent recognition (e.g., Rubin et al. [24]), and procedural support (e.g., Sundström [30]); more details are given in Johannsen [11].

A mathematical framework for interaction between traditional automation and decision-support systems is supplied by Johannsen [11] and explained with four case-studies. This mathematical framework requires much more precision and details that can only become available after a lot of further multidisciplinary research.

2.3 Man-Machine Interfaces with Graphical and Dialogue Support

The presentation level and the dialogue level describe the two distinct aspects of any man-machine interface, as shown in figure 1. The presentation level concerns the form of the information and, thus, includes the displays and controls, which are more and more often computer graphics screens and computerized control input systems, as well as some knowledge-based graphics support modules. The dialogue system is a kind of central intermediary between the human operator(s) and all levels of the whole technical system, including the decision-support systems, and deals with the contents of information flows in the man-machine communication. Loose

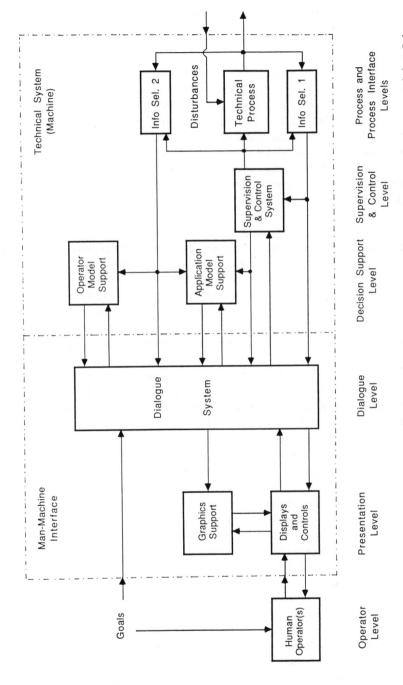

Figure 1 Extended Operator (User) Interface Management System structure for dynamic technical systems (after Johann-sen, 1992)

ends of dialogue can occur, particularly in critical situations, when an urgent request of a subsystem of the technical system interrupts the interaction of another one with the human operator [2]. Handling these loose ends of dialogue can be supported by small-scale knowledge-based dialogue assistants for the interaction with the different subsystems of the technical system.

The visual display units with colour graphics screens allow flexible and task-adapted forms of presentation. Many options for picture design exist, but the multitude of alternatives can also be misused if the designers lack enough knowledge about ergonomics and the information needs of human operators [10]. Dynamical shifts of information needs in the problem space of the two dimensions of the degree of detail and of the level of abstraction must be taken into account even more with advanced information display concepts, as already suggested by Goodstein [8]. The design of graphics support can lead to knowledge-based graphical systems that dynamically generate new picture contents by using knowledge of the application, of the operator model based on information search needs, of graphical presentation techniques, and of dialogue techniques. Such concepts require an integration of computer graphics and knowledge-based technologies [5, 6]. A similar approach for knowledge-based designer support and intelligent man-machine interfaces for process control was suggested by Tendjaoui, Kolski, and Millot [32]. The cognitive engineering approach must also be applied here to answer the question of how to present information to the human operator(s) and to design appropriate graphical decision support for the goal-directed, spatial visualization of task-meaningful units and their relations [34]. The searcher for which interaction media to use has, among others, to consider recent results on multimedia research [1].

Advanced man-machine interfaces that integrate knowledge-based task and operator models as decision-support systems with presentation and dialogue submodules were particularly investigated by Tendjaoui, Kolski, and Millot [32] and by Sundström [30], and reviewed by Johannsen [11]. Application and process knowledge as well as ergonomic knowledge and knowledge about the information search needs of the human operator(s) in different operational situations determine which information needs to be displayed, when, and how. In principle, cooperative interactions between several human operators and the knowledge-based decision-support systems are possible. However, the state of the technology has not yet advanced to the level of rigorous industrial applications.

2.4 Transformation and Use of Knowledge by Knowledge Engineers, Designers, and Operators

Human operators do not use only their own knowledge in interactions within man-machine systems. In addition, a lot of knowledge is available to them directly or indirectly through the information acquisition system, the supervision and control system, and the decision-support systems. Much of this

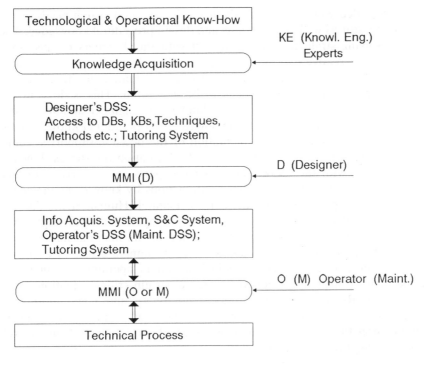

Figure 2 Stages of knowledge transfer and man-machine interaction for knowledge engineers, designers, operators, and maintenance personnel

knowledge has been supplied by the designers of these systems. The maintenance personnel find themselves in a similar situation; however, the decision-support systems, as well as the man-machine interfaces designed for them, may be different from those for the human operators. This is indicated in the lower part of figure 2. The figure also shows that in addition, a tutoring system may have been designed to teach novices how to operate or to maintain the whole technical system.

The designer creates knowledge support implemented in the different subsystems for the human operators and the maintenance personnel. Similarly, the designers themselves can be assisted by appropriate decision-support systems or, in the case of novices, by a tutoring system. This too is shown in figure 2, as well as the fact that the man-machine interface for the designer is normally completely different from those for the operator and the maintenance personnel. As mentioned above, the designers' decision-support systems may include an application model with knowledge and data about the technical process and its supervision and control system; an operator model will be provided with knowledge about the information search needs of the human operator as well as ergonomic and design procedural knowledge. The problem-solving strategies are mainly left with the designers because they determine the creative part of any design.

As is clear from figure 2, the knowledge contained in the designers' decision-support systems also needs to be generated. This is done in the knowledge acquisition process by the knowledge engineer and/or the domain experts who supply their technological and operational know-how for the conceptualization and implementation of the decision-support systems. Thus, the whole process of knowledge and information transfer as shown in figure 2 starts with the know-how of some experts and is carried through successive stages with their respective interactions by knowledge engineers, designers, and operators or maintenance personnel. Human errors that are made at any of the earlier stages may be propagated to later stages and, finally, may adversely influence the operation of the man-machine system. Reason [20] pointed out, for example, that the human errors of designers can sometimes be compensated for by the human operators but may also be the main contributing factor to turning a certain critical situation into a catastrophe. Regulations with product liability are more and more concerned with these aspects. Errors in early stages may eventually have severe consequences that call for careful avoidance and correction strategies as well as for training towards responsibility.

3. Knowledge-based Information Access by Means of Simulation and Self-learning Tools

3.1 Training Supervision and Control Tasks with Simulated Man-Machine Systems

Simulators of cars, trucks, aircraft, ships, power plants, chemical plants, and other technical processes are valuable tools for research and development as well as for training. They have a long tradition in aeronautics and astronautics, but nowadays are more often used in other application domains. The appropriate simulation fidelity is an important technical, psychological, and economic issue. Often, part-task simulators are sufficient, particularly for research. The simulation fidelity needs to be higher when professionals are to be trained in how to deal with difficult task situations. On the other hand, cheap simulators would be sufficient for less difficult tasks such as driving a car. A new technology for such inexpensive simulators for regular refresher training of drivers is needed, in my opinion, as one countermeasure to the far too many traffic accidents.

Simulated man-machine systems will be a technology that goes beyond the use of simulators. Here, the human operator's behaviour in supervision and control tasks is also simulated. This requires fairly accurate human performance models as well as models of mental workload. As Stassen et al. [29]

summarized, such models are available for lower cognitive levels of human behaviour such as those required in manual control, failure detection, and monitoring. The use of these models in closed loops with the technical system leads to completely simulated man-machine systems. They can demonstrate in real time to any one interested how good or bad human interaction with a specific technical system may be. For training purposes, play-back capabilities for repeating any situation of interest, human interaction facilities with the possibility of comparing the trainee's with the ideal or bad simulated behaviour and performance, and explanations in separate text and graphics windows are necessary. Particularly with explanation facilities, such a simulated man-machine system becomes a tutoring system or self-learning tool for the training of supervision and control tasks for specific technical systems.

The three cognitive levels of human behaviour suggested by Rasmussen [19], namely skill-based, rule-based, and knowledge-based behaviour, also have to be addressed appropriately during any training process. As Sheridan [25] proposed, computer aiding has to supply different support and needs different forms of interaction at these three levels. Demonstrations seem to be appropriate at the lowest level, whereas rules and advice are called for respectively at the two higher levels. A simple example may illustrate this. We assume a simulated man-machine system for training in car-driving tasks and select a typical driving situation on a motorway. This situation may be characterized by driving into and through a construction zone with narrower lane width and dense oncoming traffic. The demonstrations may show the outside view in real time for different selected speeds and may exemplify certain critical situations or even accidents. Rules for explaining the reasons behind the speed limits and the behavioural choices in critical situations may be shown in a separate window on the computer screen. This needs to be visible when observing the simulated outside view on the same colour graphics screen but shall not impair the main driving task. The explanations may be further supported by additional demonstrations. Only after the on-line demonstrations are terminated is the knowledge-based advice given. The human operator can take as much time as needed to interrogate the computer support in an interactive dialogue fashion. In this way, knowledge-based behaviour can be trained and, thus, through training, becomes rule-based behaviour.

Time pressure in the latter training phase and, generally, on the knowledge-based behavioural level is dangerous because it can even lead the human operator to fall back on the skill-based behavioural level with, then, a high risk of human errors [4]. The very last phase of the Chernobyl nuclear power plant accident can be explained by the choice under dramatic time pressure of the wrong skill-based behaviour instead of the necessary knowledge-based behaviour [16]. It may even be possible to demonstrate such human errors on-line in the simulation by freezing certain situations and by giving additional explanations for the wrong behaviour and for the causes behind it.

3.2 Self-learning Tools for Problem-solving Tasks

Not only supervision and control tasks but also the higher cognitive problem-solving tasks need to be trained appropriately. Unfortunately, our research results, particularly with respect to modelling problem-solving tasks, are not as well developed as our knowledge about control tasks [15]. It is even arguable whether it will ever be possible to model higher cognitive human functioning. Human creativity is a major ingredient in these functions and is mainly responsible for keeping the human being in the man-machine system. Planning tasks have not as yet been modelled sufficiently. Some models exist for human fault diagnosis tasks [23]. The more important research results contribute evaluations and concepts for experimental fault-diagnosis situations. Also, knowledge-based decision-support systems as computer aids for fault diagnosis were developed, as mentioned above (see also Johannsen and Alty [14]; Tzafestas [33]).

The idea here is to use and to integrate concepts, experimental results, models, and computer support systems for developing tools that allow self-learning and the information transfer of problem-solving tasks. The combination of such tools with the on-line simulation tools, as described in the preceding subsection, may further enhance the quality of the training system. It must be possible for human beings anywhere in the world to sit by themselves in a silent place in order to learn or just to understand basically how a particular technical system can be managed and operated, and how problems with this system can be solved once they occur. A reasonably large personal computer should be sufficient for such self-learning and information transfer purposes. Only the recent hardware revolution makes it feasible to supply enough computer power for such new training endeavours, even in geographically very remote places and in fairly limited economic situations. In the latter cases, access to the computer tools has to be organized for sharing their use in a way similar to that of using public library facilities.

Two off-line advisory systems for training maintenance personnel in fault-diagnosis capabilities will be described here as existing examples for self-learning tools in problem-solving tasks. These two systems are FAULT (Framework for Aiding the Understanding of Logical Trouble-shooting), developed by Rouse and Hunt [23] and their colleagues over several years, and ADVISOR (Advanced Video Instructor), developed by Tanaka and others [31].

The original idea of FAULT was to investigate whether generalizable capabilities for fault diagnosis exist in human beings. Therefore, an early experiment was designed with the very abstract task of diagnosing a single fault in a complex network of nodes. In a second, more realistic example, the faulty components had to be identified in an electrical network with AND and OR gates. The connections between the components also included feedbacks. Both examples are suitable for training human operators in fairly general, context-free skills. With the FAULT system, a series of further experiments was performed that were context-specific, as in real-life tasks. The chosen networks were now schematics of car, aircraft, and ship engines with their main subsystems and components. The human subjects started the fault diagnosis with rather general symptoms (e.g. the engine will not start). Then they had to check gauges, ask for information about specific components, make observations, or remove components for testing. The symptom, the status of all available gauges, and the selected actions with their costs were displayed in different windows on the computer screen. Several forms of computer aiding based on models of human problem solving were used to assist the maintenance personnel in their troubleshooting tasks. The results of the experiments showed that human problem solving, which depends on how well the human operators understand the problems, is not optimal; see Rouse and Hunt [23]. There seems to be a trend towards context-dominated problem-solving behaviour. This is, however, not completely context-specific, as can be seen from human beings' abilities to make transitions into unfamiliar problems and to cope with ill-defined and ambiguous situations.

The recognized human cognitive limitations in problem-solving tasks can be overcome by computer-aiding systems based on some problem-solving performance models. Simpler computer aids may just be structure-oriented bookkeeping tools. In FAULT, it was shown that the bookkeeping methods consistently improved human performance, even after later transfer to unaided problem solving. FAULT was also used as a self-learning tool for maintenance personnel of ship engines. Like a tutor, the tool guided the human subjects through different knowledge-acquisition stages and problem-solving phases. Nowadays, similar tools are used for individual self-learning and for quick prescribed testing of the required status of knowledge at freely selected assessment times in airlines. These tools supplement the use of aircraft simulators in pilot training.

The ADVISOR system is another learning environment for maintenance [31]. It was developed jointly by the Tokyo Electric Power Corporation and Mitsubishi Electric Corporation. An interface-centred design approach was chosen, whereby multimedia techniques were also used. The system has some pedagogical interfaces to enhance the understanding of novices and to help them manage troubles and make appropriate guesses. Useful information and suggestions are provided in a format best adapted to the current understanding of the human subject, although the system does not provide a perfect model of human problem-solving performance. A prototype was developed for the maintenance of gas insulated switch gears in the power industries.

ADVISOR is implemented on a workstation, a personal computer, a videodisc, and a TV display. A real TV picture of the interesting views on and into the equipment in the maintenance situation is shown on the TV display with superimposed text windows for explanations and touch-sensitive virtual keys for interrogations. On the neighbouring colour graphics display of the workstation, learning from examples can be supported by explaining what the cause-consequence structures for a number of alarms look like in a multiple fault situation. Thus, the novice can learn the principles behind the examples. Also, the pedagogical interfaces can be shown with several windows on the display of the workstation as supports for recognizing the current status in the whole learning space as well as the significance of each step and the intention behind it in the maintenance sequence. It is not known to this author whether the evaluation of the ADVISOR system was terminated, and what kind of results were achieved.

3.3 Self-learning Tools for Understanding Man-Machine Systems Engineering

Any man-machine system is a part or subset of a larger system with broader boundaries in the systems engineering sense. This has already been shown in figure 2 with the relationships between knowledge engineering, designing, operation, and maintenance; several man-machine systems are involved in this case. Nowadays, even broader contexts and perspectives need to be understood by different human agents dealing with man-machine systems; these include political decision makers, managers, designers, or operators. Different levels of understanding in different contexts are required. For example, environmental and economic issues, matters of legislation and liability, dependencies within a logistics network and of market developments, and many more technical and non-technical influences are to be considered in a real-life systems engineering approach towards decision support, design, and operation (see figure 3).

If knowledge about the needs and the functions of safe systems operation and careful problem solving in man-machine systems is better understood all over the world, then people will probably be more concerned about the selection and the proper use of many technical systems in their societies. The current challenges of technology transfer into developing countries and of conversion with risky systems in the industrialized countries, as mentioned already in the introductory section of this paper, may be better and more rapidly understood; possibly more competent opinions might be expressed if self-learning and information transfer tools were already available. Certainly a particularly strong need for such tools exists among high-level decision makers and public opinion moderators, namely among politicians, leaders of big and medium-sized companies and banks, and journalists (see figure 3). Their knowledge is often too limited and their time to learn something new very restricted, but they nevertheless have to make decisions quickly and as correctly as possible or to comment on actual events almost immediately. The suggested computerized tools for self-learning and information transfer will certainly complement libraries and more traditional information retrieval systems.

In a way similar to that in the prototypes mentioned in the preceding sub-

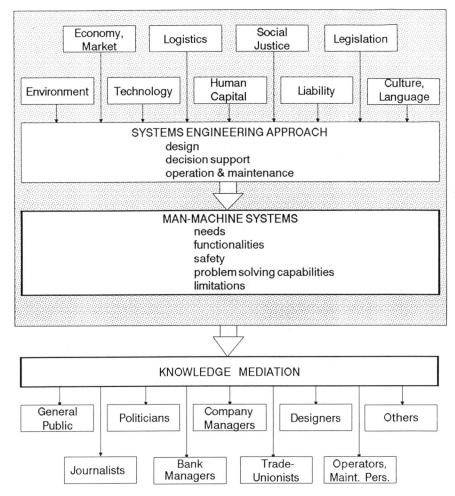

Figure 3 Understanding man-machine systems engineering

section, namely FAULT and ADVISOR, self-learning tools for understanding the broader issues of human interaction with technical systems in a
societal context have to draw on a combination of decision-support systems,
appropriate multimedia presentation and dialogue techniques in advanced
man-machine interfaces, and explanation and justification facilities. The design approach for such tools has to start from requirements and cognitive
task analyses that lead to the different decision-support functions needed for
different types of later users of the tools. The explanation and justification
facilities as well as the advanced man-machine interfaces have to be adapted
to different information transfer and learning needs that are to be satisfied
by the different decision-support systems.

3.4 Cultural Aspects of Using Simulation and Self-learning Tools

Simulation and self-learning tools, as mentioned in subsections 3.1 through 3.3, have to be used in free societies by diffferent kinds of people with different kinds of education, different cognitive abilities for understanding systems, functions and context, and different kinds of responsibilities. Thus, one may say that different cultural patterns for approaching problems and interacting with socio-technical systems exist even within one particular society. Sometimes the differences between two distinct views of the same system are cultivated unnecessarily, as often occurs with the distinction Snow [28] showed between the two cultures of science, namely one for the human and the social sciences and one for the natural and the engineering sciences. Bridging the gaps between all kinds of possible views seems to be more and more important for understanding our world and for contributing to a peaceful future.

These gaps between cultures can be much larger when we look at the differences between societies that developed quite independently from each other over many centuries. Take the Japanese and the German cultures as examples that show a completely distinct dimension of cultural differences from those within either of these two societies. Sheridan et al. [27] dealt with many facets of these cultural apects.

Although the gaps between different cultures, individual as well as societal, became much smaller through the influence of many technologies, particularly the information technologies, many unique traditions and language barriers remain. They have to be accepted and even preserved with patience, goodwill, and historical sense. However, they need also to be overcome mentally in order to allow worldwide cooperation as well as exchange of ideas and products in a successful and mutually rewarding manner. The ideas and technologies of one culture have to be appropriately adapted to another culture when a high degree of user acceptability shall be reached. This requires sensitive tutoring for systems designers in the delivering culture as well as concerned training for end-users in the receiving culture. Thus, two types of simulation and self-learning tools may well be suitable. For the designers, one type can aid the understanding of other cultures. For training purposes, the other type can support understanding of the culturally adapted technology. Multilingual explanation facilities seem to be important aids where appropriate and where necessary to enhancing understanding for the tools that have to consider cultural differences. The use of voice output and input within a multimedia environment may even allow illiterates to get acquainted with new technical systems by means of the suggested tools.

The need for investigating and for considering cultural aspects in the design and use of technologies, in particular automation and information technologies, have only recently been recognized, also by at least some engineers and computer scientists. In the International Federation of Automatic Control (IFAC), a Working Group on Cultural Aspects of Automation was formed in 1990, for example. They organized their first workshop in October 1991 [7].

4. Needs for Future Research and Socio-technical Development

The overall objectives mentioned in the introduction of this paper for using computer-aided information transfer and self-learning tools can only be achieved with further research combined with particular socio-technical developments. Future research is needed in the areas of human performance modelling and problem solving, goal and knowledge structures, cognitive limitations, knowledge acquisition, and all their contributions to the design of more cooperative interactive decision-support systems. More research is required in the development of knowledge-based explanation and justification facilities as well as in advanced man-machine interfaces with multimedia presentation and dialogue techniques. Information retrieval and tutoring techniques also have to be further developed for the particular needs of these tools.

In addition to the future research requirements just mentioned, the socio-technical development of simulated man-machine systems and self-learning tools has to be fostered for many application domains. An integrated multi-disciplinary approach towards development and training for expanding information access seems to be mandatory. Several specialized companies as well as research and development centres with an existing or planned international operational basis have to take up the new challenges of information dissemination in a more systematic way. Then we can hope that a sufficient number of people all over the world will appropriately understand human interactions with all kinds of technical systems in order to contribute through responsible decisions to the future and peace of mankind.

Of course, the most elementary needs of food, shelter, and health have to be satisfied first. This has yet to be accomplished for more than 40 per cent of the world population, due in particular to the lack of environmental conservation according to a very recent report of the World Health Organization. Only after an acceptable minimum standard of living is reached will people be concerned with seeking better information supply. However, even for solving these elementary problems, a lot of information support is required, for example in order to enforce environmental conservation. Thus, we face a vicious circle. To get out of this is also very much a matter of politics, economics, and world trade. But nothing can be accomplished without more proper use of science and technology.

REFERENCES

1. Alty, J.L. (1992). "Multimedia Technology: A Design Challenge." See p. 172, this volume.

2. Alty, J.L., and G. Johannsen (1989). "Knowledge-based Dialogue for Dynamic Systems." *Automatica* 25: 829–840.

3. Borndorff-Eccarius, S. (1990). *CAUSES – State-based Diagnosis Support Expert System*. ESPRIT-GRADIENT P857, Report no. IMAT-MMS-11. Labor. Man-Machine Systems, University of Kassel (GhK).

4. DeKeyser, V. (1986). "Technical Assistance to the Operator in Case of Incident: Some Lines of Thought." In: E. Hollnagel, G. Mancini, and D.D. Woods, eds. *Intelligent Decision Support in Process Environments*. Berlin: Springer-Verlag, pp. 229–253.

5. Elzer, P., and G. Johannsen, eds. (1988). *Concepts, Design, and Prototype Implementations for an Intelligent Graphical Editor (IGE1)*. ESPRIT-GRADIENT P857, Report no. 6. Labor. Man-Machine Systems, University of Kassel (GhK).

6. Fejes L., G. Johannsen, and G. Strätz (1992). "A Graphical Editor and Process Visualisation System for Man-Machine Interfaces of Dynamic Systems." *The Visual Computer*. Forthcoming.

7. Forslin, J., and P. Kopacek, eds. (1992). *Cultural Aspects of Automation*. Vienna: Springer-Verlag.

8. Goodstein, L.P., ed. (1985). *Computer Aided Operation of Complex Systems*. NKA Report no. LIT (85)5, Riso.

9. Hessler, C. (1989). "Use of the Task Model for Detection of Operator Errors and for Flexible Task Allocation in Flight Control." In: *Proceedings of the ESA/ESTEC Workshop on Human Factors Engineering: A Task-Oriented Approach*, Noordwijk.

10. Johannsen, G. (1990). "Design Issues of Graphics and Knowledge Support in Supervisory Control Systems." In: N. Moray, W.R. Ferrell, and W.B. Rouse, eds. *Robotics, Control and Society*. London: Taylor and Francis, pp. 150–159.

11. Johannsen, G. (1992). "Towards a New Quality of Automation in Complex Man-Machine Systems." *Automatica* 28: 355–373.

12. Johannsen, G. (1993). *Mensch-Maschine-Systeme*. Berlin: Springer-Verlag. Forthcoming.

13. Johannsen, G., S. Borndorff, and G.A. Sundström (1987). "Knowledge Elicitation and Representation for Supporting Power Plant Operators and Designers." In: *Proceedings of the First European Meeting Cognitive Science Approaches to Process Control*. Marcoussis, France.

14. Johannsen, G., and J.L. Alty (1991). "Knowledge Engineering for Industrial Expert Systems." *Automatica* 27: 97–114.

15. Johannsen, G., A.H. Levis, and H.G. Stassen (1992). "Theoretical Problems in Man-Machine Systems and Their Experimental Validation." *Proceedings of the Fifth IFAC/IFIP/IFORS/IEA Symp.: Analysis, Design and Evaluation of Man-Machine Systems*, The Hague, June 1992. Forthcoming.

16. Munipov, V.M. (1990). "Der menschliche Faktor bei Havarien in den Kernkraftwerken Tschernobyl und Three Mile Island." In: *2. internat. Kolloquium Leitwarten*. Cologne: Verlag TÜV Rheinland, 239–247.

17. Nakagawa T., and H. Ogawa (1986). "The Identification and Control, Partially Added with the Artificial Intelligence Approach." In: *IFAC Conf. SOCOCO*, Graz, pp. 121–126.

18. Rasmussen, J. (1983). "Skills, Rules and Knowledge, Signals, Signs, and Symbols, and Other Distinctions in Human Performance Models." *IEEE Trans. Systems, Man, Cybernetics* SMC-13: 257–266.

19. Rasmussen, J. (1986). *Information Processing and Human-Machine Interaction*. New York: North-Holland.
20. Reason, J. (1990). *Human Error*. Cambridge: Cambridge University Press.
21. Rouse, W.B. (1983). "Models of Human Problem Solving: Detection, Diagnosis, and Compensation for System Failures." Special Issue on Control Frontiers in Knowledge-based and Man-Machine Systems. *Automatica* 19: 613–625.
22. Rouse, W.B. (1991). *Design for Success*. New York: Wiley.
23. Rouse, W.B., and R.M. Hunt (1984). "Human Problem Solving in Fault Diagnosis Tasks." In: W.B. Rouse, ed. *Advances in Man-Machine Systems Research* 1. Greenwich, Conn: JAI Press.
24. Rubin, K.S., P.M. Jones, and C.M. Mitchell (1988). "OFMspert: Inference of Operator Intentions in Supervisory Control Using a Blackboard Architecture." *IEEE Trans. Systems, Man, Cybernetics* 18: 618–637.
25. Sheridan, T.B. (1987). "Supervisory Control." In: G. Salvendy, ed. *Handbook of Human Factors*. New York: Wiley, pp. 1243–1268.
26. Sheridan, T.B. (1992). *Telerobotics, Automation and Human Supervisory Control*. Cambridge, Mass.: MIT Press.
27. Sheridan, T.B., T. Vámos, and S. Aida (1983). "Adapting Automation to Man, Culture and Society." Special Issue on Control Frontiers in Knowledge-based and Man-Machine Systems. *Automatica* 19: 605–612.
28. Snow, C.P. (1959). *The Two Cultures*.
29. Stassen, H.G., G. Johannsen, and N. Moray (1990). "Internal Representation, Internal Model, Human Performance Model and Mental Workload." *Automatica* 26: 811–820.
30. Sundström, G.A. (1991). "Process Tracing of Decision-making: An Approach for Analysis of Human-Machine Interactions in Dynamic Environments. *Internat. J. Man-Machine Studies*, pp. 843–858.
31. Tanaka, H., S. Muto, J. Yoshizawa, S. Nishida, T. Ueda, and T. Sakaguchi (1988). "ADVISOR: A Learning Environment for Maintenance with Pedagogical Interfaces to Enhance Students' Understanding." In: H.-J. Bullinger et al., eds. *Information Technology for Organisational Systems*. Amsterdam: North-Holland, pp. 886–891.
32. Tendjaoui, M., C. Kolski, and P. Millot (1991). "An Approach towards the Design of Intelligent Man-Machine Interfaces Used in Process Control." *Internat. J. Industrial Ergonomics* 8: 345–361.
33. Tzafestas, S. (1991). "Second Generation Diagnostic Expert Systems: Requirements, Architectures and Prospects." In: R. Isermann, ed. *Fault Detection, Supervision and Safety for Technical Processes*. Preprints IFAC/IMACS Symp. vol. 2, pp. 1–6.
34. Woods, D.D. (1986). "Paradigms for Intelligent Decision Support." In: E. Hollnagel, G. Mancini, and D.D. Woods, eds. *Intelligent Decision Support in Process Environments*. Berlin: Springer-Verlag, pp. 153–173.

Human-centred Design of Information Systems

William B. Rouse

ABSTRACT

The design of information systems is considered in terms of the viability, acceptability, and validity of the information support provided. These issues are discussed in the context of several examples, including systems for bibliographic information retrieval, aircraft operations, maintenance information, sales transaction support, and design information. A variety of "lessons learned" that illustrate the impact of adopting a human-centred approach to designing information systems is summarized.

1. Introduction

Information is an essential ingredient in much that we do. We spend much time gathering, refining, and interpreting information. This process of digesting information has become increasingly complex as the store of information has become larger and more diverse.

The size of the information store makes it difficult to consume all relevant information. The increasing diversity of information sources, forms, and languages makes it difficult to identify and interpret all relevant information. Often indigestion and sometimes "information poisoning" result.

Information technology appears to provide the means whereby these problems can be overcome. Hypermedia, multimedia, natural language processing, expert systems, and CD-ROM are notable examples. Numerous commentators over the past 40 years projected that technologies such as

214

these would soon help us to deal with the information explosion. Each of these commentators has, at best, been a bit too optimistic.

One could argue that the proclamations of success were premature because the cost and/or power of computer technology, as well as related technologies, did not evolve as quickly as anticipated. While this may be true, I believe that more subtle problems have hindered progress. Put simply, enabling technologies such as those noted above may be necessary, but they are not sufficient for success.

To support this assertion, consider the following two examples. In the late 1970s and early 1980s, we undertook an effort to put hard-copy procedural information on-line. It seemed intuitively obvious that problems associated with the growing size and number of technical manuals could be lessened, or perhaps eliminated, by moving to computer-based information systems.

Two studies were performed in the context of aircraft operations manuals [5, 6]. Of particular interest here is the first study, where one condition involved putting on the computer display the exact same information, in the same format, as in the hard-copy manuals. Experimental results for this condition indicated the computer-based system was substantially inferior to the hard-copy presentation.

The problems appeared to be due to the inherently limited screen size and the distinct possibility of getting lost in the display hierarchy. Fortunately, means were devised for alleviating these problems and a derivative of the display system discussed by Rouse, Rouse, and Hammer [6] is being used in the Boeing 777 aircraft. Nevertheless, the "lesson learned" is clear – simply putting information on a computer does not necessarily make it more useful than presenting it in more traditional ways.

The second example concerns an effort in the mid-1980s to develop intelligent bibliographic information retrieval systems, primarily for use by engineers and scientists. Five studies were performed to understand how various computer aiding schemes affected users' abilities to retrieve information of value. The results of this series of studies are reviewed by Morehead and Rouse [3].

Three of the studies considered the impact of providing links among articles based on reference lists. As we expected, such links helped considerably. This led us to add further links based on citations of articles. In this way, an article was linked to both its ancestors and its descendants.

Much to our surprise, the citation links substantially degraded users' performance. Users tended to wander down citation paths long after they ceased to be productive. We modified the system to display the productivity metric of articles selected divided by articles viewed. In this way, the decrease in productivity of citation paths became evident to users and they abandoned citation paths much sooner. The "lesson learned" here is also clear – providing more links among information elements is not necessarily beneficial and may be detrimental.

Thus, the intuitively obvious benefits of enabling information technologies are not always realized. The straightforward reason is that intuition is not always right, as numerous lottery customers will attest. Rather than betting on technologies, users would be much better served if we first focused on the benefits sought, and then considered alternative means of providing these benefits.

2. Human-centred Design

The types of problems noted earlier can be avoided, and the potential of enabling information technologies can be realized, by adopting a human-centred approach to designing information systems. Human-centred design is a process of assuring that the concerns, values, and perceptions of all stakeholders in a design effort are considered and balanced [8].

Stakeholders include users, customers, maintainers, investors, and so on. Further, the designers of information systems are stakeholders in these systems. While this paper necessarily focuses on users, were we to discuss the design, development, implementation, and servicing of an actual information system, we would consider all of the stakeholders.

Human-centred design can be viewed as a process for addressing and resolving the seven issues listed in figure 1. Four of these issues (i.e., evaluation, demonstration, verification, and testing) are well known to designers of information systems and are usually addressed in a reasonable manner. These four issues are not discussed within the confines of this paper. Interested readers will find a comprehensive treatment of these issues in Rouse [8].

The top three issues in the figure (i.e., viability, acceptability, and validity) are seldom addressed with sufficient rigour by designers of information systems. Human-centred design involves pursuing all of the issues in figure 1, starting at the top. Thus, the first question asked is "What matters?" while the last question asked is "Does it run?"

Rouse [8] discusses a four-phase methodology, as well as associated methods and tools, for pursuing the seven issues in figure 1. In this paper, discussion focuses on elaborating the nature of viability, acceptability, and validity. The use of these constructs is subsequently illustrated in the context of a few applications.

Viability is concerned with benefits and costs. Contrary to the apparent beliefs of many designers of information systems, the primary benefits to users seldom include having the opportunity to use an information system. Users typically use an information system to make better-informed decisions, solve problems, order products and services, save time, and so on.

Costs may include access charges; however, such costs are often paid by third parties. For most users, costs include the difficulty and time involved in learning to use and in using the system, as well as the difficulty and time associated with using the outputs of the system. Thus, for example, one of the costs of using conventional computer-based information retrieval sys-

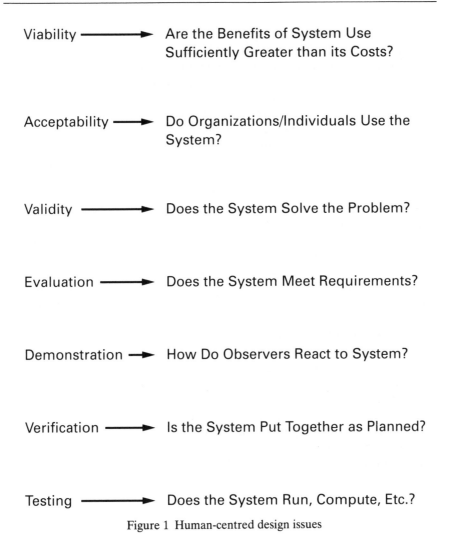

Figure 1 Human-centred design issues

tems is the difficulty and time of wading through the hundreds or thousands
of abstracts obtained, as well as locating and obtaining source documents.

Acceptability concerns the extent to which a way of doing things fits in
with individual and organizational preferences and constraints. For instance,
the hardware and software of an information system should be compatible
with other hardware and software employed by users and their organizations. A more subtle need is for usage procedures for the information system to be compatible with usage procedures for other systems used by the
same set of users. An example of preference-related acceptability concerns
would-be users' desires for colourgraphic displays despite the fact that
monochromatic alphanumeric displays would be less expensive and provide
a valid means to meeting information needs.

Validity focuses on whether or not an information system solves the users' information-seeking problems. It is quite possible for a system to meet requirements – that is, pass evaluation with flying colours – but not provide valid support. For example, an information system might rapidly retrieve and display masses of information, much of which is irrelevant, the remainder of which is only marginally understandable by the class of users for which the system was designed. While one could blame this on the quality of the databases and argue that the information system satisfies its technical requirements, it is nevertheless a fact that the system does not provide a valid solution to users' problems. One might attempt to resolve this problem by adding artificially intelligent functionality that reads and translates all of the information retrieved to assure that what users get is relevant and understandable. This would not necessarily lessen validity problems if users were skeptical of the computer's ability to perform such filtering and translation.

Note that the discussions of human-centred design in this section have only paid passing attention to display formats, dialogue structures, and so on. While these issues are important, they are *not* synonymous with the user-system interface within the human-centred design framework. Within this framework, the interface is "deeper" than the displays and keyboard. The interface includes all functionality whose goals are to enhance human abilities, overcome human limitations, and foster user acceptance [8].

Therefore, within human-centred design, one does not design an information system and then "add" a user-system interface. Instead, one begins with the user in terms of benefits, costs, etc., and progressively deepens the design. At some point, one translates the means to providing benefits into particular enabling technologies. Typically, the design of displays and input devices naturally evolves in this progression. In this way, human-centred design not only results in systems that are *usable* – it also produces systems that are *useful*.

3. Applications

In this section, three example applications are discussed: (1) maintenance information systems; (2) sales transaction systems; and (3) design information systems. The purpose of these illustrations is to show how human-centred design influences the nature of the products and systems that result.

3.1 Maintenance Information Systems

The application concerned the problem of transforming large, blueprint-size hard copy, often called C size, to small, computer-display-size images [2]. The context of interest was helicopter maintenance.

A very important element of human-centred design is initial emphasis on defining the true nature of the problem to be solved. From the point of view of the humans involved in this context, the problem of interest was helicopter maintenance, *not* reading blueprints. Thus, in terms of validity the primary concern was providing information to support maintenance activities rather than finding a way to access blueprints on a small display.

This realization led us to focus on the tasks to be done rather than on the nature of blueprints. It became clear that information is used in different ways depending on the nature of the task, i.e., problem solving vs. procedure execution. This conclusion led us to adopt Rasmussen's abstraction-aggregation hierarchy [4] as a means of organizing maintenance information.

The abstraction dimension included physical form, physical function, and generalized function depicted in terms of location diagrams, schematics, and block diagrams, respectively. The aggregation dimension included assembly, subsystem, and system-level representations. As a consequence of this approach to organizing information, it was no longer necessary to have large displays.

This system concept was evaluated in a series of five experiments. It was determined that the nature of the displays affected maintainers' activities. They performed at least as well using the new displays and overwhelmingly preferred the new displays. Further, it was determined that creation and updating of the display database would be easier with the new approach. Thus, both acceptability and viability were improved.

3.2 Sales Transaction Support

Computer-mediated sales are an increasingly prevalent approach to selling in retail stores, banks, airlines, and many other domains. Perhaps not surprisingly, there has been considerable interest in improving the user-system interface of such systems. Of particular concern, because of the high turnover among people performing such jobs, has been decreasing or possibly eliminating the need for any extensive training in the use of these systems.

We undertook two efforts in this area, one in the domain of retail sales and the other in passenger reservation systems. In both cases, we were asked to improve the usability of these systems by focusing on the user-system interface. We employed the human-centred design methodology to pursue these efforts.

In both cases, we focused initially on viability, acceptability, and validity for a period of 4–6 weeks. We discovered that usability problems, while important, were by no means the predominant concern. The benefit sought in both cases was increased sales and the cost was the time required to make sales.

It would have been quite possible to solve usability problems without enhancing viability – increasing benefits and/or decreasing costs. Focusing solely on usability would have probably increased individual user acceptance but not necessarily organizational acceptance. Finally, solving usability problems alone might have met requirements, but would not have been a valid solution to the right problem.

For both efforts, the initial focus on viability, acceptability, and validity led to an emphasis on sales support rather than solely on improved operability of computer terminals. While usability and the user-system interface still received much attention, it was given in the context of supporting the tasks that really mattered. The result was system designs that were substantially different from those originally envisioned.

3.3 Design Information Systems

The application under design information systems focused on access to and utilization of science and technology information in the context of designing aerospace systems. The motivation for this effort included a long-term interest in the value of information [7, 11], as well as a practical need to develop design information systems.

In keeping with the human-centred approach to design, we began by focusing on viability, acceptability, and validity. These issues were pursued using questionnaires, interviews, and observational techniques involving a large number of designers [9]. We found that very little science and technology information is accessed by formal means.

Why don't designers take advantage of science and technology information? One answer is that they perceive little benefit and great cost in accessing this type of information. They attach *no* benefit to using the information system *per se*.

They are concerned with making informed design decisions. They become informed by asking other people in their organization, a conclusion also reached by Allen [1]. Why do they rely on subjective opinions rather than the "hard" objective information provided by science and technology? A primary reason is that they find published research results to be applicable in general but not to their specific problems in particular. They want contextually based answers to their questions rather than generic simplifications. In other words, they question the validity of available science and technology information.

There are also acceptability problems. Almost all science and technology information is created, written, and published for consumption by scientists and technologists. Designers seldom have the specialized expertise, or the patience, to penetrate this information. They find the context and format of presentation totally unacceptable.

The essense of the designer's dilemma is depicted in figure 2. Each transformation in this diagram requires time and effort. Both time and effort increase as one moves to the right in this diagram. It is easy to see why a designer would not want 1,000 abstracts of research articles on human memory to answer a question concerning usability of radar modes. The cost of answering questions in this way far outweighs the benefits.

The above conclusions concerning designers' perceptions of viability, acceptability, and validity caused us to focus on designers' tasks and information needs rather than on the nature of science and technology information. Thus, rather than focusing on how to get designers to access and

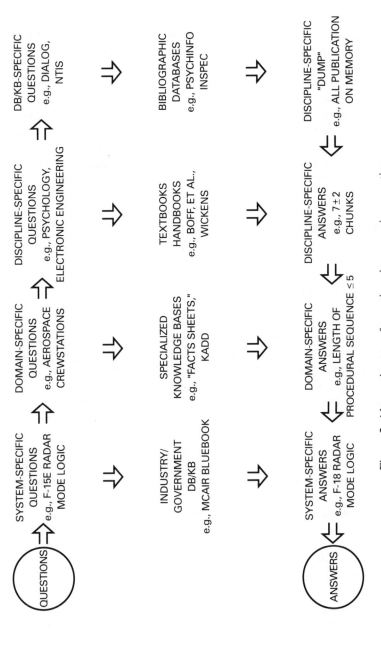

Figure 2 Alternative transformations in answering questions

utilize science and technology information, we looked at the information requirements to support design decision-making. Such requirements should drive the way in which science and technology information is created, organized, formatted, and accessed *if* this information is intended to support design.

After considering alternative representations, we concluded that information seeking in design could be represented as a process of asking questions and pursuing answers in the context of a "design space" including a set of archetypical tasks focused on attributes of the design artifact and characterized in terms of abstraction and aggregation [10]. Typical scenarios or trajectories in the design space were studied to determine information requirements in particular and support requirements in general. Using a structured analysis and design methodology [8] led to identification of hundreds of requirements and an appropriate conceptual architecture that would satisfy these requirements.

4. Lessons Learned

The human-centred design issues of viability, acceptability, and validity have been discussed in the context of several examples, including systems for bibliographic information retrieval, aircraft operations, maintenance information, sales transaction support, and design information. In this section, the lessons learned from these efforts are summarized.

First and foremost, it is essential to recognize that information access and utilization are seldom ends in themselves. The benefit sought is successful task performance, not information seeking. Thus, primary tasks of interest do *not* include operating an information system.

Regarding primary tasks, the information requirements associated with these tasks would dictate information system design. The existing organization and format of information should not, to the extent possible, constrain the nature of an information system. It should also be noted that the ways in which information requirements are satisfied are likely to vary with tasks, despite the fact that the same information content may be required for two or more tasks.

Simply putting information on a computer display is not necessarily better, and may be worse, than using other media, unless appropriate aiding is provided to enable using the information in new ways. Similarly, additional information is not necessarily better, and may be worse, without appropriate aiding to enable using the new information.

People tend to interpret the validity of information in a very context-specific manner relative to their needs at the moment. People are also more likely to find information acceptable if its format and content make it easy to understand and interpret.

Finally, by focusing on the issues of viability, acceptability, and validity within the human-centred design framework, one is much more likely to solve the right problem and solve it in an acceptable way. The result is information systems that are both *usable* and *useful*.

5. Conclusions

The information explosion continues unabated. The promise of information technology has long been heralded as a means of containing, and perhaps counteracting, this explosion. This paper has argued and illustrated with many examples that the problem is not amenable to technology panaceas. Instead, success is much more likely if the concepts, principles, methods, and tools of human-centred design are used to determine contextually relevant information requirements, as well as synthesize support systems that satisfy these requirements.

REFERENCES

1. Allen, T.J. (1977). *Managing the Flow of Technology*. Cambridge, Mass.: MIT Press.
2. Frey, P.R., W.B. Rouse, and R.D. Garris (1991). "Big Graphics and Little Screens: Designing Graphical Displays for Maintenance Tasks," *IEEE Transactions on Systems, Man, and Cybernetics* 21.
3. Morehead, D.R., and W.B. Rouse (1985). "Computer-aided Searching of Bibliographic Databases: Online Estimation of the Value of Information." *Information Processing and Management* 21: 387–399.
4. Rasmussen, J. (1986). *Information Processing and Human-Machine Interaction: An Approach to Cognitive Engineering*. New York: North Holland.
5. Rouse, S.H., and W.B. Rouse (1980). "Computer-based Manuals for Procedural Information." *IEEE Transactions on Systems, Man, and Cybernetics* 10: 506–510.
6. Rouse, S.H., W.B. Rouse, and J.M. Hammer (1982). "Design and Evaluation of an Onboard Computer-based Flight Management System for Aircraft." *IEEE Transactions on Systems, Man, and Cybernetics* 12: 451–463.
7. Rouse, W.B. (1986). "On the Value of Information in System Design: A Framework for Understanding and Aiding Designers." *Information Processing and Management* 22: 217–228.
8. Rouse, W.B. (1991). *Design for Success: A Human-centered Approach to Designing Successful Products and Systems*. New York: Wiley.
9. Rouse, W.B., W.J. Cody, and K.R. Boff (1991). "The Human Factors of System Design: Understanding and Enhancing the Role of Human Factors Engineering." *International Journal of Human Factors in Manufacturing* 1: 87–104.
10. Rouse, W.B., W.J. Cody, K.R. Boff, and P.R. Frey (1990). "Information Systems for Supporting Design of Complex Human-Machine Systems." In: C.T. Leondes, ed. *Control and Dynamic Systems*. Orlando, Fl.: Academic Press, pp. 41–100.
11. Rouse, W.B., and S.H. Rouse (1984). "Human Information Seeking and Design of Information Systems." *Information Processing and Management* 20: 129–138.

Designing Interactive Systems Based on Cognitive Theories of Human Information Processing

Norbert A. Streitz

ABSTRACT

Hypertext and hypermedia are introduced as a solution to the question of how best to produce and provide the information now required by all kinds of users. The new generation of information systems must be user-oriented and task-driven, and their design should rely on knowledge about the manner in which the human mind processes information. Extensive reference is made to the cooperative hypermedia authoring environment SEPIA.

Introduction

I start from the assumption that the world of today – and, certainly, that of tomorrow – depends more and more on the availability of the right information at the right time and in an appropriate quantity and quality. Not judging in this context whether this development is good or bad, a question remains on how to produce this information and how to provide it to clients, students, teachers, engineers, readers, and users in general. My first claim is that hypermedia systems will offer a solution to this problem. They represent the beginning of the development of a new generation of information and publication systems. It is my second claim that the development of these information systems has to follow the approach of user-oriented and task-driven system design. It is my third claim that the application of this approach has to rely heavily on knowledge about human information processing, i.e. cognitive theories. In summary, I claim the inherent capabilities of hypermedia systems will meet the requirements of cognitively adequate human-computer interfaces and the demands for task-driven provision of functionality needed to support a variety of work activities.

The goal of this contribution is to show for selected areas what is needed in order to live up to the expectations raised by the concept of hypertext and hypermedia. The paper is structured as follows: First, I introduce the concepts of hypertext and hypermedia. Second, I present five basic requirements for user-oriented and task-driven system design. Third, I describe the proposed research and development strategy in the context of the cooperative hypermedia authoring environment SEPIA. SEPIA is an example of how the design of an interactive system is heavily based on theories of human information processing, in this case on cognitive models of authoring hyperdocuments. SEPIA is part of the current reserach activities at the Integrated Publication and Information Systems Institute (IPSI) of the Gesellschaft für Mathematik und Datenverarbeitung (GMD) in Darmstadt, Germany.

1. Hypermedia Systems

1.1 Basic Concepts

First, we need to distinguish between *hypertext* and *hypermedia*. In my usage, *hypertext* uses the structural aspects. This concept is based on the idea of a non-linear organization of pieces of information ("nodes") that can be referenced and related to each other by "links" in an associative manner and constitute a network structure (directed graph including cycles). An essential feature of hypertext is the capability of having machine-supported links within and between documents. On the other hand, I use the term *hypermedia* if the nodes contain multimedia contents, e.g. sound, complex graphics, pictures, video, or animation. No doubt multimedia aspects will contribute to the attractiveness and dissemination of the innovative hypertext concept. But one has to note that multimedia provides only the technological basis, e.g. digitizing pictures and compressing them, showing video in a window, editing sound, etc. Currently, multimedia applications consist more or less of the presentation of a collection of multimedia content. What is lacking is structure, a concept of how to relate information elements to each other, how to use this information, which again has implications for the combination and presentation of information. But this is what hypermedia is all about.

I claim that hypermedia systems will provide qualitatively new means for producing, communicating, and comprehending knowledge and will radically change the conditions of the information soceity [26, 27]. Additional information will be provided as we go along. Since the expectations about the potential of hypermedia systems are very high, the role of hypermedia in a

comprehensive information environment has to be clarified. In my opinion, expectations will not be met by relying solely on the concept of hypertext and hypermedia. Rather, these are the crystallization nucleus for the development of a new generation. They must be complemented by considering and integrating existing results and future achievements in the following areas:
– ergonomic design of human-computer interaction
– database management systems
– information retrieval
– knowledge-based components
– publication and high-quality layout systems
– telecommunication and computer networks

Current hypermedia systems are only a first demonstration of the elementary principles that raise our interest in what is still to come in the future. Since there is no space to discuss the deficits of existing systems and approaches, I refer to Conklin [4], Halasz [6, 7], Russell [21], and Streitz [26, 27]. Much of the confusion about hypermedia is caused by people's attempts to define it as one application among others, as similar, e.g., to desktop publishing. From my perspective, hypermedia systems are technological examples of a set of basic but very powerful principles with a high potential for innovations that enable the definition and creation of new applications or value adding for existing applications.

1.2 Authoring versus Retrieval

Discussions about the implications of computer technology for the information society seem to address primarily problems of how to provide information (presentation, retrieval, filtering, etc.). While these are valid issues, at least an equally important aspect is largely neglected: a prerequisite for information retrieval and presentation – not only in hypermedia systems – is that this information must have been produced. Therefore, we have to provide tools for authoring and production. The quality of the next generation of hypermedia systems will depend on the extent and quality of support in authoring environments. Most of the hyperdocuments currently produced rely on the method of "turning (existing linear) text into hypertext." There may be value in turning existing paper or linear electronic documents into electronic hyperdocuments. But the results rarely show what an innovative hyperdocument could actually be like. Most of the existing tools for creating hypertexts are not well suited for this task. A review of existing guides shows that the issue of providing adequate conceptual support for authors of hyperdocuments has not been sufficiently addressed. There is another deficit in hypertext research with respect to writing. Current research is not really addressing the crucial problem that producing a non-linear document might require very different concepts of creating, revising, and composing documents and therefore different kinds of support. But it is my strong belief that the concepts of hypertext and hypermedia will only be convincing and successful if there are dedicated tools that correspond to the special characteristics inherent to the innovative potential of hyperdocuments.

2. User-oriented and Task-driven System Design

I propose that the design of interactive systems in general should adhere to the following five requirements:

(1) User-oriented design. One should take into consideration the limits and the capabilities of the human information-processing system when designing features and mechanisms of human-computer interaction. This concerns primarily the design of the user interface. It is important to have potential users participating in the design process.

(2) Task-driven design. Even a very user-friendly system is worthless if it does not provide the extent and quality of performance necessary to solve the problems that are part of accomplishing the overall task. A detailed analysis of the task and the corresponding activities is required in order to decide on the kind of support needed by the user and then provided by the computer system.

(3) Theory-model guided design. To achieve usability and utility, it is necessary to have an adequate model of the different processes and knowledge structures involved in the problem-solving activity of the user. I propose that these models be based on cognitive theories of human behaviour.

(4) Empirical-based design. The use of models and theories implies the use of empirical investigations to test and validate the theoretical assumptions and their applicability. This implies rapid preparation of prototypes for testing the ideas with users and engaging in a step-by-step design process.

(5) Technology-knowledgeable design. One has to keep in mind the role of available technology for building systems. At a given point in time, the state of the art of technology and computer science will always constrain the actual implementation. On the other hand, innovations in technology can serve as a source of inspiration for new applications in different domains. Beyond this, it has been argued [24] that it should be a goal to progress conceptually one step ahead of existing technology and provide requirements to be met by future interactive systems on the basis of cognitive and social theories of human behaviour.

These requirements are statements about the approach. We have also derived a number of design principles. Here, the principle of "cognitive compatibility" is central [24, 25]. This requires minimizing the discrepancies between the mental problem representations the user forms or has formed about the task and – on the other side of the user interface – the presentation and the function of objects and operations available to the user that are determined by the system's representations.

3. SEPIA: A Cooperative Hypermedia Authoring Environment

In this section, I will describe how my ideas that the design and development of interactive systems can and should be based on theoretical considerations and empirical results in the field of cognitive science were reflected in the research and development strategy for SEPIA (Structured Elicitation and Processing of Ideas for Authoring). SEPIA and its basic design principles were first described in Streitz et al. [28] but did not include the cooperative aspects.

3.1 The R&D Strategy for SEPIA

Our R&D strategy that addresses the cognitive processes, the product, and the social aspects of the authoring activity is characterized in figure 1, showing the relationship of the activity under investigation, the theoretical basis, and the resulting components of SEPIA. Paying attention to the process aspect requires developing and refining a model of the cognitive processes of writing and transforming these results into requirements, as in our activity space concept. Looking at hyperdocuments as a *product* with features of a new rhetoric [15, 31] results in requirements for a corresponding performance, e.g. our construction kit in the rhetorical space. To get valid requirements, we built a large hyperdocument in a separate reading environment testing our assumptions about a new rhetoric for hyprmedia [9]. Considering that most large and complex documents are prepared by a team, social cooperation models had to be defined, and SEPIA was extended from a single-author to a multiple-author environment by providing corresponding cooperation modes. Thus, detailed knowledge about the process, the product, and the social situation played equally important roles in the development of our hypermedia authoring environment.

3.2 The Authoring Activity

While publishing is communicating knowledge, authoring, and in particular writing, is knowledge production and transformation. From our point of view, writing is a complex problem-solving and design activity with multiple constraints. The final product – in terms of a hyperdocument – can be viewed as an externalized representation of internal knowledge structures that have been developed by the author. Thus, authoring tools that are especially geared to hyperdocuments offer much better facilities for conveying the message and intention of authors: authors can externalize knowledge in a format that is closer to their knowledge structures than was possible with traditional linear documents. Making additional properties of the author's knowledge structure (as part of the hyperdocument) available to the reader facilitates integration rather than delinearization and thus more comprehensive processing on the recipient's side. Documents produced with these tools

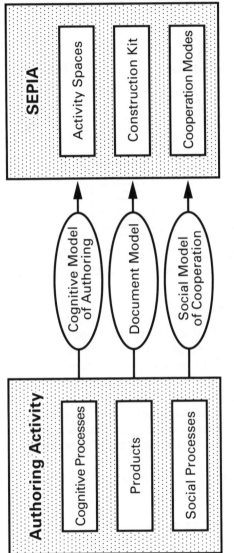

Figure 1 Research and development strategy for SEPIA

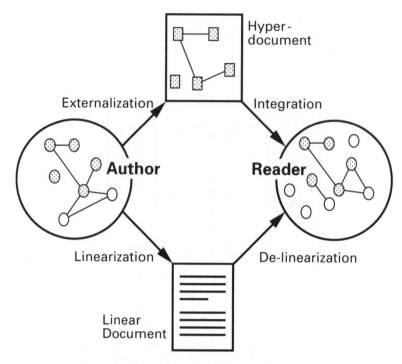

Figure 2 Communication of knowledge structures

keep authors' knowledge structures alive by preserving, e.g., their argumentation and theoretical structures, which can be used for subsequent processing. This improves reception not only by human readers but also by text analysis components for machine translation or automated abstracting. While it is difficult today to have an adequate analysis of natural language text beyond the sentence level, such hyperdocuments would contain more of the necessary structural information and facilitate the analysis process. Figure 2 shows the differences between the linear and non-linear document structures in the process of communicating knowledge.

3.3 The Role of Cognitive Theories for the Design of Authoring Systems

The construction of writing tools is mainly based on intuition and first-order task analysis. What is lacking is a sound theoretical foundation for building cognitively compatible interfaces that provide intelligent support for writing. Kintsch [13] forecasts that progress in this field will remain restricted unless a sufficient cognitive theory of writing is developed. We still do not know what is going on in authors' minds when they progress from "chaos to order," as Brown [2] has characterized this process.

Based on an analysis of the cognitive processes of writing [1, 11, 10], authoring can be characterized as a "journey of discovery." On this journey, ideas are not merely translated into text ("knowledge telling"); they are also generated and refined, because the production of words and phrases triggers new associations ("knowledge transformation"). The widely cited model of Hayes and Flower [11] emphasizes the problem-solving aspect of writing. Based on the analysis of thinking aloud protocols, it identifies three main sub-processes (planning, translating, and reviewing). A model that reflects fundamental differences between novice and expert writers has been proposed by Bereiter and Scardamalia [1]. Knowledge transformation is conceived as an interaction between two problem spaces: the "content space" and the "rhetorical space." While the content space is the space of generating and structuring the author's knowledge about the domain of the intended document, the organization of the document structure and the final editing take place in the rhetorical space. This is also where decisions are made on including, excluding, sequencing, and reformulating information. The interdependencies of extensive planning, production, and revision activities are characteristic of the writing process and lead to both an external product – the text – and an internal product – a new knowledge structure.

3.4 Supporting the Authoring Process via Activity Spaces

On the basis of research writing, we distinguish three closely related sub-problems that an author must solve to produce a document: (1) the content problem, (2) the rhetorical problem, and (3) the planning problem.

According to Newell and Simon [20] and Newell [19], the mental representation of these three problems can be described in terms of separate but *interacting problem spaces* formed by different constraints, objects, and operations in which different knowledge sources are brought to bear [8].

Applying the principle of "cognitive compatibility" [24], we can use this decomposition into sub-spaces as a basis for dedicated requirements of components of the authoring environment. We require that these (cognitive) problem spaces be "matched" in SEPIA by corresponding *activity spaces* (see figure 3). Each activity space provides specific objects and operations to facilitate the author's activities when working on these sub-problems. Since argumentation is a crucial cognitive activity that plays an important role in writing for a large number of document types, these three spaces are supplemented by a fourth called "argumentation space."

We derived specific requirements for the design for SEPIA from models of authoring. On the basis of these considerations, we determined the characteristic features of the user interface and the functions. The resulting "conceptual interface" is depicted in figure 3. Subsequently, we implemented these ideas. The result is shown in figure 4 as a "screen dump" of the SEPIA environment. Figure 4 presents details describing objects and functions of each activity space. (Note: the order of these descriptions does not mean that an author must use the spaces in that sequence.)

Objects and operations of the *content space* facilitate the development of

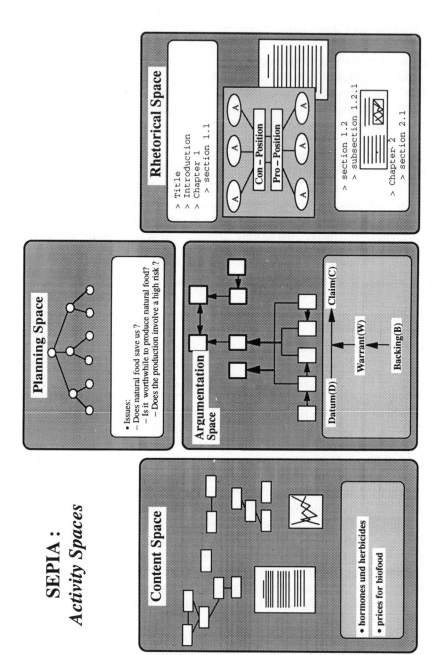

Figure 3 The conceptual user-interface of SEPIA

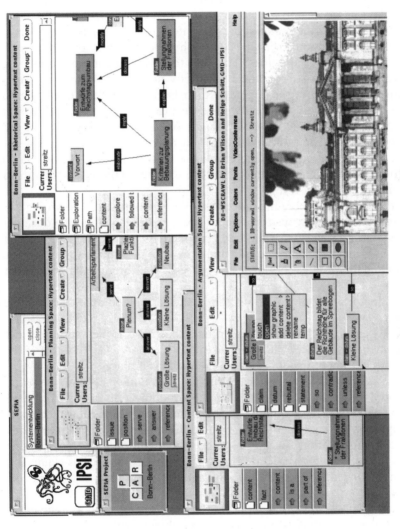

Figure 4   Screen dump of the implemented user-interface of SEPIA

a domain model. This can also involve access to background material from either internal (e.g., previous documents) or external sources (e.g., querying a database). Therefore, SEPIA uses the structuring facility of hypertext to support collecting ideas in textual and graphical nodes, their grouping in topic-related clusters by composite nodes, and connecting them via typed links.

In the *rhetorical space*, the author creates the reader-oriented, final document. To a large degree but not exclusively, it is produced on the basis of transformations of the material in the content space. This final product can be either a conventional, linear text or a hyperdocument, formed by a typical network of nodes and links. Together these document types constitute a scale reaching from strictly linear to strictly non-linear documents. Hyperdocuments can vary in the degree of their linearity between these two endpoints and can be very different with respect to their structure and presentation. Nevertheless, they all should satisfy one major requirement: in order to support comprehension and navigation on behalf of readers, they must appear as *coherent* entities. Therefore, the rhetorical space provides a construction kit consisting of objects that are explicitly tailored to fulfil this requirement.

In the *planning space*, the author is able to externalize his writing plans with respect to goals, construct issues to be concerned with in the document, and establish an agenda of the authoring activity. Consequently, this space serves as a meta-space for coordinating the activities in the other three and controlling the progress of the design process. The formulation of issues may lead to

- the identification of important topics that are elaborated in the content space;
- the generation of positions and claims that have to be supported in the argumentation space; and
- the structuring of the document in the rhetorical space.

For the development of an issue structure, the author can rely on a set of dedicated nodes and links. Here, we use a modification of the IBIS method [14] by extending the definition of the issue concept and introducing a new principle for linking issues [18, 17]. In addition, the planning space is linked to the argumentation space. Positions that are formulated as answers to issues in the planning space are transformed and re-created as claims in the argumentation space prompting the author to provide supporting documents.

The *argumentation space* supports the development of an argumentative structure by providing appropriate design objects and operations based on an extension of the well-known argumentation schema developed by Toulmin [30]. Using the argumentation space, the author can elaborate an argumentation by generating support or objections on different levels, by formulating contradictions, and by constructing argumentative chains (for details, see Streitz et al. [28]).

When "travelling through activity spaces," the author does not need to follow a predetermined route. At every point in the authoring process, he can decide which sub-space to use next. In every sub-space, he can external-ize intermediate results, generate new ideas, and revise earlier decisions. To guarantee such high flexibility, it is necessary to facilitate the interaction and smooth transformation of knowledge between the different activity spaces. In SEPIA, this is accomplished by the automatic transfer of objects between specified spaces, their reuse, and indication and control of references be-tween activity spaces.

Beyond the structural aspects, hyperdocuments are very much charac-terized by the type of media that is used. In general, all atomic (content) nodes carry *multimedia information*, including text, graphics, pictures, and sound. Currently, we are working on the integration of video as the content of a node. This way of using multimedia capabilities, i.e. as part of nodes and in combination with other media, has to be distinguished from using them for communication purposes as, e.g., in audio and video conferencing systems (see 3.5).

3.5 Supporting the Social Process: From One Author to a Group of Authors

Considering that especially large and complex documents are prepared by a *group of authors*, our system should be able to support the social processes involved in team work. For this reason, we had to develop social coopera-tion models and extend SEPIA from a single-author to a multiple-author en-vironment by providing corresponding cooperation modes.

First of all, authors have to be able to access and modify shared hyper-documents concurrently. If authors want to work on the same part, they have to be prevented from destroying each other's work. In most cases, the overall task is divided into sub-tasks that are assigned to specific authors or subgroups of authors. The authoring environment should support this divi-sion of labour. This is facilitated by the modular structure of hyperdocument networks and accomplished by introducing the notion of "composites," which allows a set of nodes that belong to one sub-task to cluster.

We distinguish between the following three modes of collaboration: *indi-vidual, loosely coupled*, and *tightly coupled* work. These modes differ in the extent of awareness each author has of the activities of his co-authors. In in-dividual work mode, authors communicate via annotations that are reached by each other subsequently. In loosely coupled mode, several authors can work on the same sub-task and manipulate nodes in the same composite but not the same nodes. In tightly coupled mode, authors share the same view of the structure of the network and – if they want to – the content of the same node. In this environment with "shared screens," they have full WYSIWIS (What You See Is What I See). Communication and coordination are facili-tated by "telepointers," which are cursors with the names of each author attached so that each one knows who is pointing at which object. But addi-tional communication channels are needed. SEPIA provides a digital audio channel (via the ethernet) for audio conferencing only, as well as a desktop audio/video conferencing device that enables (currently two, but soon up to four, authors) to see and talk to each other using a separate in-house video net. Furthermore, a shared white board called WSCRAWL [16] has been integrated for the exchange of meta or additional information. It is a group-aware, colour, pixel-oriented, shared drawing tool. Permanent telepointers support gesturing on the whiteboard. Each drawing is immediately visible on all connected screens.

The awareness of the co-authors' activities is a prerequisite for smooth ad hoc transitions from one mode of collaboration to another. Currently the transition from individual work to loosely coupled work is triggered automatically when a second author opens a composite node already "occupied" by the first author. Each author is notified by a "doorbell" sound. The first author then knows that someone has entered his composite and the second author is aware that he has entered where somebody else is working. The status linen of user names displays who is currently in this composite node. There is no limitation to the number of authors in one composite. Being in loosely coupled mode, the various awareness-producing means can cause a need for authors to have a tightly coupled session. To start a tightly coupled session, one co-author selects all or a subset of those co-authors currently in the same node to have them join the session. The system asks each of them to confirm if they agree or not. Then, the browsers of those co-authors who confirmed are shifted into tightly coupled mode. The request for confirmation is necessary because this mode constrains the freedom of each one by imposing strict WYSIWIS in this browser. This implies, e.g., if one author scrolls up in the browser of this composite, all other coupled browsers do the same and the network is moving. Or, if one author opens the content of a node, the content is displayed on all other coupled browsers. On the other hand, this is necessary if authors want to discuss details and be sure that each participant sees exactly the same thing. Authors can exit a tightly coupled session either by closing the composite node or returning to loosely coupled mode.

For more details, especially on the system architecture and the implementation of SEPIA, I refer to Streitz et al. [29]. For implementation details of the database support, see Schütt and Streitz [22], and for the group-work facilities, see Haake and Wilson [5].

4. Conclusion

I presented the hypertext and hypermedia as a powerful concept for the creation, structuring, and presentation of information. In a second step, I introduced the approach of user-oriented and task-driven system design. Building on this, I described the application of cognitive theories of human information processing for the specification of design requirements for a hypermedia authoring system. And finally, I described the user interface and the functions of the authoring environment for one author as well as multiple authors. Reviewing my three claims from the introduction, I can state that my experience in following this approach has been very positive. I do not think that another concept of information structuring would have been as appropriate as the hypertext concept. Its inherent opportunities for modularity, flexible decomposition and combination, reuse of existing material, guided tours, group-specific views, annotations, clustering of information, graphical presentation, and navigation of document structures are what is needed for the complex task of authoring large documents in teams. There is still one aspect missing. We finished building the prototype system only recently. The task ahead is to provide the system to a group of real users outside the laboratory and have them use it. Their feedback will be important for the evaluation of our design decision and their implementation. Based on this information and our own use and thinking, we will review our decisions, redesign this prototype, and produce a new version – as the idea of an iterative design presented in section 2 requires.

REFERENCES

1. Bereiter, C., and M. Scardamalia (1987). *The Psychology of Written Composition*. Hillsdale, N.J.: Lawrence Erlbaum.
2. Brown, J.S. (1968). "From Cognitive to Social Ergonomics and Beyond." In: D. Norman and S. Draper, eds. *User-centred System Design: New Perspectives on Human-Computer Interaction*. Hillsdale, N.J.: Erlbaum, pp. 457–486.
3. Card, S.K., and A. Henderson (1987). "A Multiple, Virtual-Workspace Interface to Support User Task Switching." In: J.M. Carroll and P.P. Tanner, eds. *Proceedings of the CHI und GI'87 Conference on Human Factors in Computing Systems*, Toronto. New York: ACM, pp. 53–59.
4. Conklin, J. (1987). "Hypertext: An Introduction and Survey." *IEEE Computer Magazine* 20 (9):17–41.
5. Haake, J., and B. Wilson (1992). "Supporting Collaborative Writing of Hyperdocuments in SEPIA." *Proceedings of the ACM 1992 Conference on Computer Supported Cooperative Work*, Toronto, Ontario, 1–4 November 1992.
6. Halasz, F.G. (1988). "Reflections on Notecards: Seven Issues for the Next Generation of Hypermedia Systems." *Communication of the ACM* 31:836–852.
7. Halasz, F.G. (1991). "'Seven Issues': Revisited." Final Keynote Talk at the Third ACM Conference on Hypertext (Hypertext '91), San Antonio, Texas, 15–18 December 1991.
8. Hannemann, J., and M. Thüring (In press). "Schreiben als Designproblem: Kognitive Grundlagen einer Hypertext-Autorenumgebung." In: H.P. Kring and G. Antos (Hrsg.). *Neuere Forschungen zur Textproduktion*. Trier: Wissenschaftlicher Verlag Trier.
9. Hannemann, J., M. Thüring, and N. Friedrich (1992). "Hyperdocuments as User Interfaces: Exploring a Browsing Semantic for Coherent Hyperdocuments." In: R. Cordes and N. Streitz, eds. *Hypertext und Hypermedia 1992. Informatik Aktuell*. Heidelberg: Springer, pp. 87–102.
10. Hayes, J.R. (1989). "Writing Research: The Analysis of a Very Complex Task."

In: D. Klahr and K. Kotovsky, eds. *Complex Information Processing – The Impact of Herbert Simon*. Twenty-first Carnegie-Mellon Symposium on Cognition. Hillsdale, N.J.: Lawrence Erlbaum, pp. 209–234.

11. Hayes. J.R., and L.S. Flower (1980). "Identifying the Organization of Writing Processes." In: L.W. Gregg and E.R. Steinberg, eds. *Cognitive Processes in Writing*. Hillsdale, N.J.: Lawrence Erlbaum, pp. 3–30.

12. Kant, E., and A. Newell (1984). "Problem-solving Techniques for the Design of Algorithms." *Information Processing and Management* 20 (1–2): 97–118.

13. Kintsch, W. (1987). "Foreword." In: C. Bereiter and M. Scardamalia. *The Psychology of Written Composition*. Hillsdale, N.J.: Lawrence Erlbaum, pp. 9–12.

14. Kunz, W., and H. Rittel (1970). "Issues as Elements of Information Systems." Working Paper 131. Berkeley, Calif.: University of California, Center for Planning and Development Research.

15. Landow, G.P. (1987). "Relationally Encoded Links and the Rhetoric of Hypertext." In: *Proceedings of the First ACM Workshop on Hypertext (Hypertext '87)*. University of North Carolina at Chapel Hill, 13–15 November 1987, pp. 331–343.

16. Lemke, A., N.A. Streitz, and B. Wilson (1992). *WSCRAWL: The Use of a Shared Workspace in a Desktop Conferencing System*. Technical Report: GMD-Arbeitspapiere Nr. 672.

17. McCall, R. (1991). "PHI: A Conceptual Foundation for Design Hypermedia." *Design Studies* 12: 30–41.

18. McCall, R., I. Mistrik, and W. Schuler (1981). "An Integrated Information and Communication System for Problem-solving." In: H.S. Glaeser, ed. *Data for Science and Technology – Proceedings of the Seventh International CODATA Conference 1980*. London: Pergamon, pp. 512–516.

19. Newell, A. (1980). "Reasoning, Problem-solving, and Decision Processes: The Problem Space as the Fundamental Category." In: R. Nickerson, ed. *Attention and Performance VIII*. Hillsdale, N.J.: Lawrence Erlbaum, pp. 693–718.

20. Newell, A., and H.A. Simon (1972). "Human Problem-solving." Englewood Cliffs, N.J.: Prentice-Hall.

21. Russell, D. (1990). "Hypermedia and Representation." In: P. Gloor and N. Streitz, eds. *Hypertext und Hypermedia: Von theoretischen Konzepten zur praktischen Anwendung*. Informatik-Fachberichte 249. Heidelberg: Springer, pp. 1–9.

22. Scardamalia, M., and C. Bereiter (1987). "Knowledge Telling and Knowledge Transforming in Written Composition." In: S. Rosenberg, ed. *Advances in Applied Psycholinguistics: Vol. 2. Reading, Writing, and Language Learning*. Cambridge: Cambridge University Press, pp. 142–175.

23. Schütt, H., and N. Streitz (1990). "HyperBase: A Hypermedia Engine Based on a Relational Database Management System." In: A. Rizk, N. Streitz, and J. André, eds. *Hypertext: Concepts, Systems, and Applications: Proceedings of the First European Conference on Hypertext – ECHT'90*, Versailles, France, 28–30 November 1990. Cambridge: Cambridge University Press, pp. 95–108.

24. Streitz, N.A. (1987). "Cognitive Compatibility as a Central Issue in Human-Computer Interaction: Theoretical Framework and Empirical Findings." In: G. Salvendy, ed. *Cognitive Engineering in the Design of Human-Computer Interaction and Expert Systems*. Amsterdam: Elsevier, pp. 75–82.

25. Streitz, N.A. (1988). "Mental Models and Metaphors: Implications for the De-

sign of Adaptive User-System Interfaces." In: H. Mandl and A. Lesgold, eds. *Learning Issues for Intelligent Tutoring Systems*. New York: Springer, pp. 164–186.

26. Streitz, N.A. (1990). "Hypertext: Ein innovatives Medium zur Kommunikation von Wissen." In: P. Gloor and N. Streitz, eds. *Hypertext und Hypermedia: Von theoretischen Konzepten zur praktischen Anwendung*. Informatik-Fachberichte 249. Heidelberg: Springer, pp. 10–27.

27. Streitz, N.A. (1991). "Hypertext: Bestandsaufnahme, Trends und Perspektiven." In: J. Encarncao, ed. *Telekommunikation und multimediale Anwendungen der Informatik*. Informatik Fachberichte Nr. 293. Heidelberg: Springer, pp. 543–553.

28. Streitz, N.A., J. Hannemann, and H. Thüring (1989). "From Ideas and Arguments to Hyperdocuments: Travelling through Activity Spaces." In: *Proceedings of the ACM Conference HYPERTEXT'89*, Pittsburg, Pa., 5–8 November 1989, pp. 343–364.

29. Streitz, N.A., J. Hakke, J. Hannemann, A. Lemke, W. Schuler, H. Schütt, and M. Thüring (1992). "SEPIA: A Cooperative Hypermedia Authoring Environment." In: *Proceedings of the 4th ACM Conference on Hypertext and Hypermedia (ECHT '92)*, Milan, Italy, 1–4 December 1992, pp. 11–22.

30. Thüring, M., J. Haake, and J. Hannemann (1991). "What's ELIZA Doing in the Chinese Room – Incoherent Hyperdocuments and How to Avoid Them." In: *Proceedings of the 3rd ACM Conference on Hypertext (Hypertext '91)*. San Antonio, Texas, 15–18 December 1991, pp. 161–177.

31. Toulmin, S. (1959). *The Uses of Argument*. Cambridge: Cambridge University Press.

Personal Hypermedia Systems

Mitsuo Takahashi

ABSTRACT

The paper discusses hypermedia technology applications as navigation tools, drawing on many media, for information seeking on personal computers. Stress is given to the tools designed for people without extensive technical background. The Hypertext and Multimedia softwares widely available at PC stores are surveyed, as are Personal Information Manager and Executive Information Systems softwares and Outlining aids.

1. Introduction

It is often said that we are in an era of an information flood. There is much information around us and we do need an efficient way to use it for our everyday activities. For example, we read a newspaper every morning, and there are many pages in a newspaper, even on a usual morning without any big news; but not all the pages are of interest to us: we are interested in some topics. The contents of a newspaper are the source data and the specific topics we are interested in are information. It can be said that data become information when they are evaluated by us as useful or meaningful. So, I should say that we are in an era of a data flood and that we need an efficient way to navigate and to evaluate data in order to reach the useful information.

Businesses depend on organizational structures to filter and evaluate data into information. On each level of hierarchical organization, there are people who have the experience and the professional background to analyse and evaluate data and to obtain information for their own and their colleagues' activities in the company. They use varied office equipment to collect, store, and analyse data. This includes copying machines, electronic filing machines, computers, etc. They also use diverse software to keep data in a systematic way and to retrieve, analyse, and evaluate data. Even those persons who have insufficient experience or professional knowledge can expect to get help from these intermediaries around them in a company.

The information flood is also a big problem for ordinary people, who can not expect much help from intermediaries. Even in a company setting, there may be many people who cannot expect help of this kind, as is the case in small companies or in certain departments of larger companies.

Hypermedia is an information technology that can help individuals navigate through the flood of data and reach the useful and meaningful information with little or no help from intermediaries.

I would like to survey in this paper hypermedia tools designed for people who have no technical background. I would like to survey hypermedia products for the personal computer that are available at local shops at a reasonable price. Anybody who owns a personal computer can use these hypermedia tools when they want to try by themselves. Of course, at research centres or laboratories, they are conducting advanced research for innovative hypermedia tools for the next century, but we cannot try them by ourselves today. The theme assigned to me is "Personal Hypermedia Systems," and I understand that my role is to identify those hypermedia systems for ordinary people that are commercially available at present.

2. What is Hypermedia?

Hypermedia is an extended concept of Hypertext. Hypertext is a software tool that establishes links among blocks of data stored on a computer. We can navigate through the data jungle by passing the links of data. The concept of Hypertext came from the creative work of V. Bush, T. Nelson, and D.C. Engelbart. Software products that reflect the Hypertext concept were released for Apple Macintosh in 1986 and 1987. They are Guide (developed by P. Brown at Kent University, United Kingdom, and released by Owl International, Inc.) and HyperCard (developed by B. Atkinson and first released by Apple Computer and now by Claris Corporation).

Macintosh is well known to many users for its GUI (Graphical User Interface) and was selected as a first platform for the realization of the Hypertext concept. But, in IBM-PC, which is much more widely used in business, Microsoft's Windows environment is now rapidly becoming popular, and there are also a couple of Hypertext tools for IBM-PC. Windows gives us a similar GUI environment and is also a good environment for Hypertext because it has multi-tasking, multi-window systems and data transportability among its applications. Windows has already been installed on more than 10 million PCs. Now, Guide has a Windows version, and

Toolbook (Asymetrix Inc.) gives us similar capabilities on HyperCard in a Windows environment.

Computers have been used to process numbers, text, and graphics; and now they can also process image, audio, and movies. This variety of data presentations (called media) has called for a new word, multimedia. In order to be processed by computer, images, audio, and movies must be digitized and they require much more storage space and much more speed to be processed efficiently for practical use. The amazing evolution of information technologies makes it possible to use Multimedia even on personal computers. Now, many products are available for Multimedia, like the digitizing board, the image-capturing board, the scanning machine, the data compression and decompression board(chips), the high resolution colour display, the graphics accelerator, etc. And mass storage devices like CD-ROM drives, optical storage devices, computer supported VCR, and Laser Disk, etc., are available even at the street corner PC shop.

Since the evolution of hardware for Multimedia, many softwares have been released that can process image, audio, and movies in addition to text, numbers, and graphics. For example, word-processing softwares now can show pictures and movies on a page among texts, and spreadsheet softwares can also show pictures and movies on a worksheet and can paste audio annotations on a specific part of a worksheet.

Typical Hypertext softwares like HyperCard and Guide can process Multimedia. They can store images, audio, and movies in addition to text, numbers, and graphics as a block of data and link them. We can navigate through the hierarchy of a data block composed of text, numbers, graphics and pictures, audio, and movies. The Hypertext concept is now extended to include Multimedia and there appears a new word, Hypermedia. Hypermedia is a software that can link blocks of data that are composed of text, numbers, graphics, pictures, audio, and movies on a computer storage device. We can navigate through these various data media to get meaningful information. I am interested in the use of Hypermedia by business people, but in order to get a clear understanding of Hypermedia, I would first like to survey Hypermedia software products in general.

2.1 HyperCard and Guide

HyperCard uses cards of a fixed size to store data. A card can include text, numbers, graphics, pictures, sound, and now even movies. Several cards that share the same framework and purpose make a stack called Hyperstack. We can develop our own Hyperstacks but there also exist numerous hyperstacks for specific purposes like an address book, an appointment book, etc. Cards in a stack can be linked to each other and can also be linked to cards in other stacks. There are items called objects, such as "background," "field," "button," etc., on a card and cards are linked by those objects.

Figure 1 shows a card that is going to be linked to another card by button. Button information shows where we link the object on a card. We can link

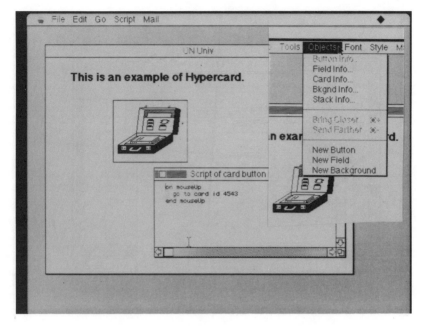

Figure 1

objects to outside equipment like CD-ROM, VCR, etc. Hypercard is a general purpose tool and is also flexible.

Figure 2 shows an example of Hyperstack named Symphony #9 from the Voyager Corporation. It allows us to enjoy listening to Beethoven's symphony with the guidance of a music expert. As you see in the figure, it shows music scores and the buttons at the right are linked to the sounds on a CD on a CD-ROM player connected to the PC by a special language script called Hypertalk and XCMD built in the buttons. It also shows a menu from which we can access notes about many items such as cello and chorus that again link to specific parts of the CD where we can listen for example to cello playing or choral singing. The buttons at the bottom are linked directly to other cards and by using arrows we can move forward or backward.

Guide is another Hypertext software for Macintosh and Windows. Guide has four types of linkage. They are Note link, Replacement link, Reference link, and External link. We open a window, where we can type any text, paste any graphics and images, or paste buttons from a button library. These texts, graphics, images, and buttons can be linked to the four types of links. Figure 3 shows the words "Note link" in a text that is going to be linked to a note link. Note link pops up another window while we click the note-linked object. Figure 4 shows the popped up window that is linked to the word by Note link. If we link a Replacement link to an object, then when we click the object, it is replaced by the linked window. Reference link takes us to another window linked to it when we click the object in a window. External

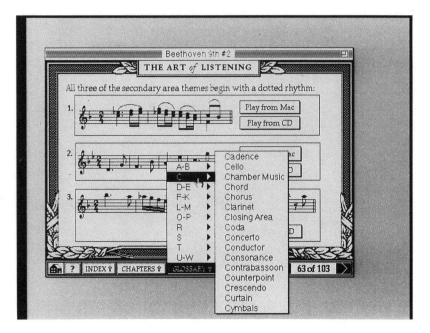

Figure 2

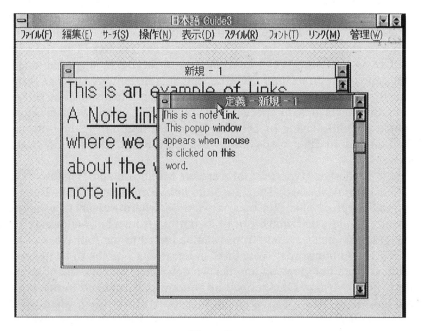

Figure 3

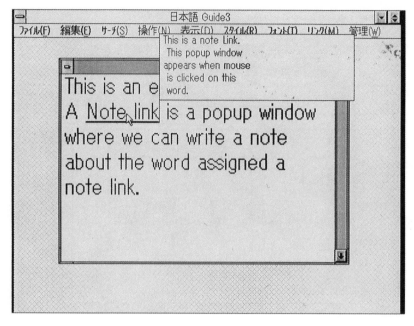

Figure 4

link invokes a script written by a special programing language called LOGiiX and performs a specific task like playing a specific part on CD-ROM or starting another software like spreadsheet software to show a graph based on specific data.

I have surveyed two original Hypertext softwares. Their most important function is linking data. Next I would like to show a little about a recent development in multimedia. It is the motion picture and QuickTime done for Macintosh.

2.2 Movie Files and QuickTime

Apple has released for consumer products a new protocol named Quick-Time. It is a protocol for a data file with a time dimension. The most revolutionary application is for movie files. Movie files have a sequence of pictures to be viewed at a specific time interval, e.g. 1/30 of a second. Any QuickTime file can be played on any QuickTime compatible machine without additional hardware. It can be cut and pasted between different softwares. QuickTime also supports various data compression schemes.

The number of QuickTime compatible softwares and hardwares is increasing rapidly. Figure 5 shows the desktop of Macintosh, where there is displayed World View (a CD-ROM software that includes many professionally prepared movie files) and a program named Screenplay (SuperMac Inc.). With ScreenPlay, we can play movie files on a window. Figure 6 shows a

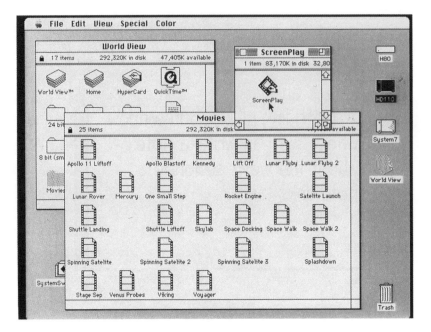

Figure 5

Figure 6

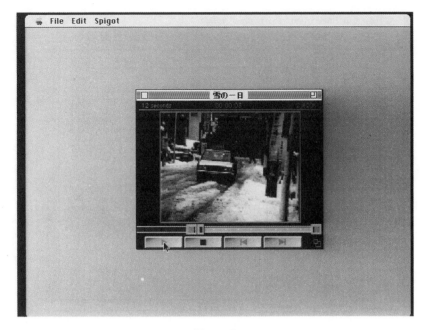

Figure 7

sample frame from Shuttle Liftoff from World View. There are several buttons at the bottom of the window by which we can play, rewind, and forward the movie.

One can buy a collection of movie files (called "movie clips") and view them using the QuickTime compatible software. HyperCard is now QuickTime compatible and Guide for Macintosh will become compatible in the near future.

Apple is now planning to make a QuickTime protocol for Windows as well. But soon we may also feel the need for our own movie files for our own purposes. Recently, there appeared several special boards by which we can easily digitize movies captured through a camcorder.

Figure 7 shows the Screenplay window playing a movie I made last January in Tokyo using a camcorder on a day when we had a lot of snow. It is a short movie that lasts only 12 seconds, but it takes 2.6MB to store the file.

After obtaining movie files, we will use them on a QuickTime compatible software. I tried to paste the movies on a presentation software called Persuasion (Aldus Corp.). I typed several topics on a slide and then imported two movie files on it. Figure 8 shows the slide and we can see movie files of "Shuttle Liftoff" and "A snowy day in Tokyo." One of the movies has a pull down menu with the Play command. We can see movies on the slide if we click the play button.

Multimedia is becoming accessible for everyone through this kind of technological development. There are already many products that have some

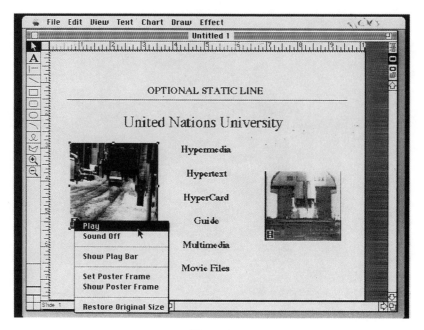

Figure 8

kind of linking function and include multimedia such as pictures and sounds. I will survey several of those Hypermedia products.

3. Hypermedia Products

The number of multimedia products with text, graphics, sounds, animation, and movies is increasing. Many of them are CD-ROM titles and there were more than 2,000 of these in 1991. Many of them have some kind of linking to navigate through.

3.1 Electronic Books

How do you read printed books? If you read novels, you will start with the first page and read through sequentially to the end. In this case, particular parts of the book do not need to be linked together. But what if you use an encyclopaedia? How do you read an encyclopaedia? You will not read it from the first page. You will first look at the table of contents and then turn to the page with the item you are interested in. Or you will look at the index at the end of the book and go to the pages to which you are referred. In an encyclopaedia, there are simple linkages of blocks of pages by contents, index, and words. The page where a specific item is explained might include references to other parts of the book or to other books. The reference to

Figure 9

other books can be thought of as links to external sources of data. Anyway, Hypertext that can include sophisticated linkings might not be very useful for software products where we can so easily anticipate how users will use the data they contain.

Now, let us examine two electronic books where we can see a simple linkage of data and multimedia such as audio and pictures. One is a CD Book for children and another is an electronic encyclopaedia. Figure 9 shows pages from Peter Rabbit (DISCIS Inc.). The left page shows text and the right page shows a picture (on the screen it is in colour). We can flip pages forward and backward using the clips at the bottom corners of the pages. Clicking any part (object) of the illustration, we can get the word that explains the object and can hear it pronounced. We can hear the pronunciation of any words on the text page and can also hear their definitions. This means that objects on the picture pages and words on text pages have links to the text and the voice. This CD Book is an electronic edition of the famous book and we need not navigate through it, but there are simple links between the objects on the pages and the data hidden within the pages.

Figure 10 shows the screen of the Grolier Encyclopedia, which is an electronic edition of the printed encyclopaedia. There are more than 30,000 articles in the book and we can browse through them by title index in a search menu at the top of the screen. But if we need to read articles about some specific word, we must have some navigation tool because there are so many items and words included in this book. The encyclopaedia has a search sys-

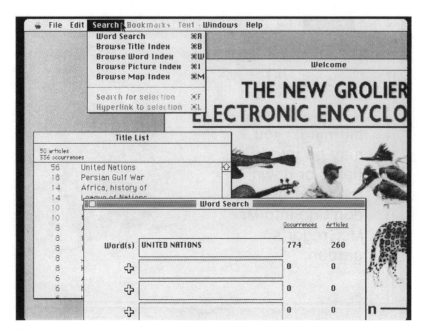

Figure 10

tem based on words. Figure 10 also shows an example of a word search, where we entered the phrase "UNITED NATIONS" and 260 articles were found.

This encyclopaedia has more than 3,000 colourful pictures and many sounds including music and bird's singing that are linked to articles. We can enjoy beautiful paintings, scenery, animal pictures, etc., and we can also enjoy listening to a symphony orchestra and bird's singing. Figure 11 shows the article about Beethoven that was retrieved by word search. The screen shows the text article, a picture, and audio. We can see menu icons in the upper corner of the article through which we can see pictures or listen to sounds. We can use arrows to go to the next occurrence of the word we want to see. We click the camera icon to see the picture that is linked to this article. We click the headset icon to listen to the audio that is also linked to this article.

These two electronic books do not have sophisticated links, but we can enjoy multimedia (pictures and sounds) and we can easily access the articles we want to read.

3.2 An Electronic Magazine

Authorware Inc. (now MacroMedia Inc.) tried to publish an electronic edition of MacWorld magazine published by the IDG group. Figure 12 is the first title screen of that magazine. Though it was a trial that included only

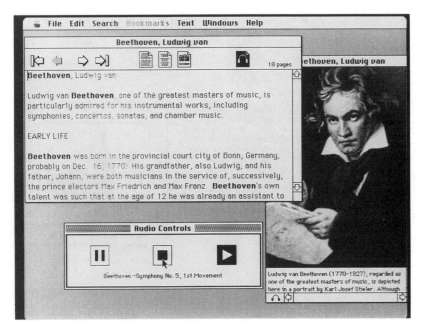

Figure 11

Figure 12

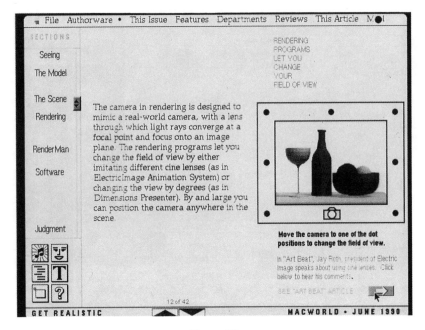

Figure 13

part of the printed copy, it shows us the future electronic magazine. Figure 13 shows one page from that magazine. At the left there is paging and we can go to a specific section or flip from one page to another. The text inside the screen has highlighted words in it and those highlighted words have links to notes and pictures. Figure 13 shows the picture revealed when we click the highlighted word, "field of view." The pictures linked to the word are animated, and if we move the camera on the picture by Mouse, then the camera view will change to a different angle. It is also linked to another article. If we click the arrow, we can, for example, go to another article that explains "ART BEAT."

This electronic magazine has text, pictures, animation, and sounds. We can put bookmarks and memo on any page of the magazine. There are also monthly magazines based on CD-ROM, like Nautilus. Many of them have some sort of linking and use multimedia.

3. How Useful is Hypermedia for Business People?

Hypermedia might be interesting from the technical point of view, but it must also have a practical application to be used in the business environment. Why do we need links of data blocks and why do we need Multimedia? Because there are so much data that we need some tool to get at the relevant information; because information limited to texts and figures is not fully satisfying to management. If they can see information in graphical form, in animation, in movies, and if they can listen to spoken information, then it might give them more satisfaction.

In order to process the flood of data, there exists a variety of database softwares. Data are stored in databases in a structured way and we can retrieve meaningful information from databases based on key subject fields. But, traditional database software needs a rigid file structure on which to store and organize the data to be retrieved. If we could place all data on a specific framework, then it would not be a big problem to have access to relevant information from the database. But the difficulties are that there is such a variety of data in the world that we cannot structure all of them in a specific framework. We need new tools in order to get meaningful information from unstructured piles of data and the linking function of Hypermedia will help us navigate through unstructured data.

There is another problem. If we have an information intermediary at hand and can ask this person to get information for us from the database, then it is rather easy for us to get information. But in order to let the intermediary know our information needs, we must discuss them, which might take rather a long time if that information is unstructured. Even so, if we do have easy access to professional help, then we are very happy; but in fact there are not many people who are able to enjoy this help.

3.1 Structuring Data by Outline Functions

When we write a long report of 50 pages, it is difficult to write from the beginning to the end. We need some framework or structure to provide an overview of the full document. Outlining is a good tool for giving structure. We can write a long text based on the structure and can then easily restructure the document. The outline function was first introduced in an IBM-PC software called Thinktank (Living Videotext Inc., now a division of Symantec Corporation).

Figure 14 shows the structure of a document with four levels of topics (the software used is Acta of Synmetry Corporation). The top level is identified by topics headed 1.0 and the next level is identified as 1.1, 1.2. The next follows as 1.1.1, 1.1.2, 1.2.1, 1.2.2, etc. We can hide lower levels of topics with the text. In figure 14, a black arrow next to an item head shows that there are hidden items under it.

For example, topic 1.1.1 has a black arrow and does have lower level topics hidden under the screen. If we can show only the topics we want to see at the moment and can hide other parts or levels of a document, it is much easier to see a specific part. Of course, if we move a topic to another place, then all the sub-topics move with it to the new location. We can also change the level of topics. For example, if we change topic 1.1.1 to a higher level, then the topic numbers will be reshuffled to reflect the new structure.

Today, many word-processing softwares have this kind of outline function. Other softwares also have outline functions. Figure 15 shows a pres-

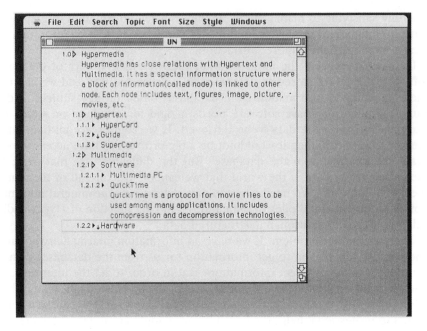

Figure 14

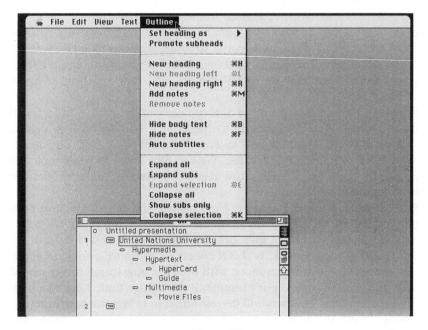

Figure 15

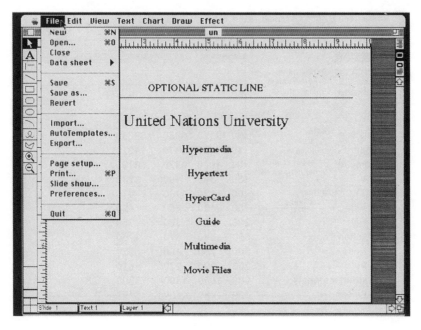

Figure 16

entation software (Persuasion by Aldus Corporation). In this figure, several topics are listed under an outline structure. If we change the screen from outline view to slide view (figure 16), the topics on the slide have a structure of topics in indentation.

It is interesting to point out that Excel (a spreadsheet software of Microsoft for Macintosh and Windows) also has the outline function. Figure 17 shows sales data on a worksheet, and there are so many items on it that we cannot see them all on one screen. In order to see another part of the worksheet, we need to scroll vertically or horizontally. But as you see in the formula menu, there is an outline function. If we use it and give a specific structure to the data based on the formula in the cells, we will get the outlined view of the worksheet. Figure 18 shows the screen after we give it a structure using the outline function. As you see on the left side of the figure, there are five levels of structure, and one can set the level of data that should be shown on the screen. If you compare these two figures, you can understand the usefulness of outlining for navigating the whole worksheet, especially when it is very large.

Two more softwares also have the outline function. Figure 19 shows a screen of Inspiration (** Inc.). Using this software, we put ideas on the screen based on various symbols. We can easily combine ideas into more meaningful ideas. In the figure, there are ideas for formulating a strategy to increase profit. There are ideas, sub-ideas, and sub-sub-ideas. We can see a hierarchy of ideas on the screen and a tool to symbolize ideas. If we change

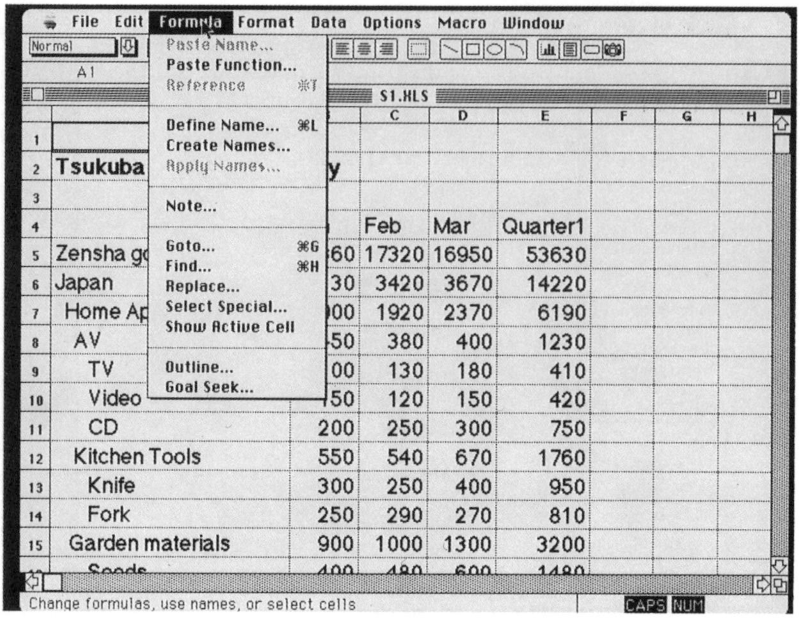

Figure 17

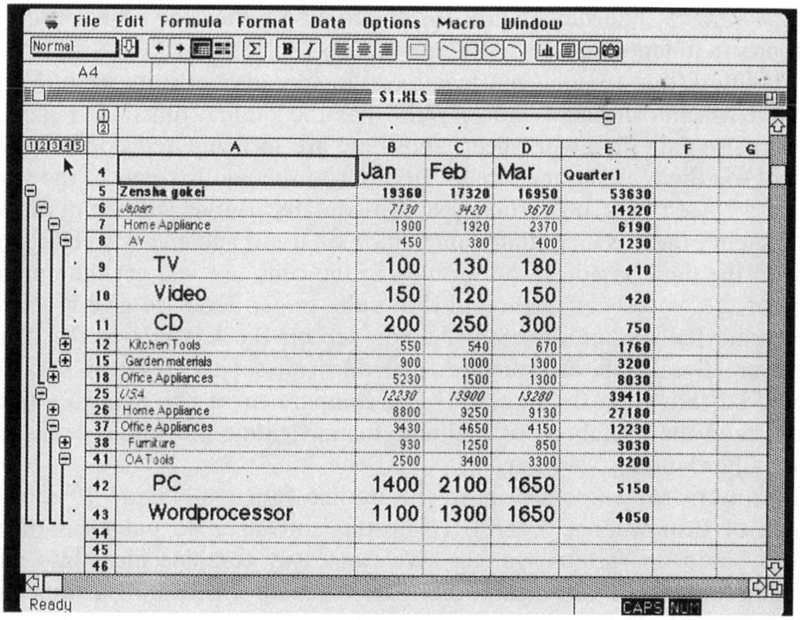

	A	B	C	D	E	F	G
4		Jan	Feb	Mar	Quarter1		
5	Zensha gokei	19360	17320	16950	53630		
6	Japan	7130	3420	3670	14220		
7	Home Appliance	1900	1920	2370	6190		
8	AV	450	380	400	1230		
9	TV	100	130	180	410		
10	Video	150	120	150	420		
11	CD	200	250	300	750		
12	Kitchen Tools	550	540	670	1760		
15	Garden materials	900	1000	1300	3200		
18	Office Appliances	5230	1500	1300	8030		
25	USA	12230	13900	13280	39410		
26	Home Appliance	8800	9250	9130	27180		
37	Office Appliances	3430	4650	4150	12230		
38	Furniture	930	1250	850	3030		
41	OA Tools	2500	3400	3300	9200		
42	PC	1400	2100	1650	5150		
43	Wordprocessor	1100	1300	1650	4050		
44							
45							
46							

Figure 18

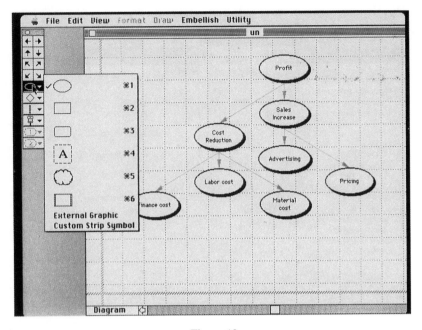

Figure 19

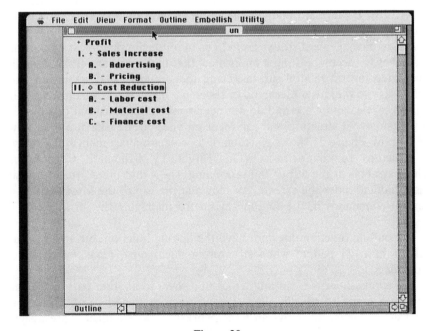

Figure 20

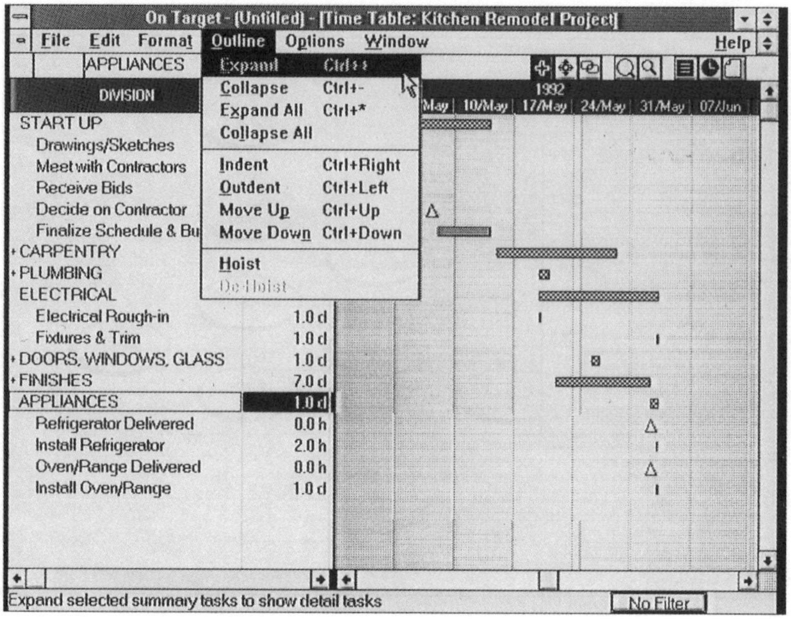

Figure 21

the style to the outline mode, then we can see the outline form of the ideas (figure 20). If there are many ideas and they have multi-level relations, it is often helpful to have them in outline form.

Project management softwares help us monitor the time, cost, and human resources for accomplishing a big project that includes many tasks. One task may have several related sub-tasks and the sub-tasks may again have sub-sub-tasks. So there is a hierarchy of tasks in a big project. When we monitor a project, we cannot see and do not need to see all the tasks at one time. So if we can give it structure and can focus on one specific part at a time, it will be helpful. Figure 21 shows a screen of a project management software with a structure of many tasks (On Target by Symantec Corporation). The + symbols at the left of the task names show that there are several sub-tasks hidden under the screen. As shown at the top of the screen, there are several commands in the Outline menu for managing the structure of the tasks.

The outline function does not have the linking function, but it is very useful for business people when structuring documents, ideas, or data on a worksheet.

In a spreadsheet, we can link a part of a sheet to another part of the sheet or link a graph to specific data. It is a macro assigned to an object on a worksheet. Figure 22 shows an Excel worksheet where macros are assigned to buttons that show graphs. If we click the Japan button, then a macro

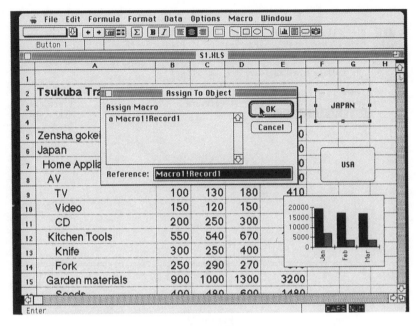

Figure 22

assigned to it shows the graph based on the specific data. Macros and buttons perform the linking functions on a worksheet.

3.2 The Use of Personal Information Manager

There are many data that cannot be placed on the structured field of a regular database. For example, topics in a newspaper may be much longer than the width of a field set in a database. This kind of data is called free format or unstructured data, and there are several softwares that handle this well. PIM (Personal Information Manager), Agenda (Lotus Development Inc.), and GrandView (Symantec Inc.) are well-known softwares in this field. Figure 23 shows a screen of Agenda, where items with the keywords are automatically linked to the words shown at the right in each category with the heading of "software," "company," etc. After getting many data that are linked to words, we can retrieve items linked with the words for which we want information.

Another useful software for free format data is Thought Pattern (** Inc.). We can easily give a tab(index) for a text and use tabs to get information from a free format database. Figure 24 shows a screen of Thought Pattern, where we select a word as a tab for this text data. The Tabs menu shown at the top includes several commands to create tabs, and if we use Cross Index, Thought Pattern automatically compares registered tabs with the words in a text data and makes links to the words appearing in the text data.

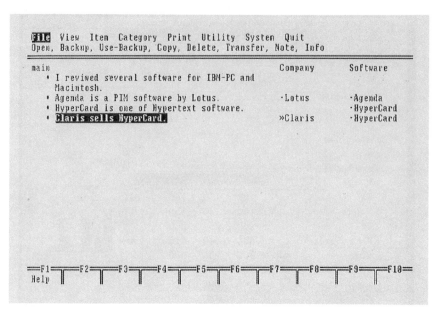

Figure 23

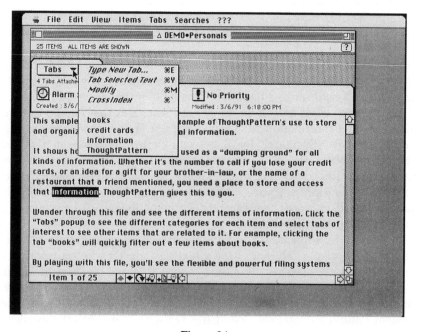

Figure 24

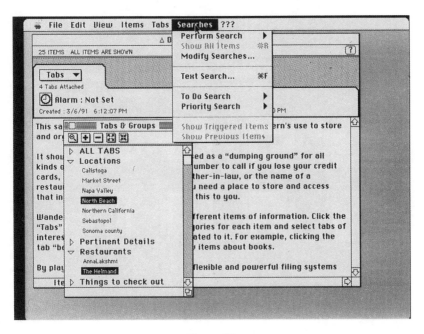

Figure 25

We can easily set tabs for retrieving data by selecting tab words in the tab lists, as shown in figure 25. There, tabs are grouped in a hierarchy, and we can use outline mode to show the level of tab we need to use for retrieval.

4. Executive Information Systems

There are softwares called Executive Information Systems (EIS). By using these, managers and their staff who are not computer professionals can get meaningful information for themselves from piles of data stored on different databases. Data can be arranged in a hierarchy from general to detailed levels. Managers can start searching information at the top of the hierarchy and go down to lower levels to make clear what information they want. This process is called "drill down."

Data may be stored in different databases. For example, sales data may be stored on a dBASE file, accounting data may be stored on a Lotus 1-2-3 file, and customer data may be stored on an Oracle file. Some files might be stored on a PC, and others might be stored on workstations or mainframes.

The EIS softwares enable us to navigate this myriad of information with minimum help from intermediaries. Figure 26 shows a screen of EIS software (Lightship by Pilot Executive Inc.). It shows icons, and there is a hand pointer on the second upper level icon. A hand on an icon shows the presence of linked information at a lower level. By clicking this icon we reach

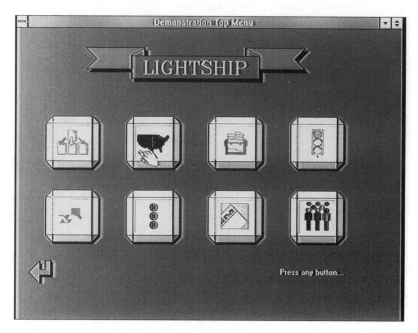

Figure 26

the lower level. Figure 27 shows the result where there is a map of the United States, which again has several icons on it. Moving the Mouse pointer to one that has linked information again changes the pointer into a hand pointer. In this figure, the Mouse pointer is on the "Profit Margin by Region" icon and we can drill down again one additional level. Figure 28 shows a graph of "profit margin" at that level.

The problem with this kind of software is the organization of data to provide meaningful links. Of course, there should be development tools built into it. Figure 29 shows a screen for the development stage. We can see several menus for this stage.

For example, the tools menu includes commands like charting, drawing, text writing, image processing, etc. We can easily make a specific screen that shows data from the database, texts, graphs, images, sounds, etc., and we can link that screen to other screens at any level of the structure.

One difficulty of EIS is how to define information needs. Information comes from a database by filtering data according to specific criteria. One of the most popular tools for defining an information need is a special language called SQL (Structured Query Language). There is an EIS software that includes many interesting functions. Figure 30 shows a screen of the EIS software (Forest & Trees, Channel Computing Inc.). There are several windows that show specific information displayed on it such as a table, a chart, or a single number. These windows are called "views," and we can define more than 2,000 views in any way we choose. Some views show in-

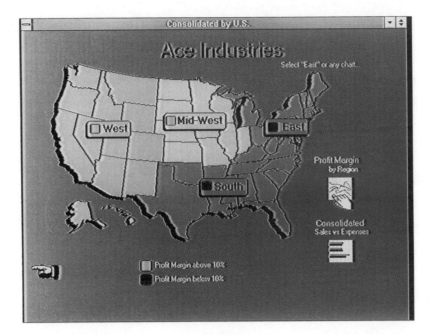

Figure 27

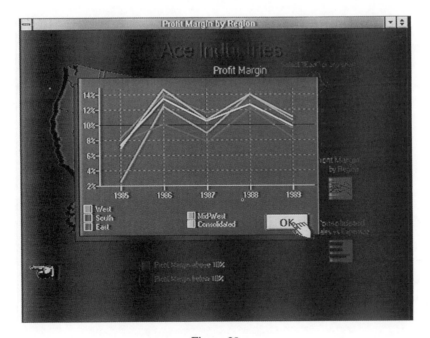

Figure 28

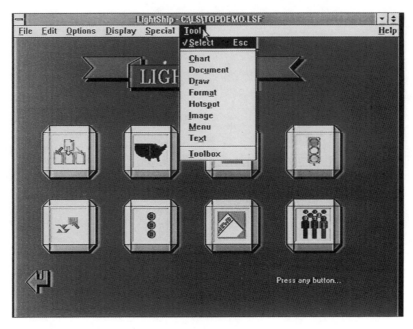

Figure 29

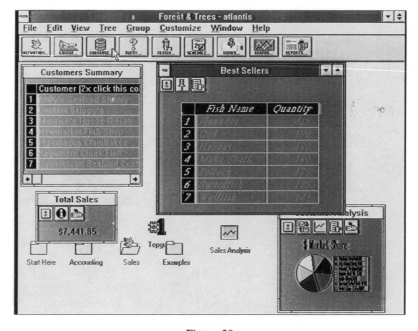

Figure 30

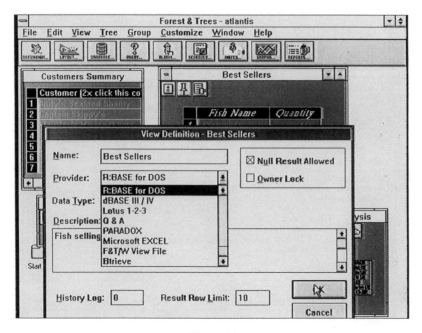

Figure 31

formation based on a specific database and some views the result of processing the numbers contained on several views.

We can use SQL to define our information needs. For example, figure 31
shows the definition of the "Best Sellers" view shown on figure 30 and displays a list of different database softwares. When we define a view, we first
select the database from which we want to get information. The Forest &
Trees shows the list of tables (files) that are created by the specific database
software, and we can select a specific table. Figure 32 shows where we can
define the information need by using SQL. This illustration shows an SQL
program to get the best-selling products based on "fish" and "sales" tables.
SQL can be used for any files listed in figure 32. There are R:BASE,
dBASE, Lotus, Excel, Btrieve, etc., and we can add drivers for mainframe
databases like DB2, FOCUS, etc. The same SQL syntax can be used for any
kind of file, and it is rather easy to write programs with the query assist
shown on the lower part of figure 32.

In Forest & Trees, there is a drill-down function by which we can navigate
the information hierarchy. Figure 33 shows the tree structure of a view
named "current ratio." This view is obtained from the two views called "current assets" and "current liabil" (liabilities). Those two views then are created from views at a still lower level. Finally, the lowest level views in the
tree get information from specific databases. We can also see many icons at
the top of the screen that may be useful to users.

One of the nice things in this software is its scheduling function. Data files

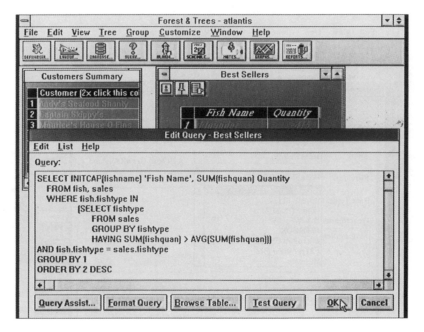

Figure 32

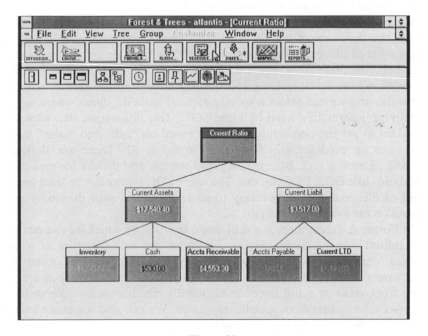

Figure 33

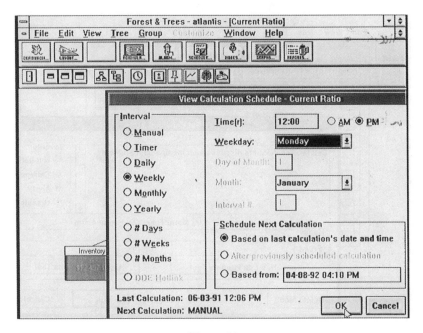

Figure 34

from which we retrieve data may be updated at different time intervals. For example, a sales file may be updated every day, an accounting file every week, and a customer file every month. Then, in order to get fresh data, we need to keep in mind the date and time of the updating schedule for all the different files, and at the scheduled time we must access the updated file to get a new result. But this procedure is tedious work. Forest & Trees helps us by automating this work. Figure 34 shows the schedule-setting window for "current ratio" view. In it, we set the schedule for every Monday at 12:00 noon. Then Forest & Trees will automatically access the data files as scheduled.

Why do we need information? Because we need to analyse it to see whether something is happening or not. If something troubling is happening, we do something to take care of that problem. For example, if we see a "current ratio" view and find it is lower than 2, we must prepare against being short of cash. But this situation may occur only once in 10 times. Almost all the time, we simply access the database to verify that the current ratio is higher than 2 and feel happy. If Forest & Trees can monitor the database and let us know when the current ratio is lower than 2, it is a great help to us. EIS should perform that kind of service.

As explained before, each view is linked to the database through SQL programming and linked to other views. Each view can be assigned a specific condition, and if the view is in that condition, it automatically signals the need for corrective action. Figure 35 shows the "set condition" screen,

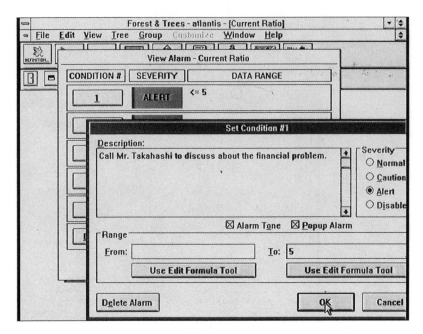

Figure 35

where we set condition in which, if the current ratio drops under 5, then Forest & Trees should pop up an alarm message on the screen to signal that the critical condition is present. We can arrange for the view to perform many more tasks.

5. Summary

Hypermedia enables us to navigate through the sea of information that is composed of text, numbers, graphics, pictures, sounds, and movies. It is interesting to experience movies, music, books, encyclopaedias, etc., interactively. The key point of Hypermedia is linking information, and it is very important for business people at the office to have access to useful information from databases. I described several Hypertext and Multimedia softwares available at PC shops. I also described PIM and EIS softwares, which have the linking feature and are helpful to business people. We also saw that outline functions are useful to help structure a document, a worksheet, an idea screen or a task list. The outline function does not have an explicit linking capability, but it does enable us to structure data and helps us to navigate the data.

Discussion

Several issues were raised by H. Yamada during the discussion. One, dealing with methodological and related socio-technical problems, was the importance of the analytical procedure vs. the prevalence of the synthetic approach in understanding man-machine systems.

G. Johannsen explained that the synthetic approach is indeed rather ambitious, but that many components have already been adequately analysed, while others have not. The emphasis was on setting a framework that includes more influencing factors than are usually considered in working with man-machine systems and especially more human agents, who need to understand the process. This means that a holistic framework should be adopted in order to recognize the kind of strategies and tools that need to be developed for better understanding and use. He said that this is done with respect to improved training for acquainting people with technical man-machine systems, dealing with knowledge-based information access by means of simulation and self-learning tools. He emphasized cultural aspects of such tools and the need for future research and socio-technical development.

Another aspect discussed was the "hostility against technology." H. Yamada said that this may also be related to the tendency of the scientist to analyse too much without integration, making science obscure and contributing to creating mistrust towards science in the general public. M. Dierkes advanced the view that at present there is less of this mistrust: the public increasingly understands that there are costs and benefits bound to the use of new scientific results and of the science-based technologies. He said that the "visions" of future technologies, as presented here, are an important social and international component in deciding on the "desirable" directions for technological development. Public discourse on such visions would help to make better selections and to avoid some of the negative side-effects that have occurred in the past. In this relation, I. Wesley-Tanaskovic stated that the adoption of the broad scope of man-machine systems, as presented here, and of the concept of "socio-technical development" was seen as a possible way to bridge the existing "gaps," especially the knowledge gap between various social groups and between nations as well.

Commenting on "human-centred design," W. Rouse said that among the various "stakeholders" interested in transfer of information, the most difficult to deal with is the aspect of the "producers," e.g. researchers and engineers. Incentives, and related structural changes, are needed to assure that they make usable and useful contributions to the world's knowledge base, without, however, constraining their creativity. New information technologies, as seen here, are means to enhance information stores, not just to fine-tune the information systems. He further stated that there are many examples of successful applications of the new information tools and telecommunications for rapid transfer of requested scientific information between the continents. This makes the scenario of the 1990s quite different from that of 20 years ago, when the first world science information systems were designed.

In relation to this aspect, M. Almada de Ascencio reiterated that scientific information systems were first thought of as being for the researchers themselves and were set up by them. Not of primary concern were other sectors, such as local government, business, and industry, which all need information for problem solving and, often, scientific and technical information as well. The proposed "human centred design," by providing more adequate information support, would seem to her to be especially useful for the developing countries building their own systems.

Speaking of hypertext systems, their potential as educational tools was questioned by J. Alty. Responding to him, N. Streitz pointed out that not every hypertext system is at the same time an educational learning environment. In order to provide the educational experience, one has to combine the hypertext/hypermedia environment with didactic strategies. In addition, he said, one has to take into account who the learner is, the subject domain, etc.

At the end, in relation to the comprehensive review of currently available Personal Hypermedia Systems, demonstrations were made of some new industrial developments by Fujitsu Ltd. of Tokyo, Japan, which could serve as examples.

Session 4

Intelligent Access to Information: Part 2

Chairperson: Meinolf Dierkes

Machine Translation

Makoto Nagao

ABSTRACT

After a quick review of the history of machine translation, the paper sketches the machine translation process. The capacities of currently available systems are indicated. The installation and use of these systems are described, as are the factors to be considered in their evaluation. The Japanese efforts and experience, as well as international experience and prospects, are outlined.

1. A Brief History of Machine Translation

An attempt at computer use for translation was proposed by Weaver around 1945, at the very beginning of digital computer development. During the 1950s, there were several research efforts on machine translation. A famous system is the Georgetown University–IBM joint project of machine translation from Russian into English. The US Government invested a significant amount of research money in this area in the late 1950s and early 1960s, mainly to develop the Russian–English system. Around 1960, 20–30 widely scattered research groups in the United States were involved with machine translation and natural language processing. However, several years of research effort did not produce really usable translation results, and therefore the US Government stopped funding machine translation research and development in 1966. After that, the boom of machine translation there faded quickly, and US researchers turned to the more basic field of computational linguistics. Europe, Japan, and the Soviet Union

273

started machine translation research in the middle of the 1950s. The Electrotechnical Laboratory in Tsukuba, Japan, demonstrated English–Japanese machine translation in 1959. Kyushu University constructed a machine translation computer that translated Japanese to English or German and the reverse. I started research on machine translation and natural language processing in 1961 when I finished my M.Sc. degree and entered research life at Kyoto University.

At that time, we had many difficulties in natural language processing by computer because we could not handle Chinese characters. We had to struggle for more than 10 years to solve these difficulties. By around 1975, we had arranged for relatively easy Chinese character input and output. But really satisfactory input/output was finally reached with the widespread usage of Japanese word processors at the beginning of the 1980s. Although we encountered many difficulties, we completed a comparatively good machine translation system called TITRAN in 1978. This translated the titles of research papers from English into Japanese. It was used on a trial basis by the researchers at Tsukuba Science City to quickly scan current research information. The success of this system encouraged the development of Japanese–English and Japanese–French title translation systems, which were completed in 1980 and 1981, respectively.

The Japanese Government, attracted by the success of this title translation, launched a national project of machine translation that aimed at the translation of abstracts of scientific and technical papers between Japanese and English. The project started in 1982 and lasted four years. I was the project leader, and four organizations, namely Kyoto University, Electrotechnical Laboratory, Tsukuba Information and Computing Center, and Japan Information Center of Science and Technology (JICST), cooperated with each other. We constructed two prototype systems, one from Japanese to English and the other from English to Japanese. After the project was completed, the Japanese to English system was further improved at JICST, and the dictionary was enlarged from 70,000 words to 540,000 words and is now in daily operation at JICST. An abstract that is composed of about 300 Japanese characters on the average is translated in 25 seconds. Some pre-editing and post-editing are done by outside companies, which require about three weeks for 1,000 abstracts. This translation speed is more than twice that of human translation. The total cost of translation by machine is about 60 per cent that of human translation. JICST is now translating about 91,000 abstracts of Japanese scientific and technical documents into English in a year. These translated abstracts are in the JICST database and can be accessed from the United States and Europe as well as from inside Japan.

This national project had a great impact on Japanese industry. Many computer companies and other information-related companies started the development of machine translation systems in the early 1980s. Nowadays, there are about 10 commercial machine translation systems, some from Japanese to English and others from English to Japanese. Furthermore, several other companies are developing machine translation systems for commercialization.

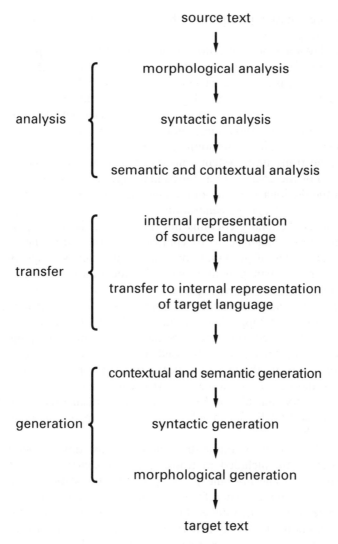

Figure 1 Machine translation process

2. System Configurations

The machine translation process is shown in figure 1. Morphological analysis determines the word form, including inflections, tense, number, part of speech, and so on. Syntactic analysis determines which word is the subject, which the object, and so on. Semantic and contextual analysis determines the correct interpretation of a sentence from the multiple results produced by the syntactic analysis. Syntactic and semantic analyses are very often combined and executed simultaneously. The result is the internal representation of a sentence. Then the internal representation in the target language is often the same as that of the source language, but sometimes a change in the internal representation is required. The generation phrase is just the reverse of the analysis process.

Present-day machine translation systems are still quite imperfect, and we must be skilled to use such imperfect systems. However, if they are used with care we can profit from machine translation in both translation speed and translation cost. The technology will make rapid progress, experience will be accumulated, and grammars and dictionaries will be improved step by step. Therefore, we should not underestimate the usefulness and the cost-effectiveness of machine translation in the near future.

A machine translation system consists of several components, as shown in figure 2. For the analysis of a source language sentence, a set of grammar rules and the dictionary of the source language are necessary. The same applies for the target-language generation.

The transfer dictionary contains information on the correspondence between the source language word (phrase) and the target language word (phrase). The rules of grammar for a language usually range from a few hundred to more than one thousand, and dictionaries include 50,000 or more ordinary words and many special terms for a subject field.

One of the obstacles in machine translation systems is text input. If electronic media such as floppy disks are the usual way of text input, there is no problem. But if printed pages are to be input, either human typing or optical character reading (OCR) has to be used. In the first case, the speed and the expense of typists are the problem. In the other, the problem is the expense of an OCR and the proofreading and error correction by humans. Japanese text input is very difficult for European or American organizations. They can use Japanese OCR, but they have to employ Japanese persons for error correction tasks.

Machine translation is automatic, but the translation output is poor in quality. Errors are included in translation and sometimes no translation comes out at all. The machine gives up on the translation when an input sentence is too complex. Therefore, post-editing is an essential task for current machine translation systems. For human translation, two-stage translation is often adopted; namely raw translation and brushing-up. Post-editing in machine translation may correspond to the brushing-up process. Some brushing-up is required even for human translation, and this is, of course, much heavier in machine translation. Post-editing is done either on printout paper or on a computer screen. The latter is more profitable and economical because no printout is required during error correction. Editing, such as the insertion of figures and tables, is also very easy on computer screens nowadays. Page formatting can be done very conveniently on the computer, and only the final results need be printed out.

Pre-editing is sometimes required. In particular, the segmentation of a long sentence into two or more shorter sentences is very often performed. Pre-editing and post-editing have a certain correlation. When a heavy or elaborate pre-editing is performed, very simple post-editing is sufficient, and vice versa. However, two persons have to be employed in this case, one for

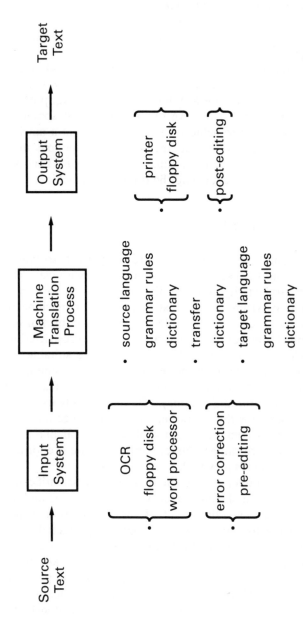

Figure 2 Machine translation system

pre-editing who knows the source language, and the other for post-editing who is a native speaker of the target language. In any case, the post-editing is unavoidable, and the overall cost of pre-editing and post-editing is generally much more than that for elaborate post-editing alone. Therefore, many machine translation users do only post-editing.

The organizations that introduce machine translation systems must not imagine a situation in which "everybody who needs copy comes to a copying machine, puts in documents, presses a button, and receives copies." Present-day machine translation systems are not that simple. Organizations must employ professionals to operate a system with pre-editing, post-editing, dictionary maintenance, and so on. An organization must sometimes change the flow of its documents to achieve the maximum from a machine translation system. Such changes include the training of people in document-writing sections to write in a clear, unambiguous style. This is ultimately profitable to both human readers and the translation machine.

3. Ability of Current Machine Translation Systems

Present-day machine translation systems are not perfect, so that we have to be very careful about the use of these systems. Otherwise we cannot achieve the anticipated productivity. One of the important factors in the use of machine translation systems is the text categories for machine translation. Headlines of news articles and the titles of papers can be translated rather easily if the subject areas of these texts are specifically restricted to a narrow domain. Technical information and news articles are suitable for machine translation without post-editing if the readers do not care about the sentence structure of translated materials and just want a rough idea of what is happening (information-gathering reading). The most widely used text category for machine translation is the operation and maintenance manuals of industrial products, such as computers, automobiles, nuclear plants, and so on. The texts themselves are rather simple, but the volume is enormous. Therefore machine translation is inevitable.

There are increasing demands for the translation of scientific and technical papers all over the world. Japan produces a lot of scientific and technical information every year, but almost all of the information is published in Japanese, and many foreigners say that Japan is protected by a language barrier. We are making great efforts to break this barrier by developing Japanese-to-English machine translation systems. There are many demands for the translation of legal texts, contract documents, and patent documents from Japanese into English. These are very hard for machines to translate, because the translation must be very accurate. The sentence structures of these documents are rather specific, usually very long and so complex that people must sometimes read them several times to grasp their precise meaning. Machines cannot handle such sophisticated sentences and generally fail at translation. Literary works and dialogue sentences are the most difficult categories of all for machines because they are mentally sophisticated, contain lots of omissions, and presuppose the implicit application by the reader of world knowledge.

We can also classify sentence categories by their difficulty from the standpoint of sentence or text types. Sentences that express only facts are very easy for machines to understand and translate. Sentences that include time relations, expectations, assumptions, conditions, etc., fall into the next category for machine translation. Then come sentences that describe the speaker's intention and mental state. These are sometimes very difficult to translate because the interpretation must include discourse information. Present-day systems translate input sentences one by one independently. They do not have the ability to see the interrelationship among adjacent sentences. Therefore, the translation of those sentences that require contextual information is very imperfect, although the minimum discourse phenomena, such as anaphora and ellipsis, are treated to a certain extent.

Other information that is required in the understanding of sentences is world knowledge, everyday-life knowledge, or common-sense knowledge. This information is particularly needed for the interpretation of dialogue, which presupposes the listener's knowledge. A new model for interpretation and translation has to be developed for such cases that includes speaker and listener models in the translation system. Intensive research is being done on this topic at ATR, the Automatic Translation Telephony Research Institute located in Kyoto, Japan. There are many sentences that include metaphors and culture-specific expressions. We still do not know how to handle these expressions, which are heavily connected to world knowledge. The study has just started.

Present-day machine translation systems are based on the compositionality principle; that is, the translation is based on a one-to-one correspondence with the words and phrases in the original input and can rarely use the information implicitly given in the sentence. In other words, the system transforms partial word sequences of an input sentence mechanically into different forms again and again before arriving at the final translation. When a given sentence is long, the process becomes very complex and very often fails. There is a statistic such that Japanese sentences composed of less than 20 characters are translated successfully more than 80 per cent of the time, but that those of more than 60 characters are translated successfully less than 30 per cent of the time, and that almost all sentences that have more than 100 characters are untranslatable by machine. However, we have recently developed a very interesting method for conquering this difficulty, and the success score for sentences of more than 100 characters has become 66 per cent.

4. Introduction and Use of Machine Translation

In previous sections it was mentioned that machine translation systems are very complex and still incomplete and that their users must be very careful in their introduction and use. Here more details will be given.

First, the projected use for machine translation must be clearly established. Then the kinds of documents must be specified, and the expected quality, speed, and cost of translation must be clarified.

When a system is introduced in an organization, it cannot achieve its full potential the next day. At least two specialists must be assigned to the possible pre-editing and to the post-editing. They need to have backgrounds in linguistics or the languages that the machine will translate. They must be trained for a few days for operations such as pre-editing, post-editing, and dictionary change and enhancement. Special terms in the subject field of the documents must be collected, and the dictionary must be enhanced. The styles of the sentences in the input and output documents must be carefully studied, and translation equivalents must be determined. The grammar must sometimes be changed to fit the document styles. The special terms used in documents in some subject fields may total more than 10,000. A substantial number of documents have to be translated on a trial basis, and the dictionary and the grammar must be improved. This test period will last at least half a year and may sometimes exceed two years, depending on the type of texts.

Users must measure the translation rate, speed, post-editing time, and estimated cost during this text period. Users must always compare the achievement of the machine with that of human translation. When the user of a system has come to the conclusion that the overall cost of running the system is no more than that of human translation, or that machine translation has become the more rapid method, then the operational stage will start. During the operational stage, the system must constantly be improved by feedback from the post-editor, who is in the best position to catch the weak points of the system.

A few users in Japan have reported a 30 per cent decrease in translation cost by machine compared to that of human translation, as well as a 30 per cent speed-up in overall translation time. The latter depends heavily on the post-editing practice. Many post-editors still prefer to work on printout pages, but post-editing on the CRT display is much more efficient and speedy. Translation speed by machine alone varies from 3,000 words to 80,000 words per hour, depending on the computer. Nowadays the shift is being made from mainframe computers to workstations, which are more convenient and cost-effective. In the latter, the raw translation speed will be between 3,000 and 10,000 words per hour. Total translation cost depends heavily on how post-editing is done; for example, on whether it is performed on printout paper or on the CRT screen, whether or not it is performed by subject specialists of the area or not, and whether a high quality of translation is required.

Machine translation is expensive, particularly when the cost of possible pre-editing and of post-editing is included. Therefore, there must be a considerable and regular volume of documents throughout the year. One vendor says that 2,000 pages per year is a break-even point for machine translation. This volume depends on many factors and varies greatly case by case.

5. Evaluation Factors of Machine Translation Systems

There is a growing interest in machine translation. Some people have an accurate understanding of the capability of present-day machine translation systems, but many others have the idea that machine translation can perform perfectly, that it will deprive translators of jobs, or at the opposite extreme, that the machine cannot do anything significant, and that a machine translation system is useless.

For the proper use of present-day systems, there is need for a kind of common standard for the evaluation of machine translation as well as detailed knowledge of how machines translate sentences. Serious discussions have just been started about establishing such a standard, but it will take a long time for wide agreement, because there are so many factors to consider. Evaluation factors will differ according to the purposes of usage. This further complicates establishment of a standard. The following are factors that must be considered in the evaluation.

(1) The cost of purchase, operation, and maintenance; training, text input, dictionary enhancement for ordinary words and for special terminology; post-editing, final text preparation, and the relation between translation quality and cost.

(2) The speed of text input, pre-processing (including the handling of tables, figures, and so on), machine translation, post-editing final text format arrangement (including tables, figures, and so on), and the relation between pre-editing and post-editing time.

(3) The quality in terms of fidelity, intelligibility, and naturalness of style.

(4) The capacity of the system for improvement in terms of dictionary change, enhancement of the system's main dictionary and user dictionary, grammar improvement, particularly for sub-language expressions in a specific domain, and the extent to which users are allowed to improve the dictionaries and the grammars, and to test the method after improvement.

(5) The capacity for system extension from one text field to another and from one pair of languages to another.

(6) The extent of effort required of users in terms of the number of operating staff required, including the pre-editor, post-editor, and dictionary maintenance person; the period of time required for testing and tuning, for adapting the organization's documentation system structure to machine translation, and for controlling source language expressions when authors create original texts.

(7) The capacities and limitations of the system for dealing with grammar, semantics, discourse – including anaphora and ellipsis – dictionary information, software speed, learning capability, interactive translation or post-editing, ambiguity resolution, and for producing only the best possible translation output or all the possible outputs for a sentence.

6. Japanese Machine Translation Systems

The Japan Information Center of Science and Technology (JICST) is translating abstracts of scientific and technical papers from Japanese to English by the improved version of the Mu machine translation system that we developed in 1985. Sentences in abstracts are comparatively long, and sentence structures tend to be very complex. Therefore, a certain degree of pre-editing is performed on the original Japanese abstracts. Post-editing is also done for the translated English abstracts. There is a complementary relation between the pre-editing and post-editing. That is, when a heavy pre-editing is done, the post-editing is light, and vice versa. JICST is now measuring what degrees of pre-editing and post-editing are best from the standpoint of overall cost.

The original system had a dictionary of about 70,000 words. Among them, about 20,000 words were common words and the rest were terminology in computer and electrial engineering. This dictionary was quite insufficient for JICST because abstracts from various scientific and technical fields were to be translated. So, JICST increased the vocabulary to 200,000 words and obtained an improved but still unsatisfactory translation rate. Finally, when the vocabulary was increased to 540,000 words, the expected translation rate was achieved.

Measurement of the translation quality at JICST is very difficult because both pre-editing and post-editing are performed to achieve a tolerable quality for general readers. JICST reports that the cost of machine translation, including pre-editing and post-editing, is about 60 per cent that of human translation, and that the speed of translation has been improved significantly.

Many private companies are involved in machine translation. Although they do not yet make a profit, they invest a lot because they consider that natural language processing will be a key technology in the future information society. The systems listed below are typical ones in Japan. Some other companies are developing small systems on personal computers. Those marked with an asterisk are R&D systems.

ALT J/E: Information Processing Laboratories, NTT*
ARGO: CSK(JE)
AS-TRANSAC: Toshiba Corp.
ATLAS-I: Fujitsu, Ltd.
ATLAS-II: Fujitsu, Ltd.
CONTRAST: ETL*
DUET-E/J: Sharp Corp.
HANTRAN: CBU
HICATS E/J: Hitachi, Ltd.
HICATS J/E: Hitachi, Ltd.
KAN-TRAN III: Carozelia Japan*
LAMB: Canon Inc.*
MELTRAN-J/E: Mitsubishi Electric Corp.

MU: Kyoto University*
MU2: JICST
PAROLE: Matsushita Electric Industrial Co., Ltd.
PENSEE, PENSEE II: Oki Industrial Co., Ltd.
PIVOT: NEC Corp.
RMT E/J: Ricoh Co., Ltd.
SHALT: IBM Japan, Ltd.
SHALT/JETS: IBM Japan, Ltd.
STAR: CATENA Resource (on Unix)
SYSTRAN: Systran
THE TRANSLATOR: CATENA Resource (on Macintosh)
Translation Word Processor SWP-7800: Sanyo Electric Co., Ltd.
Transer/EJ: Nova
UNNAMED: Nippon-Data General Corp.*
XJE: SPIRIT*

Two or three nationwide on-line machine translation services via computer network are commercially used.

7. Japanese Governmental Efforts

The Japanese Government, particularly the Ministry of International Trade and Industry (MITI), the Science and Technology Agency (STA), and recently the Ministry of Post and Telecommunications (MPT), has been interested in machine translation. The STA in cooperation with MITI supported the national project of machine translation (Mu project) during 1982–1986. MITI was the main promoter of the First Machine Translation Summit Meeting in Hakone in 1987, which was succeeded by the second in Munich in 1989 and the third in Washington, D.C., in 1991. MITI realized the importance of dictionary construction for machine translation, development of which requires a long time and a heavy investment. Individual companies independently construct their own MT dictionaries. This is a waste of money. MITI established a neutral organization for MT dictionary construction named Japan Electronic Dictionary Research Institute (EDR) in 1986. 70 per cent of the financing comes from the Government and the other 30 per cent is provided by eight major electronic companies. It constructs a kind of neutral dictionary based on concepts that are matched to lexical items and phrases of individual language. Within the project length of nine years, the EDR will establish a dictionary of 200,000 concepts and its corresponding English and Japanese dictionaries for machine use. The project is now in the improvement stage and its quality is being tested by a limited number of users. When the dictionaries are complete, they will be available to any organization in the world at a proper price.

MITI supports another project conducted by the CICC (Center for International Cooperation for Computerization). The project includes the development of a multilingual machine translation system between the

Japanese, Chinese, Malaysian, Thai, and Indonesian languages. The project started in 1987 and is continuing. They successfully demonstrated a first prototype system in 1990.

The Ministry of Post and Telecommunications supports a speech translation project, the ATR Automatic Speech Translation Telephony Research Institute, which was established five years ago. It is now located in Kansai Research Park, 30 km south of Kyoto City. The project aims at simultaneous speech translation by computer between Japanese and English. It comprises the recognition of spoken dialogue in Japanese and English, the translation of dialogue between the two languages, and the synthesis of speeches from translated sentences. At present, the dialogue domain is very narrowly restricted, and present knowledge in this domain is fully utilized. The first phase of the project is almost completed, and a demonstration of speech translation with a limited vocabulary will be given shortly. The project formulated a plan for a second stage for another 7–8 years.

8. Dictionary

Dictionaries are drawing increasing attention from more and more types of people. This is true not only of language dictionaries but also of terminological dictionaries in every field, and also of encyclopaedias. People are interested in learning the meaning of terms that appear in newspapers, books, TV programmes, and so on. In particular, they want not only the definition of a term they encounter but also that of related terms. People want on the one hand to know the specific details of a problem and, on the other, to grasp the total picture or to have a general understanding. This is one of the main reasons for consulting dictionaries. In the language area, there appeared a few new good language dictionaries, such as the Longman dictionaries of contemporary English and the COBUILD English Language dictionary. These dictionaries provoked discussion about what constitutes a good dictionary.

Dictionaries are becoming more and more important in the computer processing of natural language. Not only machine translation but also man-machine dialogue systems, information retrieval systems, and so on require good dictionaries for their own purposes. Here we have to clarify what kind of information must be included in a dictionary.

Dictionary construction requires much time as well as a big investment. It is quite difficult to change and reconstruct one. Therefore, there must be a careful design at the outset. In the past, dictionary contents were severely limited by the number of pages. This condition is no longer true for electronic dictionaries, which can memorize large amounts of information. A problem in an electronic dictionary, particularly for use by computer programs, is the representation structure as well as the kinds of information to be memorized. Humans can interpret dictionary contents in a flexible and adaptive way, but the computer lacks such flexibility. Therefore, we have to provide all detailed information explicitly for the computer.

It is difficult to construct a bilingual dictionary because a word of a source language has varieties of meanings or concepts, and for each meaning there exist many corresponding words and expressions in a target language. Very often there are cases where concepts in a language correspond to several concepts in another language, and it is not easy to fix minimum "grains" for concepts that have one-to-one correspondence to linguistic expressions in both languages.

When we consider the multilingual dictionary, we have first to identify minimum "grains" of concepts in each language and culture, and then to determine those that are common to the concepts set for all the languages concerned. In this way, a neutral multilingual dictionary depends on the language set. Bilingual dictionaries are difficult to construct, and so multilingual dictionaries of several languages are even more difficult to construct, particularly when these languages have very different cultural backgrounds. It is very hard to identify pivot concepts through which words and phrases can be exchanged between every language pair. It would be possible to establish pivot concepts for major conceptual words, but it will be very difficult for many other concepts in ordinary life. These concepts, particularly those strongly related to human life, must be explained in terms of the cultural background where the language is used. Therefore, for very detailed machine translation dictionaries of ordinary words, it is essential to construct bilingual dictionaries. It is difficult to construct multilingual dictionaries covering several languages that reflect completely different cultural backgrounds.

Multilingual dictionaries will be possible, however, for concepts in natural science and technology, which are less dependent on different cultures. Also, if high quality translation is not required, we may be able to construct multilingual dictionaries among certain sets of languages.

Another question that needs to be discussed is how to use dictionary information. Dictionaries are useful not only in machine translation but also in natural language processing systems such as dialogue systems and information retrieval systems. Internal organization of a large dictionary is important from the standpoints of speed and flexible reference to the related information. Hypertext structure will help people use complex electronic dictionaries. We have to develop more powerful ways for using electronic dictionaries.

9. State of the Art in Europe and the United States

There were several basic research projects in Europe in the 1960s, but the first system that aimed at practical use was the Systran Russian-English machine translation system at CEC Euratom, Ispra, in the late 1960s. Systran English-French was introduced to the EC Commission in Luxembourg in 1976. After long trial use and improvement of the system, it began semi-operational use at the beginning of the 1980s. Now Systran English-French is very satisfactorily used by many translators in the EC through computer network communication. For example, the translation speed is 40 pages (pages averaging 250 words) per minute, or 500,000 words per hour. On the average, minor corrections are necessary at about 10 spots on a page.

The EC Commission has several other Systran systems for language pairs, such as French-English, English-German, English-Italian, French-German, etc. The translation quality of these systems is less good than the English-French system.

The EC has to deal with nine languages in the Community. This might have meant that 72 systems of machine translation from one language to another had to be constructed. To avoid this difficulty, it seemed necessary to develop a multilingual machine translation system that could translate one language to all the others at the same time. The EC started preliminary investigation for this multilingual machine translation system in 1978 under the EUROTRA project. The official EUROTRA project was started in the early 1980s for the seven languages then used in the EC (extended to nine languages afterwards), and many researchers from every EC country participated in it. However, the R&D of multilingual machine translation systems was too difficult, and the project was stopped at the beginning of the 1990s. The EC has shifted its research from machine translation to dictionary construction and language industries.

Germany has started a speech translation project called VERBMOBILE, recently stimulated by the Japanese project of speech translation conducted by the ATR Speech Translation Telephony Research Institute. The project, which will continue for eight years, is divided into two four-year phases.

The United States had leading research groups in machine translation at its beginnings, but unfortunately the US Government abandoned R&D in machine translation and stopped governmental research funds in 1966, following issue of the well-known ALPAC report. Since then, the United States has made no significant progress in machine translation research and has adopted the research direction of computational linguistics. However, there is now growing interest in machine translation in sections of the Government, and new funding for research is expected.

In the US private sector, there are some commercial machine translation systems, such as Systran, Logos, Weidner, and recently Glovalink. Systran was based on the research results of the Georgetown machine translation project conducted in the late 1950s and early 1960s. It was improved and has been used mainly at the EC Commission since 1976 and served to increase the available translation language pairs greatly. It is one of the more successful machine translation systems in the world. Logos was first developed for English-Vietnamese, and then added as language pairs English-French and some others.

Canada is a bilingual country using English and French. The Canadian Government was interested in machine translation long ago, and supported several R&D projects. One of the most successful systems is TAUM METEO, which translates weather forecast sentences from English into French. It has been in operation on a 24-hour basis since 1978. It has no pre-editing nor post-editing. It is fully automatic. When a complex sentence comes in that the system cannot handle, it is displayed on the operator's screen and is translated and typed in immediately. The Canadian Government has a large translation bureau under the Secretary of State, and does daily translation manually. The bureau is now using the Logos English-French system on a trial basis as well as translators' workstations developed internally.

10. The International Association for Machine Translation

As mentioned before, Japan took the initiative for organizing the first Machine Translation Summit Meeting, and already three summit meetings have been held, in Hakone (Japan), Munich (Germany), and Washington, D.C. (USA). We also organized the International Forum for Translation Technology (IFTT) in 1989 at Oiso, Japan, where we proposed the organization of the International Association for Machine Translation (IAMT). Discussions were continued in Munich, and finally at the Third MT Summit in July 1991, IAMT was established. The purposes and activities of the Association are outlined:

(i) Purposes

1.1 The International Association for Machine Translation (IAMT) brings together users, developers, researchers, sponsors, and other individuals or institutional or corporate entities interested in machine translation (MT) for the purpose of promoting and fostering, by every available means, the development and active use of MT systems. The IAMT offers opportunities and occasions to exchange information, study MT technologies and applications, and discuss and establish reference criteria or standards in areas of common interest to its members.

1.2 The specific concerns of the IAMT may include:

1.2.1 Collection and compilation of information

The IAMT may serve as a repository for historical and current documentation on MT, converting the texts to machine-readable form when and as feasible.

1.2.2 Exchange of information

The IAMT may serve as a clearing-house for current information of interest to MT users and developers, including:

– For users: translation market trends, available MT systems, types of MT applications, introduction of MT systems, approaches to text file input, pre- and post-editing strategies, evaluation of MT, etc.

– For developers: theories of MT, MT technologies, improvement of MT, approaches to dictionary-building, databases and other files available for exchange, etc.

1.2.3 Dissemination of information

The IAMT may publish information of general interest in the field of machine translation, such as:

– A regular newsletter on MT
– An MT handbook, including agreed definitions of terminology
– Bibliographies

1.2.4 Standardization

The IAMT may develop and propose reference criteria and standards in such areas as:

– Common document format for MT input
– Exchange format for dictionaries
– Design of controlled language
– Evaluation of MT output and MT systems

(ii) Activities

2.1 The IAMT shall convene the biennial General Assembly of its membership. The Assembly shall preferably be held in conjunction with the MT Summit, a technical forum that is open to the general public and that shall also be convened by the Association.

The IAMT may also undertaken the following activities:

2.2 Sponsorship and support of workshops, symposia, and conferences on MT and related technologies and applications.

2.3 Organization of tutorials and training courses on MT applications and skills involved in the use of MT.

2.4 Organization of tutorials and workshops on MT technologies.

2.5 Establishment of technical committees, special interest groups, and study teams.

(iii) Membership

Membership in the IAMT shall be open to active or potential MT users, developers, and researchers, and to any other individuals or institutional or corporate entities with interest in the purposes of the Association. There shall be three categories of membership: (1) individual, (2) corporate, and (3) institutional. All of these must belong to a regional association, except where there is no suitable association in the region where a person interested in IAMT lives.

(iv) Organization

4.1 The IAMT is the federation of the Regional Associations for machine translation (one each in Europe, the Americas, and Asia).

4.2 The organs of the IAMT shall be the General Assembly, which is the supreme governing body of the Association, and the Council, which executes the decisions of the General Assembly and carries out the business of the Association.

4.3 The IAMT shall have a permanent Secretariat. The location of the Secretariat shall be decided by the General Assembly.

The Japan Association for Machine Translation (JAMT) is trying to extend its membership and to become the Asian Association for Machine Translation. The CICC is considering the linkage of Asian countries for the exchange of technology on machine translation. For example, we can imagine the establishment of a machine translation centre in each country, where R&D and services of machine translation will be performed, and these centres working together for the progress and wider usage of machine translation. The IAMT will play an important role in such activities in the future.

11. The Future of MT

(1) Machine translation software will be on a word processor or on a laptop personal computer in the near future, and machine translation will become available everywhere – at home, in the office, and at school. In such a situation, a system will be adapted quickly to individual user needs by the learning mechanism in the system.

(2) Another direction is network access to a big MT system from a home computer. Users in this category may not use MT systems regularly, and therefore particular care must be taken to develop user-friendly interfaces.

(3) Present-day MT systems have a definite limitation in translation quality because they are based on the compositionality principle. Anaphora and ellipsis cannot be handled well because the systems don't recognize relationships between sentences. Much ambitious research must be performed in the future and include that on natural language understanding and more elaborate contrastive linguistic studies.

(4) Multilingual machine translation systems must be developed not only for European languages but also for Asian languages and other languages worldwide.

(5) Machine translation must be introduced into information retrieval systems and database systems so that people all over the world can have easy access to any information sources in the world.

(6) Machine translation must be introduced for the daily news transmitted via computer network and also for electronic mail systems, which are becoming more and more popular and represent a convenient way of communication.

(7) Speech translation research must be performed. At the ATR, the Interpreting Telephony Research Institute in Japan, research has been going on since 1986 for Japanese and English. It is highly recommended that other countries start similar research projects.

(8) International cooperation is essential in machine translation research and development. Dictionaries must be exchanged, and contrastive studies of languages are to be promoted as cooperative research works among different countries. Such activities must be supported continuously for a long time, for example 10 or 15 years, because language is a very difficult objective to attack. Agreement must be reached at the governmental level and Unesco on such international cooperative research.

The New World of Computing: The Sub-language Paradigm

Bozena Henisz Thompson and Frederick B. Thompson

ABSTRACT

Following a brief discussion of obstacles to the introduction of a "telephone-computer era" (based on the use of a combination of the telephone, personal computer, workstation, and television set), the "sub-language" concept is presented, this being a form of communication that is related to natural language but is domain specific. The processes of implementing sub-languages on the computer are described. Treated thereafter are the creation and basing of sub-languages and the ways in which networking is accomplished. Finally, the relevant software development environment is discussed and suggestions made for the organization of a software and data provider industry that could stimulate the arrival of the new "telephone-computer era."

1. Prologue

We are witnessing one of history's major technological events: the advance of the telephone-computer era. Over a number of years, we have directed our research toward the solution to the problems of this coming age. The results of this research are presented here. However, it is not enough to put down some concepts from which a solution may be implied. The solution therefore is also in the form of a fully implemented system that now exists as a commercial prototype ready for product development – the New World of Computing system.

2. Obstacles to the Development of the Telephone-Computer

The next decade will see the telephone, personal computer, workstation, and television set combined into a single, ubiquitous instrument – the telephone-computer. The telephone-computer will cause a rapid, widespread acceleration in the use of information processing and telecommunications. As a result of this major development, the market for both computer hardware and software will rapidly expand in both quantity and variety. This market will soon far exceed the current market for personal computers, workstations, and large transaction processors.

Current hardware capabilities are already adequate for this development: 30 MIP processing chips, voice digitizing, image processing and communication chips, 8 megabyte main memories, and 100+ megabyte peripheral memories are quite adequate for the great bulk of processing to be done. 144 K bits per second, error free, and soon 1,500 K bits per second, telecommunications, with packet switching, are now becoming available; capabilities can be expected to stay well ahead of foreseeable needs. New technologies, such as flat screens, and parallel processing chips will all enhance this radical change in the human-computer interface.

What will the era of the telephone-computer be like? This question has been the focus of our research over a number of years. One thing is clear: in the confluence of computer technology and telecommunications technology, we are witnessing one of history's major advances in human communication.

Industry is ill-prepared for this rapid acceleration of the information technologies. There are dislocations in the current software industry that work against the full development of the telephone-computer. One major symptom of these dislocations is the high cost of software development. Industry is acutely aware of these symptoms, as evidenced by their concern with "open systems" and "software engineering" approaches. However, the roots of these problems lie elsewhere. In this paper we will identify these roots and an approach that substantially corrects them.

As industry moves into this period of accelerating change and expanding market opportunity, management – indeed the industry as a whole – needs a coherent set of concepts that can provide the perspective required for intelligent decision-making. At this point in time, there is an almost total lack of sensible, down-to-earth concepts on which to develop an understanding and a strategy for what is taking place. Lacking perspective, managers in the computer industry are preoccupied with tactical questions and short-range considerations.

The artificial intelligence paradigm has distracted us and has proven to be inadequate. The successes of UNIX, on the one hand, and the Macintosh computer interface, on the other, have led industry into espousing the minimalist philosophy of the computer as an applications independent tool kit. The situation was stated succinctly by Dr. Robert W. Lucky in his capacity as Annenberg Distinguished Lecturer, the University of Southern California, on 22 January 1990.[1] After surveying the astounding advances in tele-communications resulting from the development of digital switching and optical fibre technologies, he asked the rhetorical question: "What are we going to do with this gigabyte? To be honest with you, nobody knows."

He went on to state that there is a total lack of leadership to carry us into the emerging telephone-computer era. In parallel with the vacuum in the conceptual area, he pointed out that the legal position of the Local Area Telephone Companies has worked against any telephone company taking a leadership role.

A new paradigm is needed that puts into proper perspective the role of the computer in human communications.

3. Sub-language: A New Paradigm

We present here a simple, basic concept, that of a "sub-language." A sub-language is a form of human communication that is domain specific, appropriate to that domain, and, consequently, highly efficient. Using this concept, we put forward a new paradigm for human information processing and communication. We then use this paradigm to lay out the new world of computing that will characterize the telephone-computer era.

We have learned to impose structure on the jumble of our moment-to-moment experiences so as to create order and provide perspective. In the words of William James:

> Is not the sum of your actual experience taken at this moment and impar-tially added together an utter chaos? The strains of my voice, the lights and shades inside the room and out, the murmur of the wind, the ticking of the clock, the various organic feelings you may happen individually to possess, do these make a whole at all? . . . We break it: we break it into histories, and we break it into arts, and we break it into sciences; and then we begin to feel at home. . . . The intellectual life of a man consists almost wholly in his substitution of a conceptual order for the perceptual order in which his experience originally came.[2]

It is in terms of this conceptual order that our world becomes compre-hensible.

It is the infinitely variable expressions of language that give tangible form to our own immediate view of the world and by which we share that view with others. It should be no surprise to find human language playing the central role in leading to an understanding of information processing. The mechanisms of language are precisely the tools we need and use to express the recursive structures we impose on our experience. It is these mechan-isms of language that we universally share that form the basis of communi-cation.

What is this "natural language" that we use? The notion of natural lan-

guage has played a useful role in linguists' development of a general understanding of human communication, in the codification and maintenance of
purity of national languages, and the training of language teachers. Language, in the sense used by linguists, might more properly be thought of as
an integrated family of linguistic mechanisms, often expressible as grammar
rules. The phenomenon of information processing and communications,
although exhibiting in a given cultural community the adherence to such
syntactic forms, also has features better characterized, we believe, by the
notion of "sub-language." Current literature often refers to a person's mental awareness of the world as one's "cognitive model." But it is not a single
model; it is a large family of interrelated, comparable models – the many
alternatives that we visualize and choose among. The logician would refer to
these as the model theoretic counterpart of our sub-language; and indeed,
abstractly, the model theoretic and the linguistic representations are quite
equivalent. We do not wish to imply that we "think linguistically," as an
alternative to "thinking in terms of a cognitive model." Rather, it is simply a
more useful paradigm for the consideration of the role of the computer, a
case that we intend to make in this paper. The linguistic formulation, we
feel, grasps much more clearly the characteristics of our ongoing cognitive
processes.

In dealing with their immediate task environment, people narrow their
considerations by making judgements of relevance, value, and task effectiveness – judgements that are characteristically human. The results of these
judgements take concrete form in the sub-languages we use both in communicating within our task group and in our own internal thought processes.
A moment's introspection makes it clear that as we move about from one
task to another during a busy day we change from one sub-language to
another as our attention is drawn from one domain of activity to another.
These sub-languages differ in vocabulary, often in their cryptic syntax, and
even in the meanings of the same words. Their only commonality is their
basic linguistic structure.

Consider the sentences in figure 1 entered by the trust officer of a bank in
his ongoing dialogue with his computer. Does this look like English? It is
not. Surely it is a sub-language of English, one that has been geared to the
concerns of the trust officer. In such a defined context, the phrasing is no
longer ambiguous or indecipherable.

The concerns of a human individual engaged in a specific task environment can be characterized by that person's immediate sub-language. The
essential characteristic of a well functioning team is their common sub-
language. The stability of a given sub-language is found in the stability of
the task we undertake. When we return again and again to a task, and to
that group in which we interact in conducting that task, it is the sub-
language that codifies and externalizes our ongoing considerations. And it is
the ongoing, ubiquitous changes in that sub-language that track our
decision-making processes.

As an illustration, consider a particular work environment, say that of a

• Which of Bob Moore's equities changed by more than five percent in the last two weeks?
• How many high-tech shares does he own?
• What are our short-term projections for GM and DetEd?

Figure 1

person working as secretary for an industrial manager. One aspect of that person's environment is their typewriter. The technology of the typewriter keyboard has not changed in many years, even though a more efficient lay-out of the keys is known. The reasons for this stability are not hard to envision – the keyboard does not change because of the strong social inertia resulting from so many people having been trained on the existing one. There are many other aspects of the typing, filing, and sending of letters, reports, etc., where there are both physical and social inertia that mitigate against change in many aspects of the secretary's concern. The moment to moment sub-language of such a person is constantly shifting as a needed address must be found, a letter retrieved from a file, a phone call answered. But a part of all of these sub-languages remains essentially constant – that part related to the mechanics of typing, phoning, filing, where the physical and social inertia are high. This part can itself be characterized as a formal language. It is precisely these highly stable sub-languages that can economically be built into computer systems. Word processors are an ideal example of how the inertia of a significant part of the secretarial world can be exploited.

This illustration, concerning word processors, bears greater scrutiny. When a team is working on a project, there are at any given time a number of aspects that are undergoing change, and the rapidity of this change results in uncertainties and ambiguities in the sub-language that characterizes their interactions. Each of the participants has their own sub-language related to the common task but also containing the expertise and personal insights the individual brings to the common effort. But it is essential to the effective functioning of the team that they share completely and unambiguously and tacitly a sub-language that establishes the basis for their intercommunication. It is this that is the team's sub-language. Relative to their many interactions – the sharing of insights, the settling of differing ways to approach a problem, etc. – this team sub-language changes only slowly. And these changes result directly from interventions of the team itself. A word processor obviously does not comprehend the entire sub-language of a busy secretary. But it is a stable part of the secretary's rapidly changing sub-language.

A useful analogy is to the layout of a mountainous terrain. Suppose all sub-languages were spread out over a broad area, and that the altitude of this terrain at any point was some measure of the relevance of the associated sub-language to the task at hand. The landscape would in general consist of a tall mesa whose top was quite flat except for a small hill. As time went by, the mesa would move almost imperceptibly, while the hillock would be seen to shift about, in almost constant motion, as the concerns of the moment shifted from one situation to the next.

Consider a team, working on a common design task. If one were to ask them what they were doing, they would say "designing a. . . ." From our perspective, however, their sole task is the maintenance of the underlying team sub-language. The completion of their task is marked by their agreement that this sub-language is now ready to be passed on to those who will implement their design. Their common sub-language will, by that time, have evolved into one containing the design drawings, part tables, and specification lists in forms they know to meet the conventions established for transmittal to the industrial engineers who will further prepare them to go to tooling and the production floor. These conventions are characteristic of the hierarchy of sub-languages that characterize interactions in an organization.

There is one area where the computerization of sub-languages is already highly developed and sophisticated, namely programming languages. The stability of the Von Neumann architecture has resulted in an evolutionary development of sub-languages that exploit this stability. We will say that the computer "understands" a sub-language to mean precisely the same as when we say a computer understands a programming language. Namely, it carries out instructions in the way they were intended to be carried out, whether in the sub-language of a programmer or the sub-language of a professional when referring to work-related matters. Stable, domain-specific sub-languages, from simple word processors to the cryptic, icon-oriented yet highly sophisticated sub-language of a NASA space flight control room, can be handled by a computer as easily as any traditional programming language. Computers can be programmed to understand the highly idiosyncratic, often cryptic sub-languages of individuals and teams from all walks of life.

The academic disciplines of linguistics, foundations of mathematics, and computer science provide firm theoretical underpinnings for the study and computer implementation of sub-languages. A brief overview of these underpinnings reveals in sharp focus the full sub-language paradigm. The concept of "sub-language" is abstractly equivalent to the concepts of "recursive function" and "Turing machine." Thus our "sub-language" paradigm is equivalent to Church's thesis that human information processing can be characterized by these formalisms.

To give utmost precision to this statement, we restate this paradigm in terms of one of the above formalisms. Turing machines can be enumerated, i.e., we can speak of the ith Turing machine, $\mathcal{T}_i$. A Universal Turing machine, $\mathcal{T}_u$, is such that given any argument n, $\mathcal{T}_u(i,n) = \mathcal{T}_i(n)$; thus a universal Turing machine can simulate any Turing machine, provided it is given the proper index. We are indeed Universal Turing machines, but with a Demon $\mathcal{D}$. Having observed n, our Demon selects that Turing machine that is most informing, and $\mathcal{T}_u(\mathcal{D}(n),n)$ becomes our cognitive model. In this formalism, the sub-language paradigm is equivalent to saying that at any instant our thought processes can be characterized as a Turing machine; but the selection of what Turing machine is a non-computable process, characteristically human.

The sub-language paradigm can also be considered as the integration of two other well-established paradigms of computer science, namely object-oriented programming and compositional semantics. Approaching this relationship from the object-oriented perspective, take the object classes as "parts of speech," the "semantic categories" of Tarski's seminal paper[3] that established modern mathematical linguistics. We note the clear tie to consideration of data structures. In the terminology of compositional – or more generally procedural – semantics, a sub-language is defined by its "rules of grammar," each of which is associated with a semantic procedure. The implementation of these semantic procedures on the computer is in terms of the processes encapsulated in the object classes associated with the parts of speech occurring in the syntax rule. This point of view yields an elegant formulation of the "sub-language" notion: Let the processes and memory structures of a given object-oriented programming environment be implemented in "hardware," then the "sub-language" becomes the "machine language" of the resulting computer. In this form, sub-language is abstractly equivalent to "computer"; that is, the set of all "sub-languages" coincides with the set of all machine languages of computers.

The paradigm for human information processing can now be stated:

It is the constant re-evaluation and adjustment of the relevant view that characterizes human information processing. A succinct expression of such a view is as a formal language. When there are strong social and physical inertia in an area of broad concern, a part of the sub-languages characterizing those concerns stabilizes. For these stable areas, it is economically expedient to develop computer systems that can understand these sub-languages. When, in some area, these stabilities dissipate and others arise, it is only human intervention that can maintain their relevance and effectiveness.

A computer can "understand" a given task-specific sub-language far beyond its use as a simple query language to a database. Consider a middle-level manager in a large engineering establishment (see figure 2). He changes the estimated time of completion of one of the tasks for which he is responsible by typing instructions to his computer; the computer responds by carrying out the indicated actions. The sub-language of engineering management is thus "understood" by the computer, just as the manager would expect a staff assistant to have responded in a pre-computer era. The computer then becomes an instantaneous link, conveying the relevant implications of this simple change to wide-ranging concerns across the engineering floor. The computer is seen here in its true identity, as a powerful communications device. A brief, curt word to the computer, in the same jargon that has been developed by the management team, is enough to elicit a complex computer response, which may include composing and sending messages or controlling equipment. However, as the task develops, there will always be new instructions for the computer; therefore, there is a need to fall back on longer sentences and more complex commands, using, of course, the syntax and vocabulary of our own sub-language.

4. The Implementation of Sub-languages

How are sub-languages implemented in the computer? Members of one class of sub-languages are already implemented in computers, namely programming languages. How are they currently implemented? A compiler is written that embodies both the syntax of the language and the semantics. The compiler accepts a sentence of the language and returns a single, machine language program. When used in interactive mode, this program is then executed. That is, the abstract computer that understands the programming language consists of a hardware computer with its own machine language

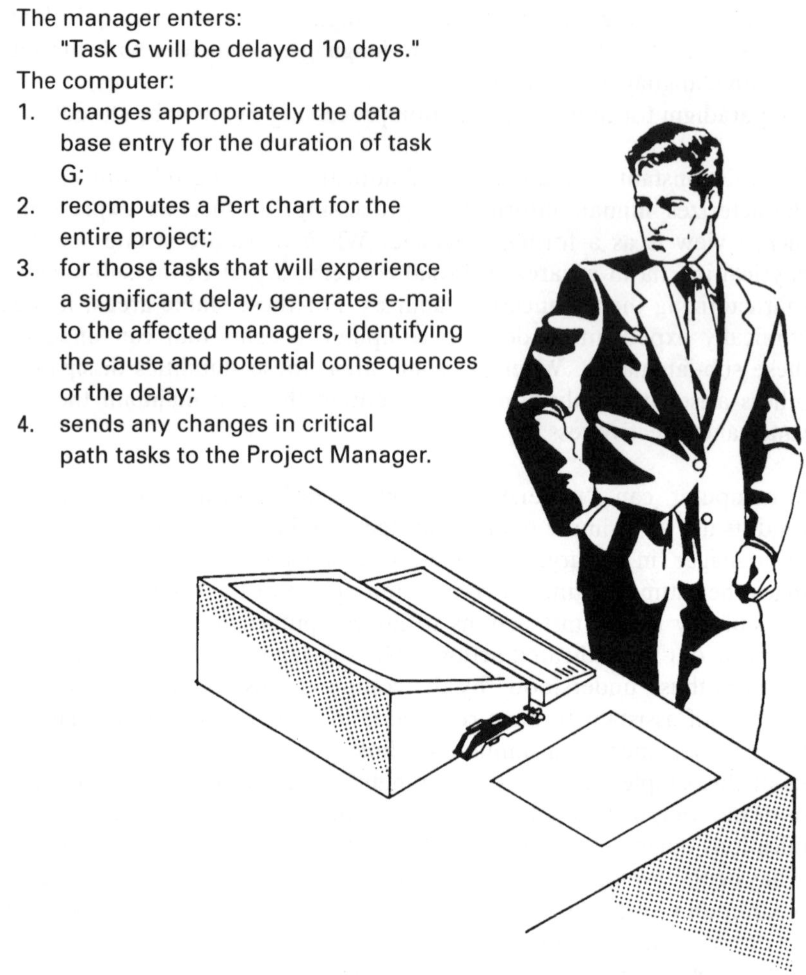

The manager enters:
   "Task G will be delayed 10 days."
The computer:
1. changes appropriately the data
    base entry for the duration of task
    G;
2. recomputes a Pert chart for the
    entire project;
3. for those tasks that will experience
    a significant delay, generates e-mail
    to the affected managers, identifying
    the cause and potential consequences
    of the delay;
4. sends any changes in critical
    path tasks to the Project Manager.

Figure 2

and the compiler, which translates the programming language into machine language. To change the programming language, one rewrites the compiler. There are compelling reasons why this is a bad technology for implementing sub-languages, including programming languages. Before discussing these reasons, we present here a radically different technology.

Let the computer be a Universal Language Processor (a hardware computer with a universal language processor replacing the compiler of a particular language, if you like). It operates in two modes:
(1) It accepts, one at a time, the rules of grammar and their associated semantic procedures that define the sub-language, building them into its internal grammar table.
(2) It accepts an input sentence, parses it according to the grammar, uses

A typical Rule of Grammar
(as understood by the computer):

RULE
>Syntax:<noun_phrase> =><adjective>""<noun_phrase>
>Semantics: POST adj_mod_proc

Figure 3

the resulting parsing graph to compose the associated semantic procedures, evaluates them, outputs the result, and cycles.

Thus it is a simple, straightforward implementation of compositional semantics.

A little insight into what is going on here will be useful in understanding
the power of this paradigm. Sub-languages are defined to the computer in
terms of grammar rules, consisting of a syntactic aspect and an associated
semantic procedure. An example of such a rule is shown in figure 3. Given
the constituents of a meaningful phrase – for example: "government" and
"contracts" – the semantic procedure goes to the two associated data files
and produces the "meaning" of the entire phrase: "government contracts."
The role of syntax is to show how words and phrases can be combined into
meaningful statements. Once the syntactic structure of a sentence is seen,
the associated semantic procedures can be composed appropriately. The
rules of grammar, along with the corresponding semantic procedures, constitute the building blocks. Each of these rules is implemented as a separate
unit. The syntax of a sentence provides the plan for combining these building blocks into the complex meaning of the entire sentence. Thus the individual semantic procedures can be efficiently composed in innumerable ways
to produce the needed answers to immediate user concerns.

(The first question that will come to the mind of a knowledgeable computer person is the effect of such an architecture on response time. Let us deal
with this immediately. In our current implementation of this architecture,
against a moderate size database concerning ships and shipping [for computational linguists, this is the well-known DARPA "blue" file.], and using a
sizable grammar, the parsing time for the following sentence: "What is the
cargo type and destination of each ship whose port of departure was some
Soviet port?" is about a tenth of a second; the through put time, including
database access, is 8 seconds. The key to these response times lies in the fact
that in such very high-level sub-languages, the object-class data structures
and processes are highly optimized, so that in processing a sentence, one is
composing a few highly optimized procedures.)

The first thing to notice are the implications of the independence of the
grammar rules – syntax and associated semantic procedures. As said above,
in building a sub-language, rules are added one at a time. These same rule-
adding utilities can obviously be used at any time to add an additional rule

or, for example, a whole family of rules implementing a new object class. It is these same utilities that implement the user's ability to extend his own sub-language by definitions.

An "insider's" problem is to determine how the great number of highly complex procedures that may all be needed at some time or another can be retained in a form that makes them available for rapid response to a query. One way that has proved particularly effective is to use "pages" in peripheral memory that are organized on the basis of semantic content. In response to a particular query, only those pages that are required are brought into main memory – whether they be database record, procedure, text, image, digitized voice, or other pages. Pages holding all manner of material are brought into the same paging area. Obviously, procedure pages require a modicum of run time binding, but since the number of paging slots is large, there is very little trashing of pages between main and peripheral memories.

The information available to the computer is organized into a network of "nodes" and "links." The "nouns" of a sub-language point to certain of the nodes in this semantic net. The syntax rules also have a geometric interpretation in terms of the semantic net; they indicate how to move from one set of nodes to another. Thus the parser composes the path from the words in the initial expression of a question to the nodes constituting the desired answer. The information about a node is kept on a database record on one or more pages of peripheral memory.

Organizing information in this way provides a highly efficient and flexible method for maintaining a rather shallow level of information organization (essentially equivalent to an entity-attribute database or relational database, plus inheritance). By linking such "database" records to more complex forms of representation (e.g., texts, pixel files, postscript files, engineering drawings) and by providing sophisticated semantic procedures that can exploit the additional complexities of these structures, the computer can give wide-ranging responses to highly complex technical questions. In the terminology of object-oriented programming, these database records constitute the object representations for the single all-encompassing object class, "noun." Any hierarchy of subclasses of objects may be created, such as "image noun," "matrix noun," "co-variance matrix noun," etc., with their associated processing procedures.

A new object class can be easily implemented as a new subclass of the "noun" object class; when an instance of the new object class is created, first its record as an instance of "noun" is created, and then a link from this record to an instance of the data structure of the new class is added. As an example, suppose one were building a new sub-language to be used by the structural engineers in an aerospace company. Suppose the company already had a major investment in files of stress data and, say, FORTRAN procedures that processed these files. The new object class, a sub-object class of "noun," would be created whose associated data structure was that of the stress data files. Syntax rules for noun phrases that engineers commonly used in referring to the stress data would be added, their corresponding semantic procedures consisting largely of calls to the relevant FORTRAN routines. Such queries as: "Plot the stress against wing tip loading for both Model A12 and Model A14 wing aileron designs" would be immediately available.

In today's highly visual world, sub-languages are seldom limited to written text. But how can this complete integration of media be implemented? Certainly the identification of the object class with its encapsulation of structure and process is a major step. Another step concerns the extended "alphabet" available to all sub-languages. All letters and characters of the usual alphabet as well as the entire extended ASCII character set; all graphic "events," such as clicks of the mouse and movements of the cursor; and all "interrupts" from internal and external sources (properly screened and identified) can be used in the input string that is fed to the language processor. (The computer, like human beings, has "fingers" for pointing and "intonation" and "gestures" it can use.) In this respect, all sub-languages have the same terminal vocabulary, namely this extended alphabet. Once this is established, grammar rules can supply the recursive, flexible link between the input string and the internal object classes. For example, one can at any time introduce a new icon, placing under it any sentence or phrase of the sub-language that is then evaluated in line whenever the icon is clicked during input of a query.

In figure 4, an airline mechanic is seen working on the radar nose-cone of a Boeing 747 aircraft. He turns to his computer for detailed technical support. He has already entered information identifying the particular aircraft he is working on and has called for a display of the nose-cone area. The computer-generated photo image of the relevant area (plus an invisible back-plane drawing outlining all significant parts) provides a highly efficient medium for communication. For example, he may type "leak" and click his mouse on the image of the place he suspects is leaking oil. The computer may respond with the spoken word: "tighten" and blink the bolt it identified in its diagnosis as the probable cause of the leak. In response to a sparsely stated but technically involved question, the mechanic receives an immediately useful response that reflects a high degree of built-in understanding.

In figure 5, a maintenance professional is entering instructions into his personal, completely mobile, telephone-computer. It eliminates any need for the usual truck-full of manuals. The professional's efficiency is greatly increased, since the computer tailors its responses to the specific installation. Astute use of hypermedia links from one data display to another quickly provides pathways to the details the professional really needs. References that establish context (e.g., "I am at . . ."), as well as pronouns and elliptic constructions (e.g., "What about the other connector?") play important roles in effective dialogue. Note that pointing to and blinking significant areas in pictures and drawings constitutes visual "pronouns" (e.g., "voltage 'there'?" or "tighten 'that'," "[show schematic icon] of 'that'").

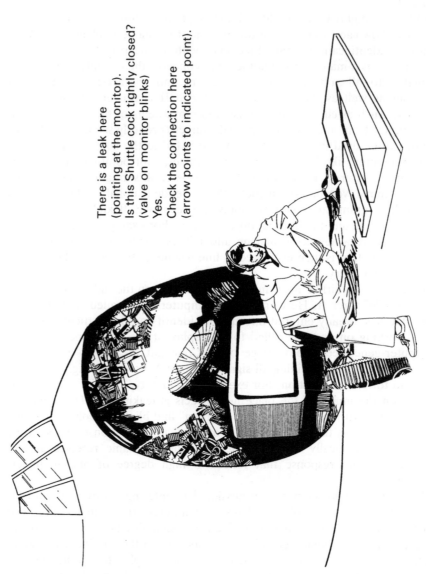

There is a leak here
(pointing at the monitor).
Is this Shuttle cock tightly closed?
(valve on monitor blinks)
Yes.
Check the connection here
(arrow points to indicated point).

Figure 4

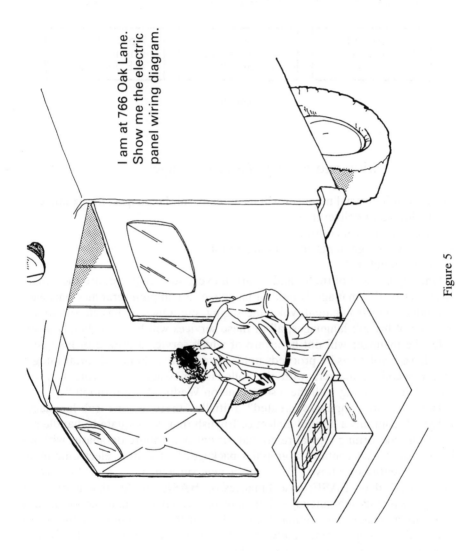

I am at 766 Oak Lane.
Show me the electric
panel wiring diagram.

Figure 5

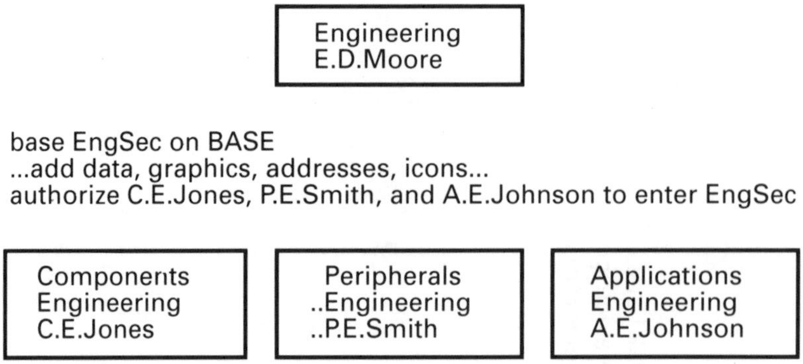

```
┌─────────────────────┐
│  Engineering        │
│  E.D.Moore          │
└─────────────────────┘
```

base EngSec on BASE
...add data, graphics, addresses, icons...
authorize C.E.Jones, P.E.Smith, and A.E.Johnson to enter EngSec

```
┌──────────────┐  ┌──────────────┐  ┌──────────────┐
│ Components   │  │ Peripherals  │  │ Applications │
│ Engineering  │  │ ..Engineering│  │ Engineering  │
│ C.E.Jones    │  │ ..P.E.Smith  │  │ A.E.Johnson  │
└──────────────┘  └──────────────┘  └──────────────┘
```

Figure 6

5. The Creation and Basing of Sub-languages

The typical industrial manager will have many sub-languages, for example:
– schedules and deliverables
– budgets and fiscal control
– personnel assignments and administration
– correspondence

Underlying each of these, and a part of every sub-language, are the general dialect of the manager's natural language, a complete graphics package, text editor, electronic mail, voice messaging, etc. Once he has chosen to use any one of his sub-languages, all of these services will be immediately available; the manager will not be aware of which service a phrase of his query may have invoked as he proceeds in his normal way: "Send this draft budget to my section managers with the following message: '. . . (voice) . . .'"; "Schedule a meeting with them sometime on Wednesday afternoon."

How are sub-languages created? Initially, there is one sub-language, BASE. It contains a limited dialect of English that is adequate to handle expressions concerning typical relational, or entity-attribute databases with inheritance. It also contains a graphics package, text editor, electronic mail, etc., as mentioned above. To create a new sub-language, say "Finances," one "bases" it on BASE: base Finances on BASE, or, for that matter, on any pre-existing sub-language that may be available. Then, choosing this new sub-language: enter Finances, one has all the capabilities of the based upon sub-language immediately available. One can then extend this new sub-language in many ways (these will be discussed below).

In figure 6, engineering manager E.D. Moore creates a sub-language to share with his three subordinate managers. Now any of the four of them can use, modify, and extend this common sub-language "EngSec." Thus they jointly maintain a common, up-to-date view of their joint activities (e.g., preliminary designs, personal schedules). This is the significance of being able to "enter."

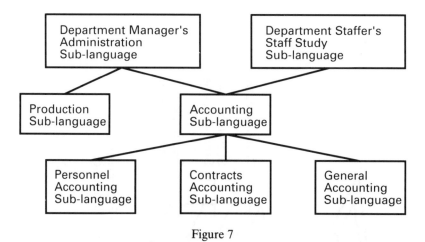

Figure 7

There is a strong asymmetric relationship between a sub-language and all of the sub-languages on which it is based, either directly or indirectly. Suppose one sub-language, "Accounting," is based on another, "Personnel Accounting": base Accounting on Personnel Accounting. Any changes in Personnel Accounting are immediately reflected in Accounting; however, Accounting can be changed in any way without affecting Personnel Accounting. This asymmetric relationship is characteristic of basing.

In figure 7, showing the accounting sub-languages, the people in the Personnel Accounting Section are the only ones who are authorized to enter the Personnel Accounting sub-language; therefore, they are the only ones who can change it; the same holds for the Contracts Accounting and General Accounting sub-languages. Accounting is based on each of these three. No one is authorized to enter Accounting; therefore no one can make any changes in it. Of course, it is automatically always up to date with the latest data from Personnel Accounting, Contracts Accounting, and General Accounting.

Appropriate managers are authorized to base on Accounting. One of the Department Manager's sub-languages is based on both Accounting and Production, and therefore always has available the very latest accounting information. The manager may well have had the application programmers add a number of grammar rules, graphic output formats, and icons, so that overviews of the complete operation are always readily available. These added facilities would only be available in this particular sub-language, but would always utilize the latest accounting and production data. A member of the manager's staff, looking into the possible change in the pricing structure for company products, could also base a staff study sub-language on Accounting, change many of the entries to values reflecting the new pricing structure, then examine the inferred results, and finally arrange appropriate graphics for a presentation (without, of course, affecting the Accounting sub-language at all). In figure 8, placing one sub-language above another indicates that the top one has been based upon the one immediately below.

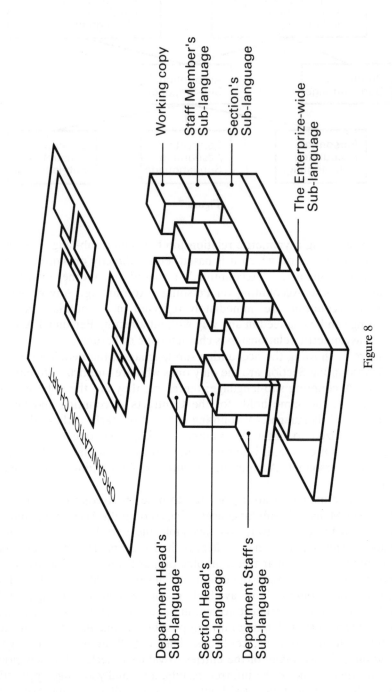

Working copy

Staff Member's
Sub-language

Section's
Sub-language

The Enterprize-wide
Sub-language

ORGANIZATION CHART

Department Head's
Sub-language

Section Head's
Sub-language

Department Staff's
Sub-language

Figure 8

6. Networking in the Telephone-Computer Era

Local area networks (LANs) are a very transient aspect of our computer environment. Once Integrated Service Digital Network (ISDN), Broad-band ISDN (BISDN), and packet switching are fully implemented and installed, file transfers will be sufficiently fast to provide current LAN services between any two telephone-computers. At that time, of course, telephone-computers will have telephone numbers, indeed will be the terminal equipment represented by one's telephone number. Simpler forms of telephone communication, such as voice, e-mail, fax, and electronic messaging, will be subsumed within a much broader spectrum of telecommunication services mediated by the telephone-computer.

When one telephone-computer calls another, pairs of windows are created; one window of each pair appears on the monitor of each computer. One pair of windows is controlled by each participant. The participant controlling a given pair of windows can enter any of his sub-languages; then any of the wide-ranging data and graphics thereby available can be displayed on both windows of the pair. If both participants are authorized to enter the given sub-language, then both can address this sub-language jointly.

For example, in figure 9, a banker calls Bob Moore, one of his clients. The banker shows Moore a bar graph of the changes in Moore's DetEd stock over the last quarter. Moore asks what happened during the previous quarter, and the banker immediately instructs the computer to bring this information forward. The banker shows Moore the implications of the changes he is advocating, and Moore's questions are quickly answered with the supporting data displayed by the computer. A change is made, and its effect is immediately seen by both parties. Once a transaction is decided upon, it is immediately implemented by means of a transaction transfer to the appropriate stock exchange.

We have already shown how a manager can create a sub-language and then authorize his subordinates to be able to enter it, thus creating a common sub-language. A sub-language on one telephone-computer can be entered from a different telephone-computer, provided the requesting person has been authorized.

The combination of basing with the single, worldwide telephone network adds up to a powerful set of capabilities. To see the significance of basing, consider the matter of virtual address space. In current practice, each computer has its own private virtual address space. Data and processes in other files must be brought into this virtual address space by bulk transfer before they can be utilized in calculations. This two-level addressing constitutes a major barrier to the expanded use of computers, in particular to the integration of information resources relevant to the changing needs of applications from the vast resources that are available. That is not how "addressable memory" is utilized in our daily life. Our addressable memory provides an encyclopaedic array of vast amounts of data and process, while only a minute fraction is involved in our immediate considerations.

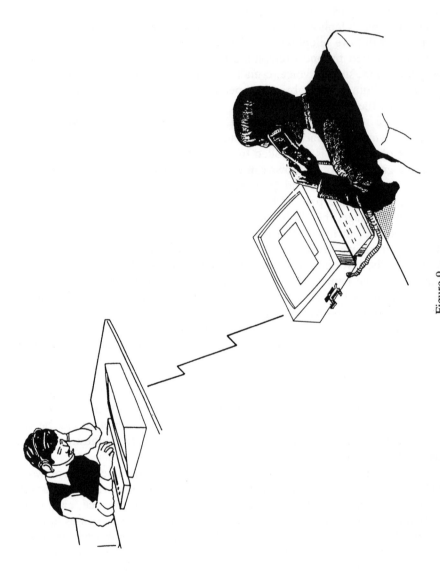

Figure 9

In the telephone-computer age, there is just one common virtual address space, down to the byte level, for all telephone-computers, for all these resources. This address space is blocked into pages, which are also the packets that move over the telephone lines. International agreements long ago have created a universal address standard, namely the telephone number. When one installs a new telephone and is assigned a new telephone number, one is thereby automatically assigned one's own slice of "virtual memory," one's own corner of the world's address space. Companies that maintain huge data and powerful processing resources have the whole world as their market because they are uniquely identified and reachable through their telephone number. It remains to provide the means of establishing the addressability to these resources. That is the role of basing.

Basing one sub-language on another establishes addressability among the data and process pages that constitute the physical manifestation of these sub-languages. Basing results in the sharing of a common address space whose key element is the addressable page. Therefore, pages whose "home" is on one station are co-mingled with pages from another. This results in a common address space (down to the byte level) across the entire hierarchy of associated sub-languages. Thus it is the sub-languages in such a hierarchy that are "networked," not computers. Such a network may be small and of short duration, servicing an immediate problem. Other networks may be large and stable, hosting an extensive hierarchy of sub-languages. Such "networks" are created, and deleted, by the simple acts of basing and unbasing.

The banker has formed such a network with the various stock exchanges and commodity markets (these being archival stations, discussed below). Since the banker shares an address space and sub-language with these markets, transactions can be completed as a natural aspect of any dialogue with a client.

Here again, in figure 10, is the maintenance professional working on the nose-cone radar of a Boeing 747 aircraft, his computer by his side. Consider the networking aspects of this maintenance situation. The computer is networked, in the strong sense of sharing address space, with the Boeing maintenance shops in Seattle, Washington. That is, a sub-language in this computer is based on the Boeing Maintenance sub-language in the Seattle shops. None of the maintenance material is in the computer being used by the maintenance person. First, the identification of the aircraft being serviced is established. In response to a call for a full-colour, annotated image of the radar nose-cone area, the pixel data sets come, via ISDN, from a single source – the high transaction rate server in Seattle. Although some processing is being done locally, all maintenance data and diagnostic analysis are being done in Seattle. The bane of having out-of-date maintenance manuals will be a thing of the past. The addressable page is both the unit of storage and the packet of telecommunication. If the maintenance professional is still puzzled, clicking the mouse on a special icon will establish an immediate conversation with a maintenance specialist at Boeing in Seattle. Both monitors will display the same material, both people can use their

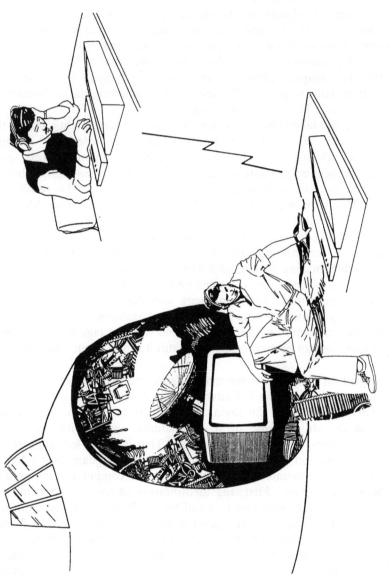

Figure 10

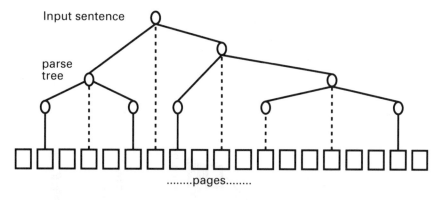

vertical links of the form ---- are to procedures
vertical links of the form ——— are to data

Figure 11

mouse to point, and both can have a voice discussion of the problem at hand.

Figure 11 illustrates the point that all of the database records, semantic procedures, utility routines, pixel data sets, etc., are stored as pages, and that only the necessary pages are brought into main memory. We see that by using basing and networking, many if not most of these pages will be drawn from distant stations at the time they are needed.

Each individual item of information in a network of sub-languages has its unique address by which it is identified. The stations whose peripheral memories are the depositories for these items are uniquely addressable by their telephone numbers. Note that it is the hierarchies of sub-languages, with their associated common address space, that are networked, and not the computers. Thus, within a single computer, a person may have many sub-languages, each in its own network. Sub-languages, not computers, are networked; sub-languages have no geographic limits.

In current database technologies, the database consists largely of links from one node of the database to another. The "links" in our semantic net database consist of page addresses. Thus for a network of more than one telephone-computer, the links within a database may refer to pages that reside anywhere in the net anywhere in the world. A database in such a network is "distributed" in an intrinsic way; the basing procedure implements this. The size of a page is 2,024 bytes.

The 64-bit page address has the following structure:

byte offset on a page	2^{11} bytes
page number	2^{21} bytes
telephone number	2^{32} bytes

Thus, up to four billion telephone-computers can be accommodated. Each telephone computer may hold two million pages, that is four billion bytes of

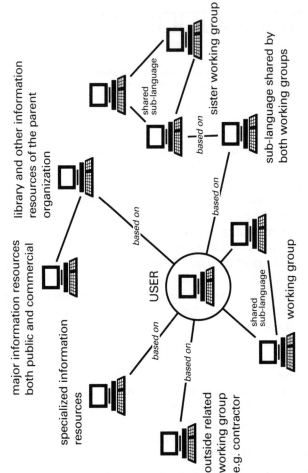

Figure 12

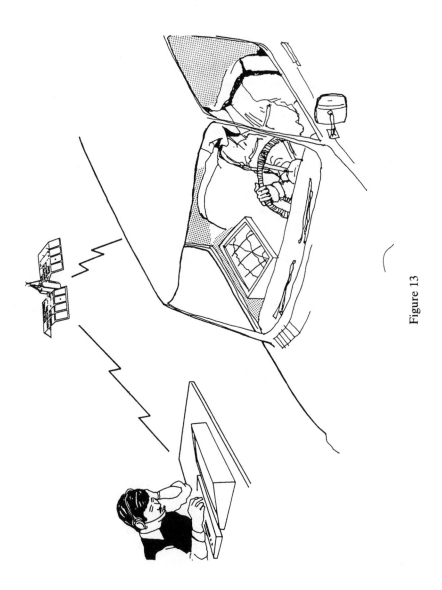

Figure 13

information. Thus the bound on directly addressable data and processes, the size of the single virtual memory, is 10^{19} bytes.

Figure 12 shows a schematic of a typical sub-language network. The inclusion in these networks of large volume servers of archival information will be typical. We have seen this need in maintenance situations. Companies like Mead Data Central in Dayton, ISML, a Houston-based supplier of scientific sub-routines, Springer-Verlag's Beilsteins Handbuch der Organischen Chemie, cookbooks and garden catalogues, the New York Stock Exchange stock closings, and the show records of the American Kennel Club will all be available, page by page as needed by thousands of users. A department store, serving several hundred thousand charge customers, will do so through a high transaction server that makes available all manner of sales material, processes incoming queries and orders, and connects customers to knowledgeable sales personnel.

Figure 13 shows a maintenance professional working for a local service company. The professional's mobile station is in at least two networks: the first, previously illustrated, is with the home maintenance station whose server has all records for the field locations served and all the maintenance information. The second is the dispatcher network. The central dispatcher can see on the map displayed before him the location of all maintenance trucks as they move about the city. When the dispatcher receives a call requesting maintenance service, he can type in the address of the caller and immediately see its location on the map, spot the nearest maintenance unit, click it with his mouse, and talk with the maintenance person directly to coordinate the new service.

7. All of the World's Information

What needs to be addressed? The obvious answer is anything that may at some time be an object of attention in some sub-language. Each object class will have its own grain and often its own special means for the identification of its relevant elements.

It is interesting to note that many, if not the majority of, objects referred to by sub-languages do not have "names" in the lexicon. Consider the marriage of Edward D. Moore and Patricia Jones Moore. Friends of the Moores often refer to their marriage; for example, "her daughter by her second marriage," but it does not have a name. "Texas Instruments FB74 transistor" may name a class of its instances as they exist in a great variety of circuits, or alternatively, of its instances in circuit drawings (as sub-drawings). In either case, they will inherit the properties (such as impedance) of the FB74 transistor class. In the latter case, one can identify the particular transistor either by "the FB74 transistor used in the monopole section of the shift register" or by pointing at the transistor and clicking the mouse while viewing the circuit diagram.

How is the addressability problem for all of the world's information solved? We identify the notion of the Archival Station. You call up a station that supplies information you wish to be accessible to one of your sub-languages. A form appears on your monitor and you are asked to fill it out. After doing so, you put your charge card in the card-reader slot in your telephone-computer. You are then free to base your sub-language on any material of interest to you and available from this source. Your initiating call to the Archival Station provides it with all it needs for billing, notification, etc. The act of basing automatically identifies authorization information necessary for security. The Archival Station only infrequently initiates a call to you. Your computer calls it, requesting a page. Since the request is sent as one of your own pages, it carries not only the requested page address but also the return address as well. Thus it carries all the information the Archival Station needs to identify you and your account and provide you with the information you need. In this manner, the Archival Station information resource sub-language can be "in" as many "networks" as there are clients who wish to have its resources available. Since clients' sub-languages are based upon this resource, it itself is protected from change. Billing services for the use of these pages are handled by the telephone company in the manner of "900" numbers today.

All telephone-computer users in each of their sub-languages have complete freedom to choose and base on whatever information resources they desire, paying for accesses to only those pages required in the course of their processing. Furthermore, this information does not come as isolated, independent displays (as, for example, in the French Minitel System). Any number of such resources may be integrated in response to a single user query in a single sub-language. This may be a sub-language a telephone-computer user has developed in conjunction with one of his personal interests, having personally selected the several information resources it has been based upon. The processes of adding such resources and of extending and modifying the sub-language and its data in many ways then become just part of normal day-to-day activities.

8. The New World of Computing Applications Development Environment

The most serious problem facing the computer industry today is the prohibitive cost of software development. Without a major improvement in this area, the telephone-computer will remain underutilized. So what is the software development environment for application programming in the era of the telephone-computer?

Each sub-language has an associated meta-sublanguage. A sub-language expresses the way a person views an area of interest; a meta-sublanguage has as its focus the associated sub-language itself. This meta-sublanguage is the proper software development environment of the applications programmer. In this environment, he can extend the vocabulary and grammar rules of his user's sub-language to encompass the idiosyncratic expressions of the user's domain, construct here the utilities needed for efficiently building and maintaining the semantics of these expressions, and create the new object classes, their data structures and processes, that bring the necessary efficiency to the computer's support of user needs. The result is a programming environment that is domain specific and highly efficient.

The same language processor that handles the "English" sentences for the client also handles the "Pascal" or "C" statements for the applications programmer. So it is a straightforward matter to give the meta-sublanguage access to the lexicon and grammar table of the sub-language. When a new rule of grammar is added, the name of the semantic procedure is put into the meta-sublanguage's lexicon, linking to both the source code file and the paged object code.

When a sub-language $\mathscr{L}2$ is based on a sub-language $\mathscr{L}1$, the basing process also creates a meta-sublanguage meta-$\mathscr{L}2$, based on meta-$\mathscr{L}1$. All sublanguages are ultimately based on BASE; thus, all meta-sublanguages are ultimately based on Meta-BASE. Meta-BASE contains a rich programming environment: trace, breakpoints, and a complete spectrum of programming and debugging tools. Further, it knows all about the associated user sublanguage. The client's sub-language's symbol map and grammar table, source code for its semantic procedures, etc., are all available to the applications programmer through the efficient syntax of the meta-sublanguage inherited through basing on Meta-BASE. In the example in figure 14, both the end-user and the application programmer are referring to the British Star, however their respective interests are markedly different.

The applications programmer maintains and extends the client's sublanguage by dealing directly with the syntax rules and associated semantic procedures of the sub-language. To add a new capability for the client, the programmer first enters the client's sub-language, types "metalanguage" to enter the meta-sublanguage, and then types "RULE." The meta-sublanguage responds with the prompt "SYNTAX" and the programmer adds the syntax for the new grammar rule. If this includes a new part of speech not recognized by the sub-language, it is automatically added to the appropriate table. The meta-sublanguage then prompts for the semantic

	Sub-language	Meta-sublanguage
language	"English,""French"	"Pascal,""C"
subject matter:	domain of user's interest	user's sub-language

user: "What is the cargo and destination of the British Star?"

applications programmer: "dump record for British Star"

Figure 14

procedure, which is then programmed directly on-line. When the procedure has been completed, the system:

(1) compiles the procedure;
(2) links the result to the resident code;
(3) puts the result on a page;
(4) puts the syntax into the grammar table linked to this page.

Note that linking is done only with the relatively small resident code. The programmer types "return" and is back in the client's sub-language, with all the client's data, grammar, and previously added extensions and can immediately try out the new rule in this actual client environment. Again, entering the meta-sublanguage, the programmer can edit the procedure and iterate. One final iteration, for removing remaining debug material, and the programmer can call the client, saying that the new capability is available and giving a concrete illustration of how it can be used on the coupled windows.

In figure 15, the Trust Officer of a bank has called the applications programmer to request a new analysis procedure, "the ABC value," to be applied to various equities. One new rule of grammar with semantic procedure defining the notion of the ABC value is all that the programmer need add. It can then immediately be used by the banker in far-ranging queries.

Just as any given sub-language can be extended by adding new syntax rules and their associated semantic procedures, its meta-sublanguage can be extended by adding syntax and semantics appropriate to the context of the applications programmer. The applications programmer can extend his own domain-specific meta-sublanguage by first typing "METARULE," and then proceeding as before when adding a "RULE." The programmer can indicate any convenient syntax for calling this new procedure, including of course the standard functional notation. A new utility procedure can be added by typing "PROC"; he will then be prompted to write, on-line, the program for the utility. For example, in the meta-accounting sub-language, the programmer may add the procedure: "update_col_totals(ledger, change_ amount)," which can subsequently be used in semantic procedures either by the programmer or by an applications programmer who has purchased the accounting package. If a programmer wants to use an abbreviation for an often used sequence of code – a "macro" in the programming sense – then "MACRO" is used in place of "RULE" or "PROC."

We see in figure 16 that the BASE sub-language has been extended to an Accounting Sub-language. The Accounting Sub-language contains all the terms and syntax that are commonly used in the accounting world. For example, it may include an object class of procedures and data structures for double-entry bookkeeping. Meta-accounting includes an extensive family of accounting utilities and a convenient syntax for using them. Thus, a meta-sublanguage becomes highly domain knowledgeable. When carefully crafted, it can have the look and feel of a specification language, yet after macro expansion and compilation, produce efficient object code.

Such a sub-language/meta-sublanguage may be marketed by a software

Programmer:
"How should it be invoked?"

Banker:
"Call it 'ABC value', like in the
ABC value of General Motors
common stock'."

••••••••••••••••

Programmer:
"Your analysis procedure is ready.
Let me show you an example of
how it works."

••••••••••••••••

Banker:
"Bar graph the ABC value of each
of Bob Moore's high-tech portfolio."

"List the last closing and current
price of each stock whose ABC
value is greater than that of DetEd."

Figure 15

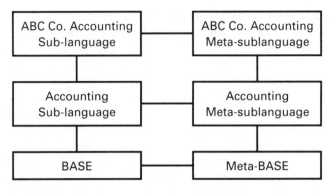

Figure 16

firm specializing in accounting. The source code for the semantic procedures will not, of course, be shipped with the package. The ABC Co. purchases this package, and their small group of application programmers tailor the ABC Co. Accounting sub-language to the special practices and jargon of their own accountants. They lose nothing of their company's pre-existing capabilities; they simply add rules that embed their own existing files and special procedures in the "accounting English" that the package provides.

Now the whole panorama of the sub-language hierarchy begins to come into focus. Each office of the enterprise has its own sub-languages, which may be related to each other through basing. Many of these sub-languages will also be based on other information resources, including archival stations. Many will have been extended to include domain-specific application packages. But this is only half the picture.

9. Toward an Efficient Organization of the Software and Data Provider Industry

A sub-language does not have to be a complete, self-contained "language." It may, for example, consist of only those syntax rules and associated semantic procedures that are the tangible implementation of a new object class. By identifying the object class by a new, otherwise unused, part of speech, the encapsulation of the processes and data of the object classes is achieved. Since this new object class is automatically made a subclass of "noun," the objects of this class satisfy the already existing syntax of "English" and thereby link to other appropriately related object classes in a natural way.

To visualize this other half of the picture, consider how easy it is to change a significant part of a sub-language, in contrast to the difficulties this would entail in current systems. One rather narrow facility that almost all sub-languages will want to include is a graphics package. A graphics package exists now, consisting of syntax rules, semantic procedures, and utility procedures, and these in turn call appropriate Postscript operators. The package only supports two-dimensional drawings. Suppose some software house that generally works close to the UNIX or DOS level produces a considerably improved graphics package that supports three-dimensional drawings. The applications programmer in any applied software shop, say one specializing in spreadsheet packages, can go into the meta-sublanguage of any appropriate sub-language and type: "delete from file:" followed by the name of the file containing the current graphics sub-language syntax, then after having mounted the new graphics system in the floppy disk drive, type: "extend from disk." The result is the replacement of one graphics package by the new, updated one. Everything else remains the same; no data are disturbed, no relinking of the whole huge system is required. And when the new spreadsheet package is purchased, the client will be able to analyse and modify their pre-existing database along any three selected attributes in a beautiful three-dimensional display.

The door is now open for a vast proliferation of software development at all levels, a layering of development levels, a multiple branching hierarchy of specialization. It is just such a hierarchy that characterizes the textile industry, the automotive maintenance industry and, indeed, today's computer hardware business. The monolithic, application-independent "tools" approach that so dominates the software industry today finds its proper niche in the lower levels within this far broader perspective.

Near the top of this hierarchy, shown in figure 17, software houses that consider themselves as specialists in their client's domain rather than in computer software can succeed in small, specialized markets because their development environment is already highly specific to their needs. The result is the drastic reduction of software development costs. This opening of the competitive market to the innovative specialist will produce sub-languages that can be fully understood by the telephone-computer and imbue it with a really penetrating understanding of the subject domain.

Let us suppose that some software house that specializes in the graphic aspects of merchandising has produced a package, including television camera, for building catalogues. A women's apparel shop has purchased this package and is seen in figure 18, preparing a Spring Sales catalogue. How is the catalogue material to be connected to the rest of the apparel shop's information system? The software house assumed correctly that video sequences and other catalogue material would be directly related to items of merchandise offered by their clients, and that the names of these items would already be in the client's lexicon as noun phrases. This was enough; what other links from these items – inventory levels, prices, etc. – are maintained by the stores is immaterial to the catalogue-building package. So when the store personnel are in the midst of preparing a catalogue, and the catalogue package asks for the description of the merchandise item to be associated with a particular graphic, the store personnel reply using their store's item number. The catalogue package procedure then links the

Well layered software development:

ABC Co. Sub-language   XYZ Co. Accounting

catelogue maker   Accounting

presentation manager   word processor   spread sheet

NEW WORLD OF COMPUTING SYSTEM

UNIX / OPEN WINDOWS   or   DOS / WINDOWS

networking   graphics chip

windows   graphics

assembler   signal processing

Figure 17

graphics to this item record in the database; it is, therefore, indirectly linked to all non-catalogue aspects of store management.

When a customer, perusing the catalogue over the telephone-computer, clicks the mouse on a graphic and types: "Send me one of those in size 14, colour blue," the following occurs:

(1) The pronoun "me" is assumed to be the person on the phone to the store's telephone-computer, who is therefore identifiable from the page address of the page sent to the store containing the request; from this the store, using a standard utility procedure, goes to the customer's telephone-computer for the number of the customer's credit card (entered by the customer in the computer slot as a part of logging in), which identifies the customer's account, credit rating, and other information.

(2) The syntax of the catalogue package, through a graphics package the software house had based on, includes the RULE: <noun_phrase> ⇒ "one of those" <click>. The semantics of this rule retrieves the item number associated by the catalogue package with this graphic.

(3) The store's order-processing package, purchased from a different software house, includes the verb "send" with the associated semantic procedure that carries out the appropriate processing of the whole transaction.

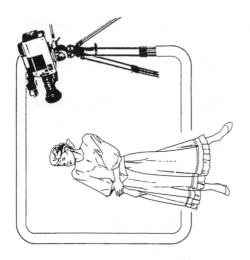

Figure 18

(4) The underlying English includes all of the other rules necessary to complete the parsing, and therefore the processing of the request. The customer's telephone-computer displays the text: "Mrs. Smith, we are pleased to send you a Calvin Klein dress, catalogue item 08249, size 14, colour blue. Your account will be billed for $124.99 plus $7.50 shipping and tax, for a total of $132.49. You can expect delivery in 3 to 5 days. Thank you for your order."

As another example, a furniture store has purchased a software plus equipment package that helps customers visualize potential purchases. In figure 19, the sales person has put the floor plan of the customer's room on the scanner (seen in the background). As the customer selects furniture items, their identifying numbers are typed in and the mouse is clicked at the appropriate site on the floor plan. The package then displays (perhaps on a giant, room-scaled screen), from any viewpoint, how the room will look. Colours, fabrics, furniture orientation, etc., can be changed and the immediate effect seen.

The scanner package, which identifies the inputs to the ultimate graphic process, would be rather easy to implement using the standard graphics package of the BASE sub-language. The output graphics pose a considerably more difficult problem. How can the cost of development of such sophisticated software be amortized? Inputs would probably be pictures of each item taken from a prescribed set of angles. The output would likely use ray tracing to get the shadows and textures just right. This package, however, would contain no other aspects, whether it be furniture, automobiles, microscopic pictures of tissue, a "blow-up" schematic of the nose-cone of an airplane, or whatever. The specialists building such packages would not need to know anything about accounting or linguistics. Their deliverable product would likely be a chip, together with grammar rules and procedures for extending a meta-sublanguage rather than a sub-language. Providing the result as meta-sublanguage syntax gives the applications programmers that use it a great deal of flexibility in the way their user interfaces can be designed and ease in using this flexibility. Thus, the resulting package can be conveniently tied in with other packages (such as the scanner package in the application cited here, a catalogue-building package, a medical lab package, or one used in developing the graphics for the Boeing maintenance package used by the airplane mechanic who was shown working on the nose-cone). This means that the high development cost of such a package could be distributed over a wide market comprised of software houses closer to applications, and labs with their own strong programming groups, each having its own distinct clientele (like the furniture store).

Many exciting capabilities are being demonstrated in academic and industrial laboratories. However, in today's software development environments, the cost is prohibitively high; and it will be a long time before these capabilities will find their way into any but military and space applications. The software development environment presented here sharply reverses that trend by supplying a simple, linguistic linking at the meta-sublanguage level.

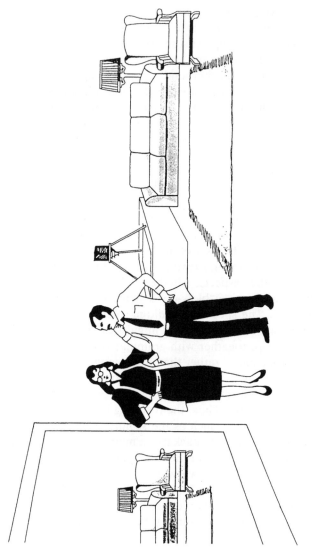

Figure 19

10. The Vision and the Realization

So here is our paradigm: It is the constant re-evaluation and adjustment of the relevant view that characterizes human information processing. A succinct expression of this view is a sub-language that is describable as a formal syntax and denotational semantics. When the basic conceptual structure of a task environment is changing only slowly, a sub-language can be established that characterizes the task and the considerations relevant to it. A computer can be programmed to understand this sub-language, providing a natural computer adjunct in accomplishing that task.

Programming languages are the proper sub-languages of system programmers, and their very limited application is the implementation of operating systems and language processors. Possibly it is only lack of a clear and relevant paradigm that has delayed even applications programmers from having an appropriate way to communicate with the computer. It is now time for the rest of us to be provided with the sub-languages appropriate to our concerns, sub-languages that are natural and efficient for our communications with the computer and for our communications with each other through the computer. Then the computer will find its appropriate niche as a medium of communication, tying people, information resources, and processing power together into the efficient focus of our appropriate sub-language. This is our vision. However, vision alone will not provide the better answer to Dr. Lucky's question. It is a sad commentary on this industry that the many visions, the promises, remain largely just that – visions and promises. The Apple Computer film "Knowledge Navigator" offers no clue for bringing that vision into being.

How do we realize our vision? How do we solve the many technological problems that lie in the path of such a development?

The problems now faced by the software industry are clear and indeed are recognized by that industry; the first of these is by far the most stringent:

(1) The high cost of software development.
(2) The integration of multiple media, of data, and of geographically dispersed people and resources into a coherent user interface.
(3) The computer's current inability to understand references to internal contents of files – to know what we are pointing at in a picture, to be able to answer questions using a table of data from a journal article, or to know to inform others of a change in an item of data.
(4) The difficulty in getting any single relevant item from all the world's information without being blocked by the enormous barriers of ambiguity, volume, and non-focused indexing.

We have faced these problems and have concluded that current software development practices, built as they are on the minimalist philosophy and the current perspective of software engineering, are not conducive to solving these problems. Object-oriented programming is a major step in the correct direction, but is insufficient because it does not deal with the adverse effects of the isolated, monolithic system that is the hallmark of the minimalist and software engineering views. Further basic changes of approach are required.

In seeking a solution to these problems, we have first put forward a clear paradigm: The sub-language is the proper focus of software development. From the vantage point of this paradigm, follow three radical changes in software system architecture:

(1) A single, grammar-driven language processor that includes language extension utilities.
(2) Segmentation at the language-processing level using a page-address structure compatible with networking.
(3) Hierarchical sub-languages sharing a common, worldwide address space to solve the distributed data and distributed processing problems.

In presenting our vision here, it has been important to us that our ideas really work, and that our words are backed by a solid, working system. The *New World of Computing System* exists. We have embodied the radical changes into a complete, rounded system. We have gone farther by extending this system to include capabilities that elucidate and amplify the basic concepts. The many technical designs, both indicated in the above presentation and implied by the illustrations, have been fully implemented in this single, integrated system. We have successfully tested and demonstrated every technological capability required to achieve every aspect of our vision. Thus, the basis has now been laid, a technical solution has been achieved for that new world of computing, the era of the telephone-computer.

11. Epilogue

The research phase of the New World of Computing System is completed. It exists today at the level of a commercial prototype. It is now ready to move into product development.

The New World of Computing System is written entirely in standard "C," except for a few hardware interface assembler procedures. It is running under UNIX[4]/OpenWindows[5] and MS_DOS[6]/Windows.[7] It currently consists of over 400,000 lines of "C" – about 3 megabytes of compiled code. Of this, only about 300 kilobytes is resident; the rest is on the System's own pages (together with data, text, etc.) and is managed by the System's own paging subsystem. The System's own pages are the packets sent across existing and future-digital telecommunication systems. This includes pages containing the digitized voice, and echoed texts and graphics, that will constitute telephone communications. Current PC and workstation hardware and ISDN telecommunication standards are completely adequate to fully support the functionality of the New World of Computing System as described in this document.

"New World of Computing" is the registered trade mark of the California Institute of Technology, which holds the copyright to the New World of

Computing System. We wish to thank AT&T/NCR for their continuing support and participation in the development of this System.

NOTES

1. Dr. Lucky is the Executive Director, Communications Sciences Research, AT&T Bell Laboratories.
2. William James, "The Will to Believe," reprinted in Kallen, ed., *The Philosophy of William James*, chap. 2. New York: Modern Libary.
3. Alfred Tarski, "Der Wahrheitsbegriff in den formalisierten Sprachen," *Studia Philosphica*, vol. 1, 1936.
4. UNIX is a trademark of AT&T Bell Laboratories.
5. OpenWindows is a registered trademark of Sun Microsystems Corp.
6. MS_DOS is a registered trademark of Microsoft Corp.
7. Windows is a trademark of Microsoft Corp.

Real-World Computing and Flexible Information Access: MITI's New Programme

Nobuyuki Otsu

ABSTRACT

MITI (Ministry of International Trade and Industry) began in 1992 a new 10-year programme called "Real-World Computing." The programme, which is to lay the theoretical foundation and to pursue the technological realization of human-like flexible information processing, is aimed at establishing the basis for flexible and advanced information technologies that are closely allied to human thought processes and capable of processing a variety of diversified information in the real world. A three-tiered research structure will attempt to establish a new theoretical foundation, investigate elemental novel functions, and develop new computing systems that can exploit parallel and distributed processing.

1. Introduction

The remarkable development of computer and communication technologies is producing an innovative change in society, not only in industrial activities but also in our way of life. It is foreseen that the variety and the quantity of information to be handled will greatly increase by the next century. Information technology will be expected to expand the information processing abilities of humans and provide everyone with flexible access to various kinds of information in the real world.

Today, the computer far surpasses human ability in certain well-defined domains, such as numerical computation, document processing, and, recently, even in logical inference. Nevertheless, it still lacks flexibility and remains far behind humans in many respects, such as pattern recognition, problem solving from incomplete information, learning capability, etc. This information-processing function seen in humans is characterized by the term "flexible information processing" or "real-world computing" in contrast to the conventional rigid information processing performed by computers that assumes complete information in a pre-assumed world or problem domain. "Intuitive" information processing, in contrast to "logical" information processing, is immature in current information technology.

In 1992, MITI (Ministry of International Trade and Industry) began a new R&D programme called "Real-World Computing" (previously called "New Information Processing Technology") as a 10-year Japanese national project to follow the Fifth Generation Computers project that officially ended in June of that year. The objective of the new programme is to lay the theoretical foundation and to pursue the technological realization of human-like flexible information processing as a new paradigm of information processing toward the highly advanced information society of the twenty-first century. The paradigm seems essential for the qualitative advancement of current information technology and for providing computers and users with flexible information access and collaboration in the real-world environment.

The programme covers a wide range of information technology: basic theory and novel functions for application to realize flexible information processing, and computational bases for massively parallel and distributed processing, including neural computing and optical computing/interconnection. Through the programme, which is open to the world, it is also intended that Japan make a greater international contribution to basic science and generic technology, which should be the common property of mankind.

The present paper, after reviewing the background, outlines the content of the Real-World Computing programme and provides details on research and development in the programme.

2. Background

2.1 Social Background

Supported by the development of computer and communication technologies, information technology is producing innovative change in society. This innovative change is not only affecting industrial activities such as new wealth and services and rationalized production and distribution systems but is also enhancing national living standards, regional progress, and educational and cultural growth.

The information network is growing from an intra-organizational network to an inter-enterprise network and, further, to one bringing together homes and individuals, and, likewise, information processing is growing beyond enterprises to include homes and individuals.

As a result, information to be processed will greatly increase not only in quantity but also in quality and variety, and a new technological basis that enables everyone to easily and efficiently take advantage of the various information resources of the network will be required for such an information network society.

Given the social background and the technological requirements for various application fields of information processing, computers should become more humanized and possess the capability to assist and collaborate with humans in the real-world environment.

2.2 Technological Background

Computers have evolved dramatically with the support of technological progress.

In the first stage, as is shown in figure 1, computers developed along the line of conventional von Neumann architectures in the fields of numerical computation, document processing and management of database storage and retrieval, since those application fields have clear algorithms and are suitable for processing by conventional computers.

The second stage of development was directed toward the manipulation of symbols and logic, casting light on the intelligent thought processes of humans such as logical (deductive) inference. The research fields of AI and knowledge engineering contributed by providing computers with the capability of handling symbolic representation of knowledge and inference rules. In 1981, Japan proposed the concept of Fifth Generation Computers as an approach to new generation computing based on logical programming. The goal was to provide computers with powerful logical inference capability and to open the door to the world of large-scale knowledge-information processing.

Today, computers have come to provide enormous computing power and far surpass human ability in solving well-defined problems, where the algorithms for solution exist and can be clearly stated in programming languages. Nevertheless, computers are still not very flexible when compared to human flexibility in information processing in the real world, where many problems are ill-defined and hard to describe in algorithmic terms. It might be called the "algorithm crisis," compared to the so-called software crisis.

2.3 Real-world Computing Paradigm

In order to cope with real-world problems and to open a new horizon in information-processing technology, it is essential to pursue research on the fundamental ways of human-like flexible information processing, by casting light on the intuitive or sub-symbolic level of human information processing, and to embody the results as new information-processing technologies for use with the developing hardware technologies.

In recent years, fundamental knowledge about the information-processing

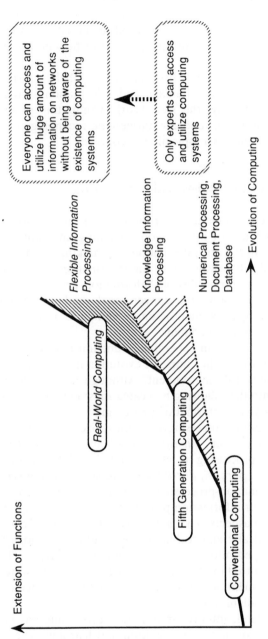

Figure 1 Extension of the functions of computing systems

capability of humans has been rapidly accumulating, and hardware technologies for massively parallel computing systems are expected to reach our hands by the beginning of the twenty-first century.

Given these circumstances, the concept of "Real-World Computing" is proposed as a new paradigm of information processing that aims at furnishing the "real-worldness" (or "flexibility") of human information processing to information systems.

In order to make the twenty-first century a globally prosperous information network society where humans and information systems collaborate more closely and flexibly, the Real-World Computing programme, which aims at developing novel fundamental generic technologies in the fields of information processing, is a topic of vital concern.

3. The Concept of Real-world Computing

In the changing environment of our daily life in the real world, we humans evaluate various kinds of information, including ambiguity or uncertainty, and acquire the necessary information for making predictions, plans, or decisions in a flexible way. Such an information-processing function is characterized by the term "flexible information processing" or "real-world computing," in contrast to the conventional rigid information processing performed by computers that assumes complete information in a pre-assumed world or problem domain.

Information processing in general is a function that has been acquired by humans and creatures in the course of evolution in surviving and adapting to the changing real-world environment. Although it is a versatile function, it can be divided into the following two categories: (1) *logical* information processing and (2) *intuitive* information processing.

Logical information processing is characterized by the words *conscious, analytical, deductive, serial, intensive, digital*, and *symbolic*. Contrarily, intuitive information processing is characterized by the words *unconscious, synthetic, inductive, parallel, distributed, analogue*, and *sub-symbolic*.

Basically, information-processing technology is intended to complement or substitute for our information-processing function through mechanization of both of the two aspects classified above. Historically, however, machines that were theoretically and technologically suitable for logical information processing developed into conventional digital computers. They progressed very rapidly, supported by the tremendous growth of hardware technologies, and serial and sequential information processing have been established as today's ruling paradigm (see figure 2).

On the other hand, intuitive information processing has been studied in the research fields of pattern recognition and learning. The original model was Perceptron, and recently neural network computing has been the case. However, intuitive information processing is still immature in current information technology. Conventional computers are far behind humans in

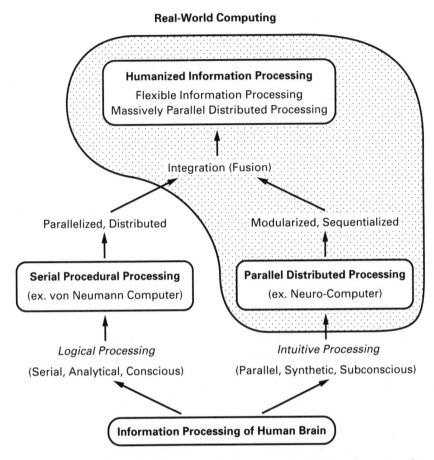

Real-World Computing

Humanized Information Processing

Flexible Information Processing
Massively Parallel Distributed Processing

Integration (Fusion)

Parallelized, Distributed

Modularized, Sequentialized

Serial Procedural Processing

(ex. von Neumann Computer)

Parallel Distributed Processing

(ex. Neuro-Computer)

Logical Processing

(Serial, Analytical, Conscious)

Intuitive Processing

(Parallel, Synthetic, Subconscious)

Information Processing of Human Brain

Figure 2 Separation and unification of the two aspects of information processing

this aspect, and their lack of flexibility is due to the unbalanced development of those two aspects.

As has been seen in the previous section, development of information systems that have human-like flexible information-processing functions and can cope with real-world problems is now one of the most important demands common to various fields, such as pattern-information processing, knowledge-information processing, intelligent robots, and friendly man-machine interface, which aim at further advanced information processing.

To develop such systems, it is, first of all, important to explore the intuitive aspect of information-processing functions of humans and embody it as a new technology. By coordinating the two aspects of intuitive and logical information processing and integrating them, Real-world Computing (hereafter denoted by RWC), which has the following novel functions, will be established as a new paradigm of information technology:

– the function to integrate a variety of complex and intricately related in-

formation containing ambiguity or uncertainty and to reach an appropriate (approximate) decision or solution within a reasonable time,
- the function to actively acquire necessary information and knowledge and to learn general knowledge inductively from examples,
- the function to adapt the system itself to users and a changing environment. Applications of RWC systems are expected to include a wide range of real-world problems, such as ill-defined (incomplete) problems like the understanding of situations in a noisy environment, large-scale problems such as the simulation of social and economic phenomena, real-time problems such as man-machine interface with virtual reality, and autonomous control of intelligent robots.

The key technical requirements for the flexibility of RWC systems are: openness, robustness, and real-time, where by "openness" is meant that the system can adaptively and autonomously change or extend itself to cope with the unexpected situations it encounters in the real world; by "robustness" is meant tough and stable behaviour of the system for distorted or fluctuating information input; and "real-time" means that the system can respond within a reasonably short time.

The reason why we humans can maintain the flexible information-processing ability characterized by the above real-worldness is that our brains incorporate distributed representation of information, massively parallel processing, learning, and self-organizing ability, and information-integration ability. Therefore, the following will be key concepts for realizing the above characteristics of RWC systems:

(1) Flexible information processing: the functional aspect of RWC, which is characterized by the admissibility and integration of ambiguous and uncertain information and the capability of adaptation and learning.

(2) Massively parallel and distributed processing: the computational aspect of RWC, characterized by processing of multi-modal, multi-variate, and strongly correlated information in a massively parallel and distributed manner.

4. Outline of RWC Programme

4.1 Purpose of Research and Development

Information systems of the twenty-first century will be based on not a single but various key technologies, such as those for massively parallel computing, optical computing, neural computing, and logic-based computing. These technologies need to be flexibly integrated into information systems in order to cope with real-world problems.

The main purpose of the RWC programme is to lay the technological foundation for the advanced information society of the twenty-first century. This programme is aimed at establishing the basis for flexible and advanced information technologies that are closely allied to humans and are capable of processing a variety of information in the real world. Such technologies seem essential for creating a cooperative relationship between humans and computers and for producing innovative and generic technologies toward the advanced information society of the twenty-first century.

The primary goal of this programme is not to develop a single computer but to contribute toward the realization of these significant but not yet established technologies. Herein, we will try to establish the theoretical foundation for these technologies, to explore their potentials, and to create some specific style of their integration. Some important real-world problems will be tackled for confirming the possibility and usefulness of the new technologies. By disclosing these experiences to the public, it is also intended that Japan will contribute to the development of the common knowledge and wealth of humankind.

In order to accomplish these fundamental and ambitious goals, it is very important and imperative to promote international and interdisciplinary cooperation in the research fields and to support collaboration among industries, national institutes, and universities.

4.2 Subjects of Research and Development

The research and developments in the RWC programme are divided into the following mutually related subjects:
- theoretical foundation,
- novel functions for application,
- computational bases.

A three-storied structure as shown in figure 3 is the fundamental framework for the organization of research and development in this programme.

4.2.1 Theoretical Foundation

The research objective is to establish a new theoretical foundation for flexible information processing. For this purpose, it is necessary to expand and generalize the conventional framework of information processing and to clarify the principle, or "soft logic," commonly underlying flexible information processing.

The research topics are:
- Flexible representation of information
- Evaluation of information and processing models
- Flexible storage and recall of information
- Integration of information and of processing modules
- Learning and self-organization
- Optimization methods

In particular, *integration* of multi-modal information (and of heterarchical processing modules) and *learning* and *self-organization* (optimization and

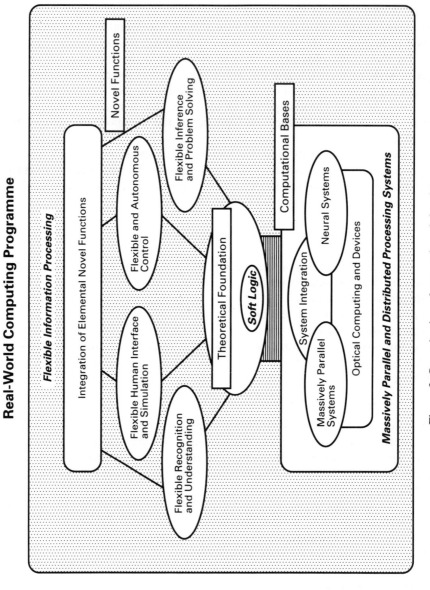

Figure 3 Organization of research and development

adaptation) are the most fundamental issues, and how to implement these in a framework of massively parallel and distributed information processing will be a key point.

4.2.2 Novel Functions for Application

Research and development on novel functions for application should be directed toward investigating elemental novel functions, which are on the whole important for realizing flexible information systems to solve a wide range of real-world problems.

Research topics are classified into the following categories.
- Flexible recognition and understanding of multi-modal information
- Flexible inference and problem solving based on flexible information base
- Flexible interactive environment for man-machine communication
- Flexible and autonomous control

Some typical application problems will be tackled for exploring the integration of these elemental technologies and for demonstrating their effectiveness.

These novel functions will be implemented on the following computational bases.

4.2.3 Computational Bases

RWC involves processing large volumes of spatio-temporally distributed information at a high speed, while taking into account their mutual interactions. As a new computational basis to support it, computing systems that can exploit parallel and distributed processing at several processing levels should be developed. For this purpose, research and development from the following perspectives is important; how to integrate these technologies will also be investigated.
- General-purpose massively parallel computing systems
- Neural systems as a kind of special-purpose system
- Optical computing systems

In the following section, more detailed contents for each research subject will be described.

5. Theoretical Foundation

The objective of research is to lay the theoretical foundation for the technological realization of human-like flexible information processing as a new paradigm of information processing.

So far, much theoretical research has been done in research fields related to flexible and parallel distributed information processing. Such fields are, for example, pattern recognition, multi-variate data analysis, probabilistic and statistical inference, fuzzy logic, neuro-computing, machine learning, regularization, and various optimization methods, and so on.

In order to provide a theoretical foundation for flexible information processing, it is important not only to continue in-depth study in these research areas but also to clarify the theoretical framework of "soft logic" commonly underlying these fields and to aim at constructing a new unified theoretical base for dynamic RWC. Here, probabilistic and statistical formulation of problems and non-linear dynamics in conjunction with learning and self-organization will be a key approach.

It will be necessary to expand and generalize the conventional framework of information processing in all aspects of information representation, processing, and evaluation, and to systematize basic theories and elementary novel functions for application on the flexible framework.

The following are conceived of as the most fundamental issues of common concern:
- integration of multi-modal information (and of heterarchical processing modules);
- learning and self-organization (optimization and adaptation).

Implementing these in a generalized and flexible framework of massively parallel and distributed information processing will be a key point in developing novel functions for application such as flexible recognition, inference, and control.

It will also be important to learn and get inspiration from nature, namely to take into account new findings in scientific research into brains, the evolution process of creatures, and ecological systems.

In what follows we will investigate the topics of theoretical research that will serve as the theoretical foundation for the novel functions in various applications and possibly for new computing architectures.

5.1 Flexible Representation of Information

In order to treat various kinds of information in the real world (such as images, speech sounds, and languages) and to construct heterarchical flexible systems, it is first of all necessary to establish a flexible framework for information representation. The framework should be flexible enough not only to represent various kinds of information in a unified manner but also to represent certainty of information. It should also be suitable and efficient for implementing associative memory and learning/self-organization procedures.

Important research topics include distributed representation (patterns) vs. concentrated representation (symbols), spatial representation vs. temporal representation, probabilistic representation, topological representation, hierarchical representation, and the representation of information and knowledge as constraints, etc.

5.2 Evaluation of Information and Processing Models

For actively interacting with the real world and learning or self-organizing from experience, RWC systems should have a flexible and systematic framework (criterion) for evaluating the importance of various kinds of input information, the output information that is a result of processing, and the processing models themselves.

One important issue will be to extend the evaluation framework from the conventional hard type, which mainly deals with "true/false," to more flexible and quantitative types, such as information criteria (e.g. AIC and MDL) and energy, which can serve as objective functions for optimization and regularization, including information-integration processes and processing models themselves.

5.3 Flexible Storage and Recall of Information

The highly sophisticated function of the human brain's memory is a key to the flexible information processing of humans. RWC systems should have such flexible memory functions to store and associatively recall various kinds of information. Thus the research should aim at establishing the theoretical analysis of associative memory and developing new efficient mechanisms for flexible associative memory.

Important research topics are association using probabilistic reasoning, association using structural similarity, associative memory for storing time series, and associative memory using non-linear dynamical systems, etc.

5.4 Integration of Information and of Processing Modules

Information processing such as inference, prediction, and planning can be considered as an integration process of various kinds of information and knowledge. It is important to carry out the theoretical research to analyse these information-integration processes and develop a new flexible way of controlling the processes.

Multi-variate data analysis methods will provide the methods of information integration, although most of those are limited to the linear transformations. Non-linear extension of those methods will be important. Neural network models are considered as giving a kind of non-linear extension. Regularization theory will provide a method for incorporating various kinds of constraints.

Cooperative processing by a huge number of processing modules is also an important issue of information integration. Research on the integration of processing modules will be important. Investigating the information-integration process of human cognitive systems will also be important as a source of inspiration.

5.5 Learning and Self-organization

In addition to information integration, learning or self-organization is one of the most important topics of flexible information processing. It will play an important role in constructing adaptive autonomous systems, complex heterarchical systems or flexible databases. Thus, it is necessary to construct computational theory of learning/self-organization for exploring the "learnability" of various concepts or structures and to develop novel, efficient algorithms.

Important research topics are learning from uncertain information, learning of probabilistic knowledge, learning of heterarchically structured knowledge, learning algorithms using active information acquisition, methods for incorporating existing knowledge with learning process, and selection of learning models, etc.

5.6 Optimization Methods

Information-integration processes can be formalized as optimization processes. Learning and self-organization can also be considered as optimization procedure. Solving these optimization problems usually demands a huge amount of computation. In order to surmount this difficulty, developing approximately correct optimization methods that can be executed effectively on massively parallel systems will be an important area of research.

Important research topics include probabilistic optimization methods such as simulated annealing, genetic algorithms, ecological algorithms, evolutionary algorithms, optimization using non-linear dynamical systems such as neural network models, and other non-linear optimization techniques, etc.

6. Novel Functions for Application

The objective of research and development is to investigate elemental novel functions in the various application fields, closely cooperating with the theoretical foundations, and to embody those in respective application fields or integrate them to demonstrate new flexible information systems.

RWC systems support should approximate various human activities by acquiring and processing various kinds of information in the real world such as images, speech sounds, tactile sensations, and so on, which are massive and modal and moreover subjected to incompleteness and uncertainty by nature. Thus, the RWC systems require novel functions with flexibility of various kinds. Terms such as robustness, openness, and real-time reflect different attributes of flexibility. Therefore, the key issue for novel functions is how to realize such flexibility.

The novelty of the functions should emerge from new concepts of theory or algorithm suitable to RWC. The bottlenecks of conventional information processing should be alleviated through new kinds of flexible functions such as integration of symbol and pattern and learning/self-organization. Merely combining conventional technologies or making ad-hoc systems for specified tasks is not what is desired.

Since flexible information processing intends to expand the abilities of in-

formation processing beyond the limitations of the conventional one, the range of expected application fields is quite wide. These fields can be divided into the following categories:
- Flexible recognition and understanding of various kinds of information, such as pattern information – images, speech sounds, and symbolic information like natural languages;
- Flexible inference and problem solving based on flexible information bases that admit direct treatment of information and have capabilities of learning and self-organization;
- Flexible human interface and simulation for realizing mutual interaction between humans and the real world;
- Flexible control and autonomous systems interacting with the real-world environment.

The following are conceived of as important directions to pursue for realizing integrated systems of novel functions:
- Real-world, adaptable, autonomous systems;
- Information-integrating interactive systems.

The former means cooperation with the real world. These systems will be flexible systems that can autonomously understand and control the environment through active and adaptive interaction with the real world and work for the purpose of partial replacement of human activities in the real world. Here it will be necessary to cope with the uncertain, incomplete, and changeable characteristics of the real world. Necessary novel functions are understanding of environment, modelling of the real world, planning for action sequences, and optimal control for adaptation, etc.

The latter refers to cooperation with humans. These systems will be flexible systems that support and enhance human intelligent capabilities such as problem solving and information creation through enlarged communication channels between humans and systems. Here it will be necessary to both understand and integrate varied information flexibly to assist humans in solving problems and creating new information. Necessary novel functions are question and answer by spoken natural language, understanding of intentions from various types of information produced by humans, realizing intelligent and interactive assistance for retrieving and presenting valuable information from a large amount of data in database, intelligent simulation to create new information findings and forecast transient states in the real world, integration methods for combining human factors, and a computational model of the real world, etc.

The fundamental novel functions should be evaluated from the viewpoint of how they contribute to realizing these two systems and from the viewpoint of how they make breakthroughs in their own technological fields, related to the attributes of RWC, namely, robustness, openness, and real-time.

In the following, we will investigate research topics that seem necessary for the realization of both systems.

6.1 Flexible Recognition and Understanding

We can easily and flexibly recognize and understand various kinds of information in the real world, such as images and speech sounds. This ability is a typical example of intuitive information processing and will be indispensable for enabling RWC systems to possess autonomy and to undertake smooth communication with humans. It will also form the base for inference and problem solving at higher levels.

Many efforts have been made in research to mechanize these functions, but the state of the art is still far behind the level of human ability. The difficulty of realizing recognition/understanding in the real-world environment lies in the incompleteness, uncertainty, and ambiguity inherent to natural patterns and further in the ill-defined and ill-posed nature of the problems themselves.

Therefore, a new framework for solution is needed that admits these difficulties and can integratively include various constraints (knowledge) and also subjective value judgements, for example flexible frameworks such as optimization, constraint satisfaction, Bayesian inference, hidden Markov models, learning/self-organization from examples, or a new paradigm of parallel computation unifying these.

Important research topics are as follows:

(1) For image understanding, the formulation of processing modules for early visual information (such as colour, shape, movement) and integration mechanisms for these, interactive segmentation and understanding of scene and motion images on the basis of flexible modelling of objects and the world environment, understanding of facial expression and gestures to infer human intentions, etc.

(2) For speech understanding, noise suppression and robust recognition under noisy environments, speaker-independent understanding of conversational speech, formulation of heterarchical processing modules from the signal level to the linguistic level and their integration mechanisms for these, flexible mechanisms for syntactic and semantic analyses, etc.

(3) For natural language understanding, robust parsers applicable to large volumes of real-world coded sentences containing incompleteness, flexible methods to manage such sentence data as databases or electronic dictionaries and to utilize those for understanding, algorithms for extracting conceptual information, computational models for understanding situations and for integrating knowledge units to treat dialogue, etc.

6.2 Flexible Inference and Problem Solving

Conventional AI realizes intelligence by reducing human intelligence, such as inference and problem solving, to problems of logical operations and retrieval in knowledge representation by symbols. Such methods are efficient for restricted or rigorously abstracted problem domains, but for the variety of problem solving required in the real world, they face many difficulties, such as the limitation of symbolic representation and manipulation, combinatorial explosion and knowledge acquisition.

In the real world, much information and knowledge contains inconsistency, incompleteness, and uncertainty. There are many cases where the problems to be solved themselves cannot be described completely, and problem formulation is required based on a user's incomplete statements. In addition, real-time problem solving is also an important factor. Therefore, a new scheme of flexible inference and problem solving is necessary to cope with these problems. This will be attained by expanding the conventional framework to a more flexible one.

Important research topics will include flexible information representation integrating symbols and patterns, flexible knowledge acquisition from real-world information (learning of probabilistic structures or causal relationships), stochastic inference and analogy based on soft logic that integrates probability theory and fuzzy logic, constraint satisfaction or network dynamics to solve ill-posed problems or problems with many tacit assumptions as optimization problems, massively parallel processing for the fast solution, and modelling of the processes of inference and problem solving in humans, etc.

6.3 Flexible Information Bases

Flexible information bases to handle various kinds of information and knowledge in the real world are important in the sense that they will support intellectual activities of humans in the forthcoming era of high utilization of information networks and also in the sense that they will form a basis for novel functions such as flexible understanding and recognition, inference, problem solving, and action control.

So far, several types of databases have been proposed and developed, such as relational, object-oriented, deductive, and knowledge bases. However, there are still many problems with the conventional databases and knowledge bases, which rely mainly on symbolic and logical representation and retrieval.

Two new technologies should be focused on here. One is the flexible representation and memorization of the variety of real-world information within a unified framework for an inner information model ensuring tractability. The other is the flexible and semantically correct retrieval of the required items corresponding to the intentions of users.

Related important research topics include flexible information representation and data structures able to reflect the topological characteristics of objects and suitable for treating hierarchies of information; self-organization functions to cope with large-scale real-world information; learning of correspondences and cooperative relationships between various kinds of information; evaluation of information value and feature extraction for information abstraction or automatic indexing; detection of lacking information and active information acquisition; inference of the user's intention from incomplete or ambiguous requirements and completion through dialogue; and high-speed retrieval by massively parallel processing, etc.

6.4 Flexible Human Interface and Simulation

Novel ways of using RWC systems must be developed in parallel with the development of the individual novel functions mentioned above. This new information processing will provide a new environment that enables broad, cooperative relationships between humans and computers and enlarges human intelligence activities. Users will be able to interact with the computing systems by such natural means as spoken language, gestures, and facial expressions, and will be able to receive information by means of real-time 3-D images, etc. This means that users can concentrate their intelligence efforts on more creative activities, freed from learning the exhausting skills required for communication with conventional computers through narrow channels.

Related research topics to realize such an information processing environment will include broad-band multi-modal interface (linguistic and visual communication, including recognition of human body actions, facial expressions, etc.), information display system with virtual reality, cognitive and behaviour models including sensor fusion to understand human intentions.

RWC will also realize novel simulation technologies to solve very complicated and difficult problems. This will assist human thinking, creative activity, and decision-making by means of visualizing phenomena that have no inherent visual appearance. Instead of performing expensive and time-consuming physical experiments that may not always be precise, the computing system will provide users with powerful tools to simulate very large-scale complex systems and to predict their behaviour in real-time. Prediction of untapped phenomena and future events, such as the global environment and weather forecasting, will be an important application of RWC.

Related research topics are learning/adaptation-type simulation, prediction and control of complex/chaotic time series, large-scale simulation and decision-making support for solving environmental, economic, and traffic problems, etc.

6.5 Flexible and Autonomous Control

Research needs to focus on the development of the technologies required for the realization of flexible, autonomous, coordinated systems operating in the real world in real time. Robots are typical examples of such systems, and applications include aids for the elderly or physically handicapped.

Such systems will be composed of various functional modules that interact with each other in perception, decision-making, and action control. On the other hand, the real world where the system operates is subject to incompleteness and ambiguity in the available information, to dynamic change in the physical environment and limitations on available time and space.

Therefore, the important problems to be solved are how to integrate these various functions and how to control the interactions between the functional modules in order to achieve desired goal states under such real-world constraints.

Related research topics are flexible modelling of the environment, task and control, active and distributed sensing and sensory integration, on-site planning and distributed cooperative searching, structurization and coordination of multi-agents (for sensing, planning, and action) for real-time skilful manipulation of objects, and maintenance of consistency between the internal world model and the dynamic real world, etc.

7. Computational Bases

7.1 Massively Parallel Systems

RWC requires a computation framework that can process various kinds of information and integrate them flexibly. A system that implements an RWC application is likely to consist of many modules that can exploit parallel and distributed processing at several levels both within and between modules. Several parallel computation paradigms have been proposed, including concurrent object oriented, data flow, data parallel, neural network, probability-based information processing, etc. RWC will probably be realized by some combination of these paradigms. These paradigms are naturally adopted to a massively parallel system and they require a huge amount of computation to solve practical problems within a reasonable amount of time.

These observations show that a massively parallel system for RWC is necessary to support the computational power, and it must also be general purpose to efficiently execute the multi-paradigms. The massively parallel system should be flexible itself, adapting itself to application environments for optimal performance while minimizing the workload on the users.

The research and development will include the following topics.

7.1.1 Massively Parallel Architectures

The following are fundamental technologies that should be pursued in the development of general-purpose massively parallel systems:
1. Model. Flexible execution models, which can be bases of general-purpose architectures, should have the ability to fill the gaps between the language models and hardware. Flexibility that allows a mapping of a virtual computer onto actual processing elements should also be pursued.
2. Architecture. The massively parallel system should be based on a general-purpose architecture that supports various paradigms efficiently. It is also important to study hardware architecture in consideration of future device technology and packaging technology for an efficient implementation of the massively parallel system.
3. Interconnection Network. The interconnection network should provide

high-speed communication that is comparable to computation speed. It should also provide support for dynamic load distribution, global synchronization, and global priority control. In implementing the high-speed interconnection network system, not only silicon technologies but also optical technologies should be considered.

4. Robustness/Reliability. Hardware-oriented robustness that can tolerate expected component failures in massively parallel systems should be examined. System components should have self-checking and self-repairing features. The total system should have a maintenance architecture or facilities to maintain system reliability.

In the first half of this programme, a prototype system that consists of 10^4 processing elements will be designed and developed as a platform. The platform system is used for tools of software development and a research platform for novel functions. Fundamental research on massively parallel model and architecture are concurrently undertaken. In the second half, a massively parallel system is expected to have the order of 10^6 processing elements. It will have the ability to execute various kinds of RWC applications at real-time speed. The architecture will be based on the new massively parallel computing model to be studied in this programme.

7.1.2 Operating System for Massively Parallel Systems

The operating system for a massively parallel system should be designed to support the execution of various processes (parallel programs) concurrently with high throughput and to build a user-friendly software environment that hides hardware details and makes parallel programming easier. The research topics are as follows:

1. Hierarchical Structure. To realize a functionally distributed management system for flexible processor management, the operating system may require a hierarchical structure. Hierarchical structure makes the system scalable. Some efficient mechanisms to control activities and its hierarchical structure and the reduction of overhead for controlling parallelism or executing critical sections should be considered.

2. Network Management. Advanced intelligent routing, addressing, synchronization, deadlock prevention, flow control, and failure preclusion should be incorporated into a flexible network management system.

3. Resource Management and Load Distribution. In the massively parallel system, the elimination of synchronization overhead, access contention, and communication overhead will become more serious issues. To overcome these problems, the operating system should be able to collect management information autonomously and undertake statistical management or adaptive management. Memory management and virtual systems for several resources should also be pursued for efficient scheduling and load distribution.

4. Fault Tolerance. In the massively parallel system, resource management should be done in a manner that allows for expected component failure rates. Therefore, it is necessary to handle the failure preclusion system as

a normal process. Multi-route processing will also be required for tackling failures.

7.1.3 Languages for Massively Parallel Systems

The language for the massively parallel system must be able to describe the coordinated operations of a number of processes. The problem is how to extract the available parallelism in the problem domain, and to be able to execute it with as much parallelism as the underlying system can provide. Various compilation techniques and run-time implementation techniques scalable to nearly 1 million processors should be studied. The following items should be considered:

1. Language Model. Language model is a description model for flexible programming languages of massively parallel systems. The model must be simple and be sufficiently close to the underlying architectures so as not to restrict their computing power, and at the same time provide powerful means of abstraction to promote software programmability, portability, and reusability. In the research for a language model, fundamental research on supporting flexible language, model of describing coordinating and cooperating actions, inheritance, and reflections should be pursued.

2. High-level Languages for Massively Parallel Systems. The primary goals of high-level languages for massively parallel systems should be ease of programming and the ability to describe computation on the scale of 1 million processors. One viable candidate will be an appropriate amalgamation of concurrent object-oriented, functional, and declarative constraint-based approaches. Currently available object-oriented models are not intended to process more than 1 million processes, so the following extensions will be needed: introduction of a description system permitting hierarchical decomposition of complexity; diversification of message propagation systems; introduction of reflective functions for adapting and evolving objects; and declarative description of object relationships.

7.1.4 Environment for System Development and Programming

1. Programming Environment. The need to develop a programming environment that can support multi-paradigm programming is expected. Tools for debugging, graphically monitoring and analysing load balance, communication characteristics, etc., will also be required. Since these functions may need hardware support, the architectural design should take these requirements into consideration.

2. System Development Support Environment. The requirements of an environment supporting the development of the massively parallel system include two features different from conventional ones. One is the support for the interconnection network development, where the overall functions such as robustness, dynamic load distribution and global synchronization mechanisms, and performance of the interconnection network

should be evaluated, in advance, by system-level simulation. The other is support for the architecture development of processing elements, where a set of basic functions for processing elements should be determined through a functional assessment of the various subsystems, including the interconnection network.

7.2 Neural Systems

In recent years, neural networks based on the model of the brain have been receiving attention for their capabilities of learning/self-organization and many types of flexible information processing. However, these networks are still limited to small-scale applications because the neural models used are very simple and the learning is mostly based on the back-propagation technique and requires a large amount of computing time. Usually neural networks are simulated on conventional computers, and the simulation speed is very slow, especially on large networks. Therefore, how to realize high-speed processing on a neural network is an important subject, and it will be desirable to have special hardware.

In the RWC programme, the possibilities for large-scale neural networks will be explored in order to create flexible information-processing systems that can operate in the real world. The research and development will include: research on new models, hardware architectures, and software environments; development of a prototype system on the scale of 10,000 processing (neuron) units to provide a platform for research on neural models and applications. Later, a final system is to be developed, which is expected to be on the scale of 1 million neuron processing units. In the final stage, the neural system will be integrated with the massively parallel system to make flexible information processing a reality.

7.2.1 Neural Models

A flexible information system for RWC can be implemented using a large neural network that changes its own structure adaptively through interaction with the real-world environment. Realizing such a large neural network will require research on new models:

1. Neuron Unit Models. Simple neuron unit models have so far been used with success in limited areas. However, more advanced applications will demand more sophisticated neuron models. To begin with, the possibilities of already proposed models, such as the chaos neuron model, the complex number neuron model, and the neuron logic model, must be evaluated. At the same time, research must be done on new neuron models.
2. Modularization and Hierarchization. In the learning process of a large neural network through interaction with the real world, new knowledge should be acquired without destroying existing knowledge, and information should be efficiently retrievable. This implies the necessity of modularization and hierarchization of knowledge. Related important research

topics are: learning mechanisms using centralized or distributed control for the purpose of realizing modularization, hierarchical structuralization and functional differentiation of a large-scale neural network, evaluation criteria for this and interaction among modules, etc.

3. Learning and Self-organization. Layered or hierarchical neural networks are effective in spatial pattern recognition, while recurrent neural networks are effective for recognition and generation of temporal patterns, and also for application to optimization problems. Since they will play a more important role in the future, it is vital to undertake research on the methods of learning and self-organization for recurrent neural networks. Another important research topic is the topology and size of a neural network, which are the most critical parameters for generalization capability of the network in learning by examples. A network should be large enough for learning and small enough for generalization.

4. Associative Memory. Association is one of the basic functions created by neural networks. Spatial or temporal patterns are memorized distributively and recalled on the principle of best match. It is necessary to theoretically clarify the principles of this association function and to work out an engineering mechanism to implement it. Related research topics are memory capacity, topological structure of memory, etc.

5. New Analog Computing Principles. Information processing by a neural system is based on the analog non-linear dynamics of the system. New principles of analog computing in neural systems, including chaos dynamics, must be clarified from this point of view.

6. Integration of Different Paradigms. Research must be done on models for integration of different paradigms. For example, the integration of a neural network and logical processing and the integration of pattern processing and symbol processing will be required for implementation of a neural system. The representation of input/output information in a neural network is important in that it functions as an interface when different paradigms are integrated, and in that it substantially affects the processing performance of the network. Therefore, theoretical and experimental research on input/output representation will be an important issue.

7.2.2 Neural System Hardware

Real-world application of neural networks might require a large network, on a scale of 1 million neurons. Such a large neural network may be modularized and consist of sub-neural networks, each of 1,000 neurons that are fully interconnected. The hardware of the neural system must support such a large network at high speed. The target processing speed is 10 TCUPS (Tera Connections Updates Per Second). In the design of neural system hardware, general-purpose and scalable mechanisms should also be incorporated so as to allow a wide variety of neural network models, because at present it is not clear which model is the best and various other new models are likely to be unveiled in the future.

Hardware for neural systems can be classified into the following three types: neuro-accelerators, VLSI neuro-chips, and engineering implementation of neural networks.

A neuro-accelerator consisting of special-purpose parallel processors should be developed for neural network processing. A large number of architectures for this have been proposed. A typical architecture consists of hundreds of processors and achieves 1 GCUPS.

A VLSI neuro-chip is the hardware for implementing neuron unit(s). The domain of neuro-chip architecture is so wide that it ranges from digital circuit chip to analog circuit chip. Digital circuit neuro-chips have various advantages such as high noise tolerance, high processing accuracy, and direct applicability of the ordinary computer manufacturing technology. They are suitable for stable operation in a large system. In addition, it is easy to implement other variations using the pulse-density model or the like so as to produce new neuro-chip possibilities. On the other hand, analog circuit neuro-chips make it possible to reduce the hardware volume because it has fewer operation circuits. This is advantageous for developing large-scale networks. Moreover, analog circuits have the potential for implementing dynamic and complex neural networks such as the chaos neural network. It is also possible to consider the use of digital-analog hybrid neuro-chips, which combine the strengths of both digital and analog circuits.

The engineering implementation of a neural network is the third approach in which the functions of the neural network are implemented through hardware logic without the use of neuron unit hardware.

It is difficult to compare these approaches, since each approach is unique in terms of learning capability, scalability, and so on.

In a neural system, all neuron units exchange their activation values. Therefore, the interconnection network architecture is an important point in design. Methods of time or frequency multiplexing are possible solutions to this problem. Related important technologies include wafer scale integration, three dimensional architecture, and optical interconnection. CAD and silicon compilers are considered to be important design tools.

7.2.3 Neural System Software

A variety of neural software systems are required for the research and development of neural systems:

1. Simulation System. A flexible, general-purpose neural simulator for large-scale neural networks would be a powerful tool. The requirements for such a simulator are high-speed processing, machine independence, extensibility, convenient user interface, and a variety of utility routines. It is also desirable that such a simulator has mathematical analysis tools to describe and analyse the convergence or cognitive performance of individual networks.

2. Neural Network Language. Neural network processing should be described using a high-level language. The design of such a language demands the following research and development: expression of ambiguous

information, description of best-match operations, integration with logical programming, and integration with simulators.

3. Operating System. When the number of hardware neurons is smaller than the number of units in a neural network, a virtual mechanism to fill in this gap will be important. Related research topics are: mechanisms for mapping the neural network onto the hardware, scheduling of resources, etc.

7.2.4 Integration with a Massively Parallel System

It is highly probable that a neural system will be one of the processors for such specific purposes as associative memory, pattern recognition, and combinatorial optimization. This means that the neural system will be combined or integrated with other computing systems. The forms of integration range from close connection to loose connection. An example of close connection is neural systems connected as associative memories to the processing elements of a massively parallel system, and an example of loose connection is a massively parallel system throwing problems such as optimization to the neural system.

7.3 Optical Computing Systems

Light is expected to be a new information medium, because of its extended transmission capacity and massively parallel processing capability. Optics will provide new device technology as well as new architectures and algorithms in the RWC programme that aims at flexible information processing using massively parallel distributed processing. Research topics will be classified into the following categories.

7.3.1 Optical Interconnection

Optical interconnection merges the advanced electronics technology represented by VLSI with optical communication technology and thus eliminates information transmission problems in electronic systems such as propagation delay, line-to-line cross-talk, space factors of wiring and mounting, and large power consumption.

By using high-density multiplexing technologies in the area of time, space, and wavelength, optical interconnection device, architecture, and design technology will offer high-speed and large-capacity optical interconnection having such flexible functions as reconfigurability and self-routing. These are also key technologies for realizing optical neural systems or optical digital systems.

In order to develop optical interconnection, the following issues are important:

– Ultrafast (sub-picosecond) optical interconnection devices for large-capacity interconnection by using time division multiplexing (TDM)
– Space-parallel/functional optical interconnection devices for reconfigur-

able high-speed interconnection by using space division multiplexing (SDM)
- Wavelength-parallel/functional optical interconnection devices for large-capacity/reconfigurable interconnection by using wavelength division multiplexing (WDM) and wavelength-selective self-routing technology
- Passive optical interconnection elements including micro-optics and diffractive elements for developing optical components having advantages of both stability and high-density optical interconnections
- Advanced opto-electronic integrated devices and circuits combining different material systems/functions for compact and smart devices for the next generation following the aforementioned devices
- Research on interchip and intra-chip optical interconnection for high-speed and flexible optical interconnection networks between processing elements, between processors and memories, and between memories
- Modularization of opto-electronic devices and passive optical elements for integration and miniaturization of optical interconnection components

7.3.2 Optical Neural Systems

Optical neural systems aim at realizing real-time processing of images and other spatially distributed information or spectral information through learning and associative processing, using massive and flexible interconnectivity of light.

In order to develop such systems, the following issues are important:
1. Optical Neural Models
    - Models for direct input and processing of 2-D/3-D image information by neural networks
    - Novel-type theoretical models using physical phenomena of light, such as bistability, chaos, and phase-conjugation, etc.
    - Expandable modular models consisting of a number of unit modules
    - Models for implementing optical analog devices that are low in accuracy but excellent in large-scale configuration at high speed
2. Optical Neural Devices
    - Large-scale optical array devices that can vary synaptic connection weights according to electric/optical learning signals
    - Optical neural devices for direct image recognition and processing and for extracting features of input images
    - Modularization and standardization of optical neuro-chips
3. Optical Neural Systems
    - Design technology for the distribution of functions, hierarchization of the system, and realization of accurate processing through system integration with digital computer
    - Learning methods for acquiring knowledge from training signals, storing them as structured knowledge, and technologies for increasing learning speed
    - Human friendly I/O interface technologies for direct processing of

multimedia information and also for image database allowing direct search of images by key image

7.3.3 Optical Digital Systems

Optical digital systems aim at realizing massively parallel and accurate processing of images and other spatially distributed information or spectral information with logical computation principles using massive and flexible connectivity of light.

To develop such systems, the following components are required:

1. Optical Logic Devices
   - High-speed binary/multi-valued optical devices and their two dimensional integration with low power consumption
   - Space-parallel optical logic devices with encoding signals in the form of spatially coded pattern
   - Wavelength-parallel optical logic devices with encoding signals in the form of a combination of light with different wavelength
   - Passive optical devices for micro-optics, planar optics, diffractive optics, and high-precision optics
2. Optical Logic Circuit
   - Reconfigurable optical interconnection between optical logic devices
   - Functional modularization of optical logic elements, such as parallel optical registers, parallel optical memories, optical crossbar switches, and optical I/O units
   - CAD technology for 3-D optical circuit design
3. Optical Digital Systems
   - Architecture and design for general/special purpose, optical parallel computers with highly accurate and flexible processing capabilities, based on explicit logical algorithms
   - I/O interface for high-speed data exchange between optical logic circuits and electronic systems
   - Technologies for implementing and integrating different functional optical modules
   - Programming language and compiler for optical parallel digital systems compatible with those for electronic computers, and compilers

7.3.4 Environment for System Development

Optical computing technologies are based on the presumption of using newly developed optical devices, and modularization of opto-electronic devices is also an important goal in the RWC programme. Optical contributions to the highly parallel and massively distributed systems shall be verified with such modules.

OEICs will be key devices for optical interconnection, optical neural systems, and optical digital systems. Development of OEICs should be based on the common platform of processing and module technologies in order to retain compatibility with the system.

The subjects that need to be investigated are: advanced OEIC processing technology, opto-electronic module technology, and standardization and CAD technology.

8. Research Organization and Plan

8.1 Basic Policy

The primary goal of this programme is not to develop a single computer but to explore the possibilities of elemental technologies that are significant and as yet unestablished. In order to accomplish this challenging and very fundamental goal, the programme is to be managed under the following fundamental policy:

1. Formation of flexible research organization. Research themes are appropriately allotted so that common-base (such as computational bases) or system integration-oriented research is performed in the central laboratory while individual or elemental research is performed in the distributed laboratories, and an organic and flexible link between both parts is secured.
2. Introduction of competitive principles. The programme introduces competitive principles in the first stage, taking various approaches, and selects the research themes to be investigated in the second stage on the basis of the results of evaluation after the initial five years.
3. Interdisciplinary and international cooperation. The programme promotes interdisciplinary and international cooperation in order to fulfil the basic aims, supporting joint research with national institutes like the ETL and universities, etc., and inviting subcontractual applications from domestic/overseas research organizations such as universities, etc.
4. Publication of research achievements. The progress and results of research and development are to be reported and publicized at domestic/ foreign conferences, etc., and by actively holding symposia and workshops as well.
5. Establishment of infrastructure for research activities. A high-speed computer network is established as the infrastructure for internationally distributed research, and formation of a flexible research organization as well as exchanges of research results are supported.

8.2 Organization Scheme

MITI selects about a dozen Japanese companies, including almost all the major ones in electronics, which form the RWC Partnership. The RWC Partnership will found its own central laboratory (RWC Research Center) near the Electrotechnical Laboratory (ETL) in Tsukuba City, expecting close cooperation with the ETL and receiving researchers from the laboratories of each company.

The ETL, which belongs to MITI and has been playing an important role in concept formation of the programme, will continue to support and lead the programme by sending some researchers into the main positions at the RWC Research Center and also carrying out its own basic, leading research for RWC.

As for the previous Fifth Generation Computer project, MITI will provide a similar amount of total budget (about US$500 million for 10 years). The main part of the budget will be allocated to the RWC Partnership (approximately half to the RWC Research Center and the other half distributed to laboratories of companies), about 10 per cent to the ETL and domestic universities, and about 15 per cent to foreign research institutes to promote international cooperation.

There is a modality for foreign researchers to participate in the RWC programme. Foreign companies and non-academic organizations will be permitted to directly join the RWC Partnership, while foreign universities will be able to participate either as subcontractors or by collaborating in joint research with the ETL and RWC Research Center. In the case of joint research, basically there is no budget flow except for the information exchange.

8.3 Time Schedule

A two-year preliminary study was undertaken in 1989 and 1990 under the research committee on the New Information Processing Technology (NIPT) and several working groups, which included the participation of more than 100 researchers in various fields from universities, national institutes like the ETL, and companies. The final report was published in March 1991. FY 1991 was devoted to the feasibility study under the new name of Real-world Computing (RWC) and toward making a master plan for the RWC programme in May 1992. Under these activities, we organized three workshops (Dec. 1990, Nov. 1991, and Mar. 1992) and one international symposium (Mar. 1991), which were open to foreign countries.

The RWC programme is starting in 1992. The RWC Partnership will be established by August, and the RWC Research Center will open in October. A call for subcontracts will be announced overseas this autumn, and the deadline will be the end of 1992. The applications will be reviewed and selected in January and contracts will be prepared in April of 1993. The chance to join the RWC Partnership or to apply for subcontracts will be kept open after this first opportunity.

REFERENCES

1. Otsu N. (1989). "Toward Soft Logic for the Foundation of Flexible Information Processing." *Bull. ETL* 53 (10): 75–95.
2. *Report of the Research Committee on New Information Processing Technology*

(1991). Industrial Electronics Division, Machinery and Information Industries Bureau, MITI.

3. *The Master Plan of the RWC Program* (1992). Industrial Electronics Division, Machinery and Information Industries Bureau, MITI.

Discussion

Introducing the discussion, M. Dierkes stressed that the second part of Session 4 dealt with two main issues related to the interface between information technology and human culture: first, the translation of information into the huge variety of world languages and the specific role of "sub-languages" in different professional fields and segments of society; second, the development of corresponding technologies that are less rigid than conventional information-processing by computers and thus make possible the handling of the more ambiguous information typical of human thinking and communication. He continued by saying that the vision of everyone communicating in their native language as they are instantaneously translated into other languages by highly "intelligent" machines seems to be socially and culturally desirable. Whether this is technologically feasible in the foreseeable future and whether it is economically viable seem to him key questions to be addressed. A third issue, he claimed, is the problem of access to these tools, especially in the case of languages spoken by small segments of the world population and by people who lack sufficient financial resources to develop the relevant technologies.

The rationale of the great number of commercial products dealing with translations from English into Japanese and vice versa was questioned by N. Streitz, especially in view of the fact that the basic linguistic principles are public knowledge. D. Lide suggested that one might look for "a neutral interchange language" to which each natural language would be translated. Comments were made by G. Johannsen on the policy of Japan's MITI to foster competition in domestic research projects but to support cooperation within Japan when it comes to international competition. He said that this might be an example to be followed by others, especially the developing countries, calling for pooling resources and sharing costs of R&D.

The point was raised by M. Dierkes that efficient machine "translation" of languages and sub-languages would be an important element in preserving cultural diversity and facilitating communication in today's multipolar world. He stressed that there is a great need for future research to improve available technologies to the point that they are able to "intuitively" process information.

Finally, the need for international cooperation in this field was strongly supported by all participants: The cost of developing dictionaries, the necessity of understanding the cultural framework, the huge variety of sub-languages representative of social and professional "subcultures" or spheres of life, the cost of hardware and software production, the research required to go beyond the bilingual machine, and the need for more "intuitive" information processing are claimed to be just some of the technological, economic, and social aspects calling for international cooperation.

Session 5

From New Technologies to New Modalities of Cooperation

Chairperson: Charles Cooper

Systems Management for Information Technology Development

Andrew P. Sage

ABSTRACT

I present a systems management approach to information technology development. While lack of an appropriate systems engineering process for the management of information technologies will not necessarily turn out to be an impenetrable barrier that prevents a technology from ultimately being developed, it can make such developments occur at a much slower rate than might otherwise be possible. Also, the final development costs may be much higher than if an efficient and effective process were followed. This may have significant harmful effects on a given organization or nation, in terms of impaired ability to compete with others as well as on society, and the cumulative effect of this can adversely effect productivity in an increasingly competitive world. There are many new scientific and technological modalities that support enhanced productivity for all. International cooperation is much needed to enhance delivery of these. Contemporary developments in information technology can potentially do much, if properly managed, to support needed actions and efforts. I describe systems management approaches for information technology development here.

1. Introduction

There is much contemporary interest in the use of science and technology to aid humankind. Often, this is related to such issues as technology transfer or infusion, innovation, entrepreneurship, and other efforts associated with the effective and appropriate development of technologies for increased com-

361

petitiveness and satisfaction of market-place demand for societal better-ment. Much of the effort to date centres on the transformation of already developed innovative technologies into viable commercial products [41].

Often, basic research and development will have been accomplished in other organizations, and even other countries, and there may be a lack of an adequate technological base to immediately assimilate a new technology by units desirous of developing a product based on such R&D. This may occur because of a lack of financial resources and expertise to develop these sys-tems, or to extend the initial developments such as to enable integration with existing systems. An objective in the use of a systems engineering approach to information technology development is to enable a focusing of research and development and associated infrastructure concerns, such as to enhance the potential for productivity, including commercialization of the resulting technological products or services.

This paper reports on one effort in this direction. It is an introductory paper and describes an overview, framework, and architecture for systems management of information technology development that could lead to im-plementation of a support system for technology developments and en-hanced use of contemporary science and technology for international better-ment. This material is based upon work supported by the National Science Foundation under Grant EET-8820124. The Government of the United States has certain rights in this material.

Identification of an operational set of critical success factors and use of them in a successful phased study of systems management of information technology development for enhanced access to science and technology is aided by a systems engineering, or systems management, process that will provide:
- the ability to quickly identify ideas and potential technologies that are worth pursuing, and that are not worth pursuing;
- the ability to identify a reasonably short, reasonably low-cost sequence of activities that will result in a cost-effective and societally desirable imple-mentation of a product, process, or service;
- the ability to identify specific projects that will potentially allow such im-plementation and provide for detailed implementation efforts; and
- the ability to identify impediments or barriers to successful information technology project implementations that will not likely allow success, and either remove the barrier or provide for a mechanism for disengagement from the potentially unsatisfactory technology implementation.

Accomplishing this will require a quick-response, action-oriented group atti-tude, an awareness of practices and future perspectives that affect the tech-nology under consideration, the systems management of this technology, and the market-place potential for the technology and the products and services that result from its use. To be sure, there are other challenges and critical success factors associated with the overall process of information technology implementations in developing nations.

The US National Research Council sponsored a 1987 study that identified eight critical success factors [30]:

(1) integration of technology into the overall strategic objectives of the firm;
(2) ability to get into and out of technologies faster and more efficiently;
(3) accessing and evaluating technologies more effectively;
(4) accomplishing technology transfer in an optimum manner;
(5) reducing new-product development time;
(6) managing large, complex, interdisciplinary and inter-organizational projects and systems;
(7) managing the organization's internal use of technology; and
(8) optimal leveraging of the effectiveness of technical professionals.

These critical success factors are associated with the entire life cycle of system development. While they were developed specifically for the United States, these prescriptions appear sufficiently generic that they are universally applicable. They can, therefore, be used as some of the attributes to evaluate information technologies proposed for development. There are, of course, many other attributes that affect technologies, including information technologies, in developing nations. These include technology transfer issues [44], national and international standardization issues [6], and issues that affect information technology development forecasting, planning, and management [1, 13, 17, 27, 64, 36].

A phased life cycle methodology for systems management [58] is especially important due to the rapid shrinking of the time between initial technology conceptualization and subsequent product emergence. The major causes of this shrinkage would appear to be the increased intensity and significance of international competitiveness and the technological changes made possible by information technology itself – such as computer-aided design, manufacturing, and production methods. One result of these two primary factors, and a host of secondary ones as well, is a shortening of the life cycle of the typical product development process. Another result is the ever increasing importance of information and knowledge as driving forces in competitive strategy selection.

The usual listing of the three primary factors of production includes capital, labour, and materials. However, information is an increasingly important driving factor in our economy, and, in particular, information technologies, including computer-aided design and production methods, are a major force in shrinking the time between technology conceptualization and product emergence. I have, therefore, included information as a fourth primary factor of production and have indicated this in figure 1. Many would argue, as does Thurow [63], that human resources and information dominate raw materials in importance in present economies. It seems clear that information now needs to be a separately identified factor that is explicitly included in the usual listing of the three primary factors of production, capital, labour, and raw materials.

Only now is this need beginning to be recognized [34, 40]. Among rel-

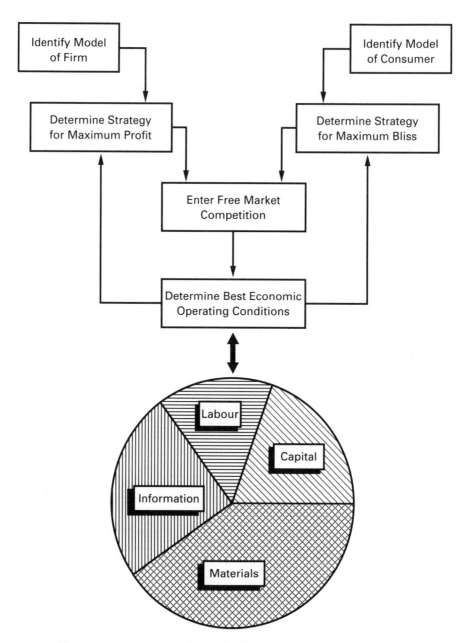

Figure 1 Four primary factors of production for economic rationality

evant research that explores this new phenomenon of the increased role of information in the production process are Cohen and Zysman [9], who suggest that American industry has not adapted to flexible manufacturing systems as quickly as it might have; Kaplan [24], who faults US management accounting systems for failing to adapt to new production patterns; Dertrouzos, Lester, and Solow [11], who discuss needed efforts for the United States to regain the productive edge; Hayes, Wheelwright, and Clark [20], who are very concerned with improving the manufacturing process through infrastructure improvements that involve the vital resources of humans, information, and leadership vision; Zuboff [66], who is very concerned with using information technology to empower people with process knowledge such that they are capable of crucial and collaborative judgement as contrasted with simply automation of production tasks; a series of reprints [19] that discusses many aspects of information technology management; and a series of papers edited by Bainbridge and Ruiz-Quintanilla [4] that explore the impact of information technologies on human work and the need for appropriate training and aiding supports to assist humans in using information technology–based systems.

The use of computers by management and in organizations as decision support systems or executive support systems is the subject of three recent efforts [45, 46, 27]. In the decision and control trilogy of strategic planning, management control, and task control, computer-based information systems are especially useful in the management control function [3]. It is especially necessary to be able to valuate potential investments in information technology, and four recent works provide detailed commentary [61, 34, 62, 7] on this subject. There are many legal implications to information technology implementations, especially when systems integration and systems management considerations are involved [5], as is invariably the case. The implications of information technology innovations on human performance are a subject of much current interest [39, 4]. Information technology has the potential for support in a variety of organizations and for a variety of purposes [2, 20]. To achieve effective support, it will be essential to manage information technology developments, to integrate information technology and institutions and organizations and, thereby, to enable appropriate design through information technology.

There are many ways in which the critical ingredients of a systems management approach to innovation and emerging engineering technologies could be described. As just mentioned, they could be described as production, capital, raw materials, and knowledge. The steps to be accomplished in each phase of an emerging engineering technology effort involve the interaction of:
- problem-solving steps;
- knowledge of technologies, and the characteristics of humans, organizations, and the environment;
- learning over time about these; and
- environmental interaction and systems management, including crisis management strategies,

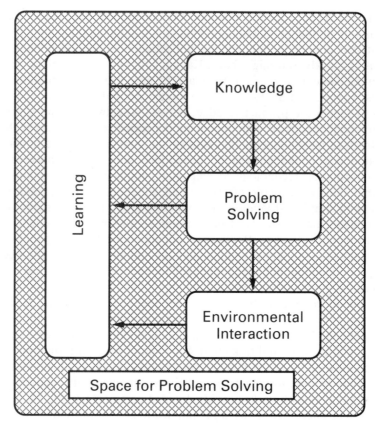

Figure 2 Critical ingredients in systems management of technologies for development or transfer

as shown in figure 2. We are surely describing a dynamic process that involves the interaction of many variables. An important issue for information technology transfers to, and associated development in, developing nations is the use of systems management and integration procedures that effectively and efficiently cope with this process in ways that support continued productivity and competitiveness through advantageous use of science and technology. A major challenge in this is that the knowledge base will be different across various units. An appropriate set of development strategies and tactics must incorporate this reality, as well as the difference in various environmental variables.

2. A Gateway Strategy for Information Technology Developments

There are a number of key strategy elements affecting information technology developments, including possible technology transfer. These include:

- technological and societal need for a technology;
- available technology base;
- research and development process management strategies;
- market and other external factors; and
- standards, including technological regulations and legislation.

There are a variety of ways in which we might conceptualize a model to describe the resulting flow of technological innovations. In figure 3, we envision several primary gateways that control the development and flow of technology, from either a push or pull standpoint. The "Gateway Concept" suggests that a technology, to reach a mature stage in which it yields useful products or services, must pass through three fundamental gateways: the technology gateway, the management gateway, and the societal gateway. These can be easily expanded into a larger number of gateways, as shown in the figure. Passing through the technology gateway requires research ability, innovation, technical merit, and a technical champion [52, 26]. Feasible scientific innovations, the available technology base to support development, and technology research and development efforts are employed, often and primarily in a *technology push* fashion.

The management gateway includes both systems management and enterprise management. Systems management is fundamentally concerned with strategic and tactical efforts associated with the multi-phased life cycle process needed to bring about a trustworthy and high-quality product. It is a systems engineering function concerned with technology management, systems integration, process and product standards, configuration management, and strategic and operational level quality assurance and management. Enterprise management is concerned with finance and accounting, organizational development, marketing, sales, and other efforts needed to bring about a successful interface between the organization and its environment.

The society or consumer gateway, sometimes referred to as a "demand pull" gateway, includes societal and market-place needs, consumer/user receptiveness, and general economic conditions. The gateway concept provides a uniquely appropriate overview of the process of technology development, and I will rely heavily on it as I develop a specific systems engineering approach to information technology development.

As noted in figure 3, the "push of technology" is basically scientific in nature in that it includes development of all feasible scientific discoveries. These are limited, however, by technological capabilities and systems management capabilities. When the resulting *technological systems design* and *management systems design* needs are satisfied, there is really only a *push from feasible technological innovations*. The pull of society, or the market-place, is basically the pull of the Maslow hierarchy of needs [21], as indicated in the figure.

Essentially all studies show that few, if any, successful products emerge only because of technology push. This leads to potentially major pitfalls in pursuing research and development from primarily a university-based *technology push* perspective, and without reference to application-oriented *technology pull* needs of an enlightened industry perspective. This is under-

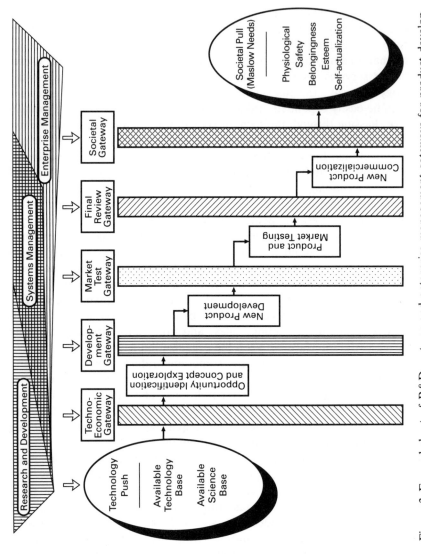

Figure 3 Expanded set of R&D systems and enterprise management gateways for product development

scored by the conclusions of a recent study by industrial R&D leaders. The study, *Industrial Perspectives on Innovation and Interactions with Universities*, which summarized opinions of 17 top industrial research officials, as stated in the *Washington Post* of 26 April 1991, concludes that "Many university officials erroneously believe that the discovery of new ideas represents the most significant step in the process of innovation, and that universities are the key source. [In fact,] industry is the primary source for innovation. Universities play only a limited role in this realm." This suggests a major role for joint industry-university and joint industry-government-university interactions in pursuit of technology development and educational goals. This is especially important relative to information technology developments.

It has been postulated [16, 42] that the growing universality of science and research now makes successful innovation more frequently driven by market pull than by technological push. There are, of course, approaches that encourage and stimulate technology push [60]. Usually these approaches involve enhanced communications and attention to scanning and targeting of potential market areas for the new technology. The several gateways – a science gateway, a technological gateway, a systems management for development gateway, an enterprise management gateway, and a societal gateway – determine what can ultimately flow through to society in the form of a realized innovation or an emerged technology. The major factors affecting, or filtering or monitoring, the flow through these gateways are also shown in figure 3. For a developing nation, the nature of each of these gateways is doubtlessly different in character from those found in a developed nation. Nevertheless, each of the ingredients is surely present.

Any conceptual diagram such as this is necessarily incomplete. What is shown in figure 3 does not adequately represent the dynamics of the process, or the distributed nature of the process, or the many feedback loops involved. It is very clear that technological development alters societal values, certainly the more pragmatic ones, and this in turn acts to change the nature of the societal pull for technological products and services. The process of technological innovation is distributed in time and space. Innovation is clearly a function of a given organization's technological and market expertise relative to a particular technology innovation or development area.

In order for there to be motivation for development or transfer of a technology, there must be a perceived need for the technology; or at least a felt need to accomplish development or transfer of it. As a consequence of this, we need to envision and consider technological needs, systems management needs, and societal needs associated with information technology development and transfer issues. These are the principle gateways we show in figure 3. If there is a technological need for a product, simply because one does not now exist, there will be no technological barrier to development of the product if there is adequate venture capital, and sufficient technological capability as provided through systems management expertise, to enable the development. If there is also societal need, then it becomes possible for a technology to actually emerge into the market-place. Again, there are dynamics involved. It is possible, for example, for a technology to be developed because of a real technological demand and a perceived market or societal demand. Such a technology will not generally be successful, at least initially. It may turn out that there is a later demand for the technology, perhaps in a somewhat modified form.

For a new technology to be developed or transferred, there must be an available technology base that supports development or transfer of the particular technology in question. Existing large investments in production facilities will enhance the propensity of developing innovations that enhance the effectiveness of this investment, for example. There can, and often will, be potentially inhibiting effects as well. The existence of a large investment in one form of technology may well impede the propensity to allocate resources to an entirely new approach that could make the old approach obsolete. At least initially, this might be viewed as a very significant impediment to technology development or transfer in a developing nation. There may well be, for example, manual methods of production that would initially become obsolete due to introduction of a new technology. Of course, a longer-term view of the development situation might show that the initially displaced workers could, upon retraining, enter the workforce more productively at a higher skill level. The criticality of human resources is a major facet of contemporary system, including organizational system, design and development approaches.

More often, however, the fact that there exists one satisfactory way to do something provides an intellectual bias that impedes thinking about new methods of approach. Therefore, a successful technology developer must be motivated and prepared to demonstrate that a new and potentially innovative approach is *better* in some societally acceptable ways.

There should be an appropriate systems management, or technology management, and integration process that support the identification of potentially efficient and effective technologies. *Should* is a key word here. The lack of an appropriate process will not necessarily turn out to be an impenetrable barrier that prevents a technology from ultimately being developed or transferred; but, it can make such developments occur at a much slower rate than might otherwise be possible. Also, the final development or transfer costs may be much higher than if an efficient and effective process were followed. This may have significant harmful effects on a given organization, in terms of impaired ability to compete or reduced quality of life.

Also, there will generally exist a set of standards and regulations, of technological, legal, and political natures, that will act to focus and constrain potential new technological innovations, including development and transfer of information technologies. These may be enhancing or inhibiting factors that depend upon the type of standards that have been identified and the way in which these standards have been implemented and interpreted.

There must also exist a set of market and other external conditions that are suitable for the emergence of an appropriate technology. Otherwise the technology will not get through the *societal gateway*. These external conditions could be rapidly occurring and of crisis proportions; such as might be due to severe drought, disease, or huge trade deficits. Alternately, they could be of a much slower time-scale. The *societal gateway* for developing nations is an area that will very strongly influence appropriate technologies for development in or transfer to a given country. It is an error to assume that this societal gateway is invariant across international boundaries. In my subsequent efforts in this paper, I will also make use of this gateway concept in suggesting a systems management strategy for information technology development.

3. Knowledge Facets for Systems Integration and Information Technology Development

It is certainly not possible to represent the all-encompassing nature of the information technology development and transfer process in a single simple figure, such as figure 3. There are a variety of outlooks on knowledge use for information technology development, including information processing technologies, that support access to science and technology for development purposes.

The thrust of this paper is that systems engineering, as a pragmatic multidisciplinary approach that is oriented to the real needs of society, is an appropriate systems management and integration technology and can be used to both study and manage the information technology development and transfer process. This is especially so since technology development and transfer processes are fundamentally multi-disciplinary ones of large scale and scope.

There are several perspectives from which knowledge may be applied to technological developments [54]. Fundamentally, these evolve from the three different types of knowledge: *knowledge principles, knowledge practices*, and *knowledge perspectives* that need to be utilized, appropriately, to enable development of operational technologies. Each is needed and if appropriate provision for these three types of knowledge is not made, efforts at development or transfer of technologies will likely fail.

Knowledge principles represent the formal reasoning-based scientific approaches that lead to the development of new knowledge. Knowledge practices represent the application of existing wisdom, often in the form of experiential-based skills and standardized rules, to the development of a new technological product that is based upon an existing product. Notions of standards for technology development and architectures that support open systems integration are especially important ones here. Finally, knowledge perspectives concern future-oriented issues that determine the relative importance of implementing potentially competing new technologies. Thus, they necessarily also involve the blend of practices and principles that should be brought to bear in resolving future-oriented technology development issues.

New systems are seldom developed, sold, or deployed in a vacuum. Usually they are improved versions of, or additions to, an existing system. The new system will normally evolve from an existing system in the generic fashion shown in figure 4. The new system may be delivered as a result of some contracted effort with an external systems engineering contractor, or it may be developed in-house. Systems integration is the process through which a number of products and services, both hardware and software, are specified and assembled into a complete system that will achieve the intended functionality.

There is an inherent relationship between systems integration engineering and standards. This exists because both system users and system developers, and purveyors or marketers, have a common need for standards that are system independent and specific developer independent. The term "open systems architecture" is now used to describe any of several generic approaches, the intent of which is to produce "open systems" that are inherently inter-operable and connectable without the need for retrofit and redesign.

An appropriate open systems architecture standard must be explicitly defined such that anyone desirous of using it can use it for implementation purposes, and must satisfy other desirable attributes of standards. Systems integration, which is fundamentally concerned with the technological and management issues needed to bring about functional operability of systems, is very concerned with these issues also, although perhaps from a slightly different viewpoint. The overall tasks of a systems integrator include:

- *System definition*: identifying user requirements and technological specifications for a system, including needs for systems integration to insure compatibility with existing and possible future systems;
- *System design and development*: including the identification of an appropriate architecture or preliminary conceptual design for the system, in turn including appropriate interfaces to existing systems, evaluating the performance of the system, potentially modifying the system architecture for better performance and enhanced inter-operability, and thereby establishing an effective open architecture for the system to be developed; and
- *System test, integration, and maintenance*: to insure that the operational system is cost effective and of high quality.

These are just the phases of the systems engineering life cycle [58], modified slightly to explicitly recognize the role of the systems integrator and the concomitant need for an open systems architecture. The confluence of systems integration, open systems architectures, and standards for these *may* be expected to lead to an open systems environment that would:

- reduce the system acquisition costs;
- reduce system integration costs;
- protect current investments in hardware and software;

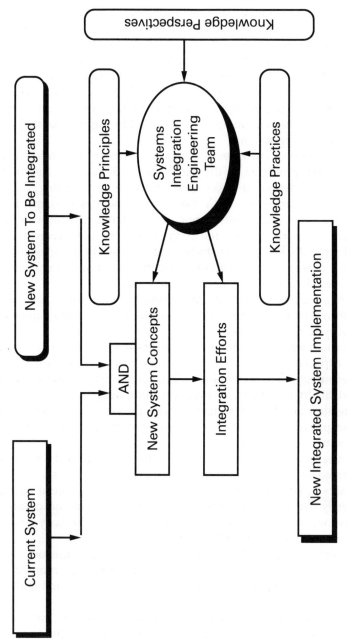

Figure 4 The systems integration engineering process

- allow increased independence in acquiring new systems, and in modifying and maintaining existing systems, and through this process
- maximize the quality and effectiveness of integrated system products in resolving user and customer issues and problems.

Of central concern in a systems integration effort is the system-level *information architecture*. The mission areas for a system will normally vary from case to case. The primary tasks of the systems integration team are design of the overall system architecture and integration of sub-systems into this architecture. The first of these tasks calls for top-down systems engineering, while the second requires management and technical direction of contract work and bottom-up approaches to achieving interfacing and inter-operability of existing systems.

Systems integration engineering requires attention to both technology and management problems on the part of both the implementation and integration teams. Technical tasks generally include assessing the impact of architectural changes on both the system under development and its stakeholders. The systems integration team should also provide systems management support relative to technical system management matters. These will generally involve cost studies and configuration management studies. Figure 5 represents the conceptual incorporation of systems integration within an overall systems engineering framework.

There are a number of motivations for systems integration. A recent study [10] identified five primary motivations for further investments in and investigations of systems integration:

(1) Experiences with information technologies have not been in accord with initial expectations. One major finding in many studies is that information technology may well enhance individual productivity, but often has great difficulty enhancing organizational productivity.
(2) Propagation of information technologies throughout nations and organizations produces the need for inter-operability and connectivity across equipment and applications.
(3) The installed base of information technology products will grow to accommodate both new technologies and new capabilities.
(4) Advances in information technology and growing appreciation for what ultimately can be accomplished will necessarily promote organizations and nations to search both for new information technology applications and new sources of competitive advantage.
(5) In the increasingly global economy of today and tomorrow, organizations and nations must rely on information technology in order to manage and coordinate their operations and to stay informed and globally competitive.

These considerations are not independent of one another, and there are others. The emergence of new information technologies, and the enhancement of those that exist today, is almost certain to occur. Information technology is so very closely related to the large-scale spectacle of global interdependence among economies, other technologies that it supports, and even

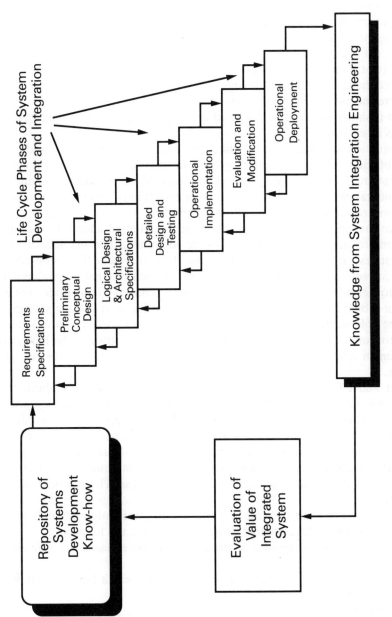

Figure 5 Evolving nature of systems integration engineering for development

nations, that its importance can only be expected to increase. These realities together with the need for continued vigilant attention to systems management for product and process quality, security, and the assurance of equity for all comprise the major challenges for systems engineering in the twenty-first century. At a pragmatic level, systems integration is very important in assuring these ends.

The first efforts in systems integration should generally be to obtain, from a variety of sources, identification of
- where the information technology systems user is,
- where the user group needs to be, and
- how it should get there relative to development strategies, including systems integration.

This situation assessment effort will define the needs, constraints, and alterables of the information technology system to be developed. This must, or at least surely should, be done with knowledge of the organizational and environmental variables extant. Of course, a preliminary version of this assessment was required to bring the effort to this stage.

Once the situation assessment is complete, potential additions and modifications to the existing system must be identified. The impacts of these alternatives on the resulting system are then analysed. This step should allow for some adjustment of parameters for each alternative implementation, to permit optimization of performance. These systems under consideration should be either immediately inter-operable with existing systems or at least able to be integrated with some degree of effort. There is a major need for an evaluation methodology to validate the software, the hardware, the human interfaces, and the trustworthiness and quality of the resulting system.

Cost and effectiveness indices will be determined for each alternative. These will be included in planning documents for systems augmentations. These documents will identify potential integration opportunities within the existing environment of computer hardware, software, communications, and physical plant. Also included with each evaluation should be an analysis of risk factors affecting each alternative. Risk has numerous facets or characteristics that affect cost, schedule, and trustworthiness. These should be fully explored.

Invariably, the goals of those working on an information technology systems acquisition or development project will include the following:
(1) To identify new technology approaches that will enhance functionality of the new system.
(2) To identify significant "cost drivers" that represent a high percentage of total costs of the system.
(3) To identify methods that will reduce costs while simultaneously retaining benefits and on-time delivery of the operational system.
(4) To field a quality system, within the constraints set by schedule and price, that is of high quality, and trustworthy in terms of satisfaction of customer needs.

These objectives apply to the overall information technology systems development effort in general and to systems integration in particular.

Operational deployment of a system, and related system integration concerns, is an iterative and evolving process. Systems that once fit well into a complete system may not do so at some future point. This evolving system development concept is shown in figure 5. The iteration and feedback are essential to ensuring continuing functionality of the system.

Systems integration has four fundamental dimensions [31]:

(1) *Integration Technology*, which supports transfer of data across different subsystems. This process includes file transfer protocols, document protocols, and remote procedure calls. Automatic data transfer, common database structures for different applications, and process-to-process communications through well-defined functional interfaces and interaction protocols are examples of how integration technology is accomplished. Some form of integration technology is generally necessary for overall systems integration, but is never sufficient to insure it.

(2) *Integration Architecture*, which structures subsystem design to insure easy and secure data sharing across subsystems. Storage of common data in databases requires functional inter-operability if the data are to be shared. Accomplishing distributed data storage through the use of an integration architecture that has direct access to data or functional access by activating other systems is a need.

(3) *Semantic Integration*, which insures either that the same concepts mean the same thing in different portions of the system or that there exists a translation mechanism that will resolve semantic inconsistencies so as to allow information exchange across systems. These inconsistencies will invariably exist when the different subsystems of a DSS are procured from different vendors.

(4) *User Integration*, which enables a system user to concentrate on the tasks to be accomplished and not the specific details of the technological system being integrated. This will generally require easy access to different applications and systems, uniform user interfaces, consistent data, and consistent use of semantic concepts.

The three perspectives on knowledge discussed earlier in this section – knowledge practices, principles, and perspectives – proactively relate to these four integration aspects, as indicated in figure 6. One major objective of any overall system acquisition effort should be to reduce implementation risks and enhance trustworthiness of the resulting system. Whether this should be accomplished through the efforts of a system integration contractor exclusively or through a more general engineering effort that might include production of new hardware and software is clearly a matter for judgement and choice based on the particular issues at hand and the particular and unique capabilities of the developing unit.

Systems integration should be capable of efficiently and effectively coping with future user needs for hardware and software acquisition. There should be an intentional linkage between information technology systems and organ-

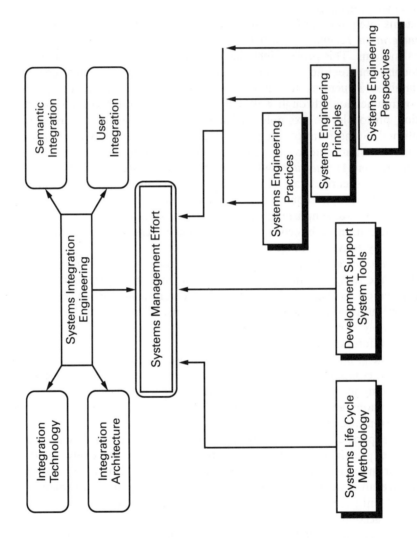

Figure 6 Principal knowledge ingredients in systems integration engineering

izational systems. Conceptual architectures and frameworks for open systems architectures and integration [48, 47] are particularly important in this regard. These are needed in order to accommodate the identification of requirements for, and the subsequent development and implementation of, an integrated system that is responsive to contemporary needs, especially for systems that can function in contemporary *high-velocity* environments.

This requires an approach that recognizes that a systems integrator and a systems user will have different perspectives on development of the system. It is only natural that a system user will be almost exclusively interested in the ultimate *product* and its trustworthiness. On the other hand, a systems engineer or systems integrator can be expected to have very strong interests in the *process* that is undertaken to insure delivery of the product. Through this interest in process, there will naturally occur an interest in product, as well. Ultimately, products or systems will be valued in terms of their ability to resolve issues or problems, and this is what the ultimate customers for a system desire. To accommodate each of these perspectives, we need a strategic level approach to quality assurance and management – one that produces *total quality management*, a subject of very contemporary interest.

Walton [65] is among those who have studied the necessary efforts to successfully integrate information technology into organizations. He indicates that positive economic and human outcomes will occur through the use of information technology developments that are patterned after a three-phase process of context generation, information technology system design, and implementation of the information technology system. These phases follow the general systems engineering life cycle phases of definition, development, and deployment and are illustrated in figures 5 and 7. There are three major manoeuvres at each of these phases: creating alignments, fostering commitments and supporting ownership, and developing mastery in the evolving information technology process. Walton suggests that anticipatory, or proactive, development of the organization will occur when organizational considerations and information technology are integrated at the strategic planning level for information technology implementation. Simultaneous development of both the organization and the information technology process may occur when the information technology process is introduced without organizational strategic planning considerations. Reactive adaption of the organization after implementation of a system will be needed when a technology push solution is imposed. Generally, this represents an imposed solution to a needed improvement. It may represent a solution looking for a problem. Walton provides case-studies of each of these implementation strategies.

Keen [25] is also very concerned with shaping the future through information technology as a means to an organizational end. He identifies four realities that normatively should guide information technology and organizational planning. The first of these is that a restructuring of the cost base for organizations will be needed in the 1990s due to the ubiquity of declining economic return margins across industries and countries. Also, total quality and enhanced service must be considered as basic ingredients of an organi-

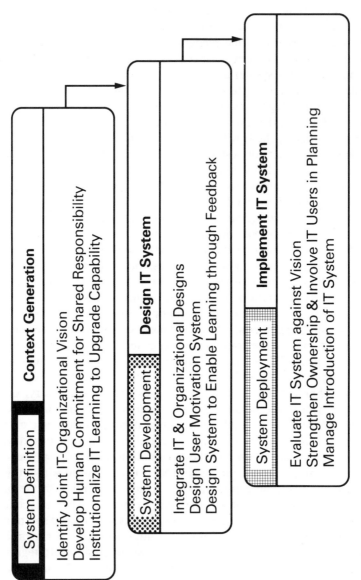

Figure 7 Three phases in the Walton IT development life cycle and phased efforts in creating alignment, enhancing human commitment, and enabling IT system user competence

zation's operations in the 1990s. Eroding prices and falling profit margins necessitate the implementation of *yield management* systems, which ensure a competitive advantage through time-sensitive information for planning, pricing, and deployment of resources. These are systems associated with such contemporary efforts as on-line customer service to minimize customer waiting time, airline pricing mechanisms that ensure maximum profits in the face of widely differing seat prices, and just-in-time manufacturing that attempts to reduce product development time and time to market. Keen envisions seven facets of organizational design through the use of information technology:
- competitive positioning;
- geographic positioning;
- organizational design and redesign;
- human resource, or human capital, redeployment;
- managing information economics;
- organizational positioning and repositioning for information technology usage; and
- strong alignments of business and technology.

The major pragmatic objectives in this effort are to enable information technology system design so as to integrate the work of people and machines in support of organizational objectives.

There are many other studies of this type. A recent paper by Orlikowski and Baroudi [33] provides a very good overview of overarching theoretical perspectives and philosophical assumptions regarding information technology and organizational phenomena. Orlikowski [32] examines the degree of change facilitation in the forms of organizing and control that is brought about by information technology deployment in the workplace. Her findings indicate that present uses reinforce established forms of organizing and that they are facilitated in strengthening and cohesion of prevalent control mechanisms. Her conjecture is that present information technology deployments tend to reflect a strong commitment to the present organizing structure and functions and that the resulting integrated information technology system reflects existing forms of knowledge and influence. Organizational and institutional learning and a variety of perspectives on rationality [29, 8] remain major ingredients in this.

A fundamental goal of strategic planning and associated systems management is to develop a balanced perspective on system development so as to insure the success of the development venture. Figure 8 illustrates some of the concerns of systems management, as expressed in the components of a systems engineering management plan (SEMP) for technology development. It is a task of systems management to develop each of the 12 elements shown in this figure, and to accomplish it in such a fashion that the system customer is satisfied with the system product in terms of its ability to resolve problems. We see, then, the need for integration of problem, product, and process as essential to systems integration.

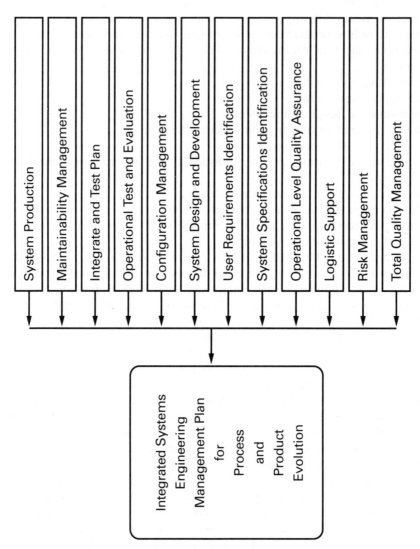

System Production

Maintainability Management

Integrate and Test Plan

Operational Test and Evaluation

Configuration Management

System Design and Development

User Requirements Identification

System Specifications Identification

Operational Level Quality Assurance

Logistic Support

Risk Management

Total Quality Management

Integrated Systems Engineering Management Plan for Process and Product Evolution

Figure 8 The need for process integration at the level of the systems engineering management plan

4. A Newness Matrix Approach to Information Technology Development

It is possible to characterize existing conditions in a developing nation along several dimensions relative to development or transfer, and implementation of a new information technology. Two questions seem to be of primary importance relative to judgement and choice as concerns exploitation of a potential technology development and/or transfer venture. They can be expressed in slightly different form for individuals, groups, and organizations. In generic form, they are:

(1) Which new technology markets should a unit enter?
(2) How should the unit enter these technology markets so as to maximize the likelihood of success and the reward to be obtained from success, and at the same time to control the risk of failure and the losses to be suffered in the event of a failure? Entry may occur through internal development of a new technology or through a technology transfer process. There are a number of related infrastructure questions and questions of system integration associated with either approach.

A potential new technology can be nurtured by one unit, which may vary in scope from one company to one nation, through the use of a combination of the following two basic approaches:

(1) internal development of the technology, or
(2) venture funding of others and subsequent acquisition, or transfer, of the technology.

There are many ways through which the questions just posed could be resolved. In part, the appropriate development strategy depends upon an analysis of four related questions:

(1) How new and different is the technology for the unit in question?
(2) How new and different is the market for the technology for the unit in question?
(3) How familiar is the unit in question with technological development needs?
(4) How familiar is the market for the technology to the unit in question?

The responses to these questions lead to a 16-cell selection matrix, shown in figure 9, that determines the extent to which a specific unit might be able to use various types of knowledge in order to determine solutions to the many potential problems that may arrive in making a potential new technology operational. The terms "base technology" and "base market" are used to describe technologies and markets with which a unit is presently concerned. Roberts and Berry [43] have described appropriate entry strategies for the nine cells that are most supportive of success in development of a new technological product or service. These are the nine cells associated with the three left columns and top three rows in figure 6. The 16-cell matrix is appropriately called a "newness matrix."

"Newness" is the key concern in the newness matrix entries in figure 9, which indicates a basic 16-cell model of experiential familiarity with techno-

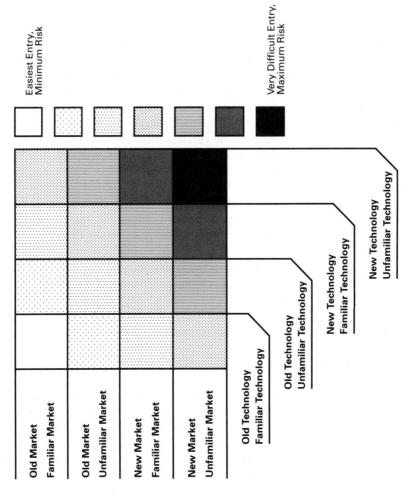

Figure 9 Experiential familiarity with market and technology effect ease of entry and risk

logy and market. Certainly, much about many information technologies will be *new* to a developing nation. But a potential innovation may be a *new technology* or a *new market*, in general – or for a specific company within a developing nation. Also, there are questions of existing technologies with which a new technology must be integrated, and experiential familiarity with the system integration process can also be expected to vary. Expanding on these concepts so as to be able to indicate generic costs and effectiveness indices, including success and failure possibilities, is a goal of a new technology identification study.

The newness matrix, illustrated in figure 9, suggests an approach for analysing the risks, hazards, and uncertainties associated with introducing a new technology. Such uncertainties, both in the market-place and in the technology itself, constitute, in the terminology of business, the risk factor. Risk, of course, is of fundamental importance as a decision-making influencer in both systems management and enterprise management.

The newness matrix is particularly relevant in the early stages of a technology's development, where there are numerous uncertainties. For example, Florida and Keeney [15] have argued that American industry has tended to rely on major technological breakthroughs rather than incremental improvements in technology as the major mechanism for technological progress, with substantial competitive disadvantages as a result. The newness matrix approach attempts to focus attention on just the types of problems that this, perhaps prevalent, *breakthrough* mentality might overlook.

Newness, or uncertainty, in markets may be due to any of the following problem areas:

- *New uses*. There will always be uncertainty where a new use, or function, is being offered, even if there are only relatively minor changes in the technology. For example, the widespread adoption of personal computers might seem to involve only minor changes in software technology for mainframes; but, this led initially to widespread user apathy and an entirely new and beneficial approach to development.
- *User skepticism* about *improved* performance characteristics. Many technologies are developed with the notion that they will substitute for existing technologies by providing more effective performance at a modest, or at least acceptable, increase in price. However, the ultimate consumer and purchaser may not be particularly impressed with the performance improvement. For example, supersonic commercial air travel has proven far less popular than developers of the Concorde had initially hoped. High definition television (HDTV) might well prove similarly disappointing, as consumers may find that intermediate forms of enhanced definition TV are quite acceptable for their needs.
- *Requirement for human behaviour adjustment* by the user. The most imaginative and potentially useful new technology can fail because users cannot, or will not, adjust their behaviour to meet the needs of the technology. A promising innovation, video phones, may flounder because people do not want callers peering at them in their home or office, but

may be reluctant to turn the video off after responding to call. As a general guideline, technologies should and must serve humans. Humans will generally not serve technologies, nor should they be expected to do so.

- *Competitive technologies*. Competitive technologies are volatile and operate in high-velocity environments. This results in very significant uncertainties. For example, efforts by commercial earth satellite-based transmission firms to boost their share of the telecommunications market necessarily must confront exciting changes brought about by fibre optics and cellular radio communications. Initially, this may make a marketing strategy for any particular technology highly uncertain.
- *Unpredictable technological developments*. Scientific or engineering breakthroughs can add enormous uncertainties to markets.
- *Legal barriers*. Regulatory and standardization requirements can add considerable uncertainty to the technology adoption process. While these may be very beneficial, there is no reason to assume that they are always beneficial.

The other axis of the newness matrix is technology uncertainty. This may be due to any of the following factors:

- *Innovativeness of technology*. Almost without exception, potentially more innovative technologies will be initially associated with greater risks and uncertainties than less innovative technologies. A need in this regard is to be able to identify what is genuinely a technological innovation and what is simply an extension of existing technology. While formal knowledge will usually be needed to deal with totally new technologies, there will exist known-to-work approaches that allow one to cope with extensions of existing technologies.
- *Number of constituent technologies*. Uncertainty may well increase geometrically, rather than arithmetically, with the number of technologies involved in an innovation. For example, successful development of HDTV requires integration of three emerging technologies: flat-screen video displays, digital video transmission, and very high-speed processing of digital video data. Success needs to be obtained in all three, and this results in current substantial technological uncertainty.
- *Manufacturing difficulties*.
- *Institutional changes*. Required to bring about process improvements such as to lead to high quality and trustworthy products [14].

Another taxonometric dimension for consideration is the type of unit involved in a possible emerging engineering technology effort, and the nature of the technology itself. Horwitch and Prahalad [22] have identified three ideal organizational modes, and we can easily add a fourth that concerns the individual innovative researcher, such that we have:

(1) the technological innovation process practiced by the individual researcher in an academic research, or potentially industrial research, laboratory environment;
(2) the technological innovation processes found in small, high technology-oriented firms;

(3) the technological innovation processes that occur in large corporations with multi-products and multi-markets;

(4) those processes found in conglomerates, multi-organizations and transnational multisector enterprises.

The types of technologies most suitable for potential development and/or transfer investigation in each of these four modes of operation will be different, as will the appropriate risk behaviour. It would seem reasonable to augment this model to allow consideration of other modes, such as those due to individual entrepreneurs and government development assistance. Also, the dimensions of the taxonomy could be enlarged through consideration of the roles, potentially very different roles, of the technology developer in organizations of four generic sizes; individual, small to mid-size, large, and multinational. Other desirable augmentations of importance involve the gateways through which a system development must necessarily pass, and the phased life cycles that involve research and development, systems management, and enterprise management.

Of much importance also will be the type of *coordination structures*, or patterns of information flow and decision-making among a set of agents who accomplish various activities in order to achieve objectives associated with technology development or transfer. The study of Malone and Smith [28] illustrates that, both for human organizations and computer systems, these structures are very important in determining production costs, coordination costs, and system vulnerability to crises of various types. Decentralized markets, functional hierarchies, product hierarchies, and centralized markets are the four fundamental structures, with functional hierarchies and centralized markets being further characterized as small-scale or large-scale. The historical evolution over time of these is in the order listed. Since market pull is generally the dominant force in the long-term success of technological innovations, it is appropriate to devote abundant attention to establishing coordination structures and associated perspectives that will enable successful development of a selected technology. It would be particularly interesting to associate the following different patterns of information flow and coordination:

– goals to be achieved, or products or services to be produced (*products*);

– people who perform various tasks (*task processors*);

– people who decide which tasks should be done (*task managers*);

– people who decide which task processors should perform individual tasks (*functional managers*); and

– communications between people (*information or message patterns*),

with different approaches to development and/or transfer of technologies so as to obtain the most appropriate relationships between organizational communications and coordination, and the development and operational implementation of specific engineering technologies in a developing nation. Figure 10 indicates some general relations among these five elements. Again, it illustrates that we have a dynamic process and that the evolution of the process variables over time is a very important consideration.

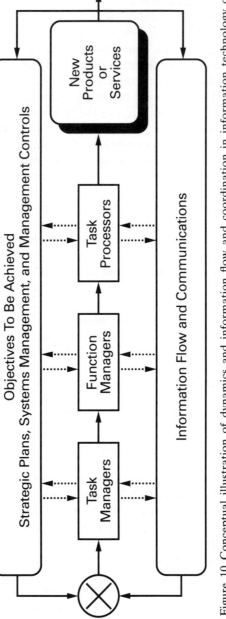

Figure 10 Conceptual illustration of dynamics and information flow and coordination in information technology development

5. Phased Life Cycles for System Acquisition

These discussions lead us rather naturally to a set of systems engineering life cycles for acquisition of a large system. First, I will consider initial development of an emerging technology. Then, I will examine systems management, or systems engineering, of a product. Finally I will look at some notions of enterprise management.

One possible life cycle for emerging technology identification and assessment, and preliminary implementation, is comprised of two major phases, and can be further characterized by seven phases:

(1) *Technology Identification and Assessment*

    (a) *scouting* and identification of requirements specifications for candidate engineering technologies;

    (b) *authoritative information documentation* concerning technological, economic, and societal need for and feasibility of the technologies;

    (c) *assessment* and evaluation of the technologies;

    (d) *selection* of appropriate technologies for initial development and implementation;

(2) *Preliminary Operational Implementation*

    (a) *tracking* of the progress of development and implementation concerning all aspects of the candidate engineering technology;

    (b) *supporting* the operational implementation of the technology in ways that are meaningful to the technology itself and the results obtained in the earlier phases of the process; and

    (c) *disengaging* from studies that prove to be productive and that have been successfully transferred, or that indicate productivity or risk potentials beyond critical thresholds.

These seven phases are also formally the seven phases [58, 55] that can, with slight modifications in the activities for each phase, be used to identify and nurture emerging technologies that may ultimately be nurtured to the point where they have potential for additional efforts that might lead to new products and services. These seven phases are used in the literature of the National Science Foundation's *Emerging Engineering Technology* Research Initiative, and were first suggested to this author by Dr. Nicholas DeClaris. The critical attributes of potential development technologies should be identified as part of the initial phases of the process. These should be identified in the form of indicators that will enable early identification of a potentially successful technology. Among these attributes are: innovativeness, timeliness, cost-effectiveness, and profitability of the products, concepts, or services of the technology under consideration. Identification of productive environments for potential technology development and transfer candidates is also a need. It is not difficult to characterize the appropriate environment as one in which a highly motivated group of people are free to pursue potentially unusual ideas, as well as not so unusual ideas. The environment should be one that recognizes and rewards success and that also recognizes that there will be some failures. It must be a sense-of-urgency environment, in that the utility of any need, idea, or actual product is temporal.

The critical attributes of a technology development and transfer process should be identified in the form of indicators that will enable early identification of a potentially successful technology. Among these attributes are: innovativeness, timeliness, cost-effectiveness, and profitability of the technology products, concepts, or services. In our discussions thus far, we have emphasized the early phases of the process. There is, of course, a major need to be concerned with all seven of the phases and to support the usual product life cycle through these. I have emphasized the early phases in the discussion here only because quality products here are so critical to success of the overall process.

The latter phases of the technology development and transfer process and effective support to the entire life cycle of product development are especially important today due to the rapid shrinking of the time between technology conceptualization and subsequent product emergence. The major causes of this shrinkage would appear to be the increased intensity and significance of international competitiveness and the technological changes made possible by information technology – such as computer-aided design and manufacturing methods. One result of these two primary factors, and a host of secondary ones as well, is a shortening of the life cycle of the typical product process, or service. Another result is the ever increasing importance of knowledge as a driving force in competitive strategies. These statements appear essentially invariant, although the specific interpretation and associated implementation may well differ across particular developing and developed economies.

Figure 11 presents an interrelated sequence of two life cycles that comprise an emerging technology R&D life cycle and a systems management life cycle. To this could be added an enterprise management for product evolution life cycle, or any of several others. Recent works by Rouse [49, 50] illustrate the interrelatedness of these quite well and the need for innovation in all of them. A particular need at this time is for efforts that involve the careful integration of the R&D life cycle of emerging technology with product development life cycles for maximum competitive advantage, and a number of contemporary publications are concerned with various aspects of the many associated issues. Roussel, Saad, and Erickson [51] are particularly concerned with concepts for R&D management as a strategic competitive weapon and needed linkages between R&D management and corporate strategy. Raelin [38] is particularly concerned with corporate cultures, professional cultures, and social cultures; and the implications of these for various management aspects relating to salaried professionals. A work edited by Goodman and Sproull [18] is broadly concerned with the ways in which contemporary technology blurs existing organizational structures and functions and changes the nature of work in an organization. It also illustrates how and why potentially more valuable technologies may not be used effec-

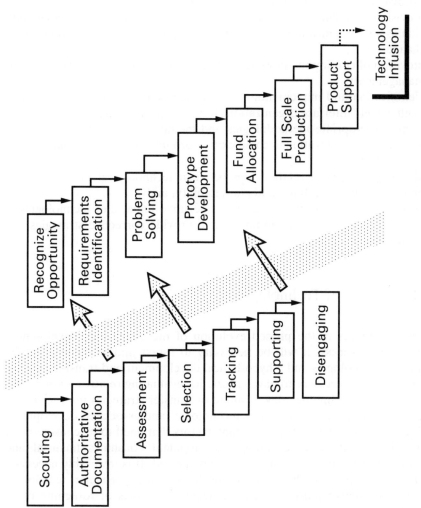

Figure 11 Technology development/transfer support to the general technological innovation process

tively, and how and why inferior technologies sucked at a lower ultimate performance limit.

Edosomwan [13] is especially concerned with the close relationship between technological innovation and management and the use of technologies in the workplace and associated product development. Specific guidelines and a framework for technology management, utilization, and forecasting are provided. The major suggestions include:

- provision of facilities for human resource development to enable individuals and organizations to contend with rapid technological and societal change;
- provision of continuous support for R&D efforts through effective systems management;
- identification and implementation of strategies and management controls to effectively interface organizations to their external environments;
- furnish appropriate methodologies and methods to manage technical resource and organizational complexities brought about by rapid technological change;
- identify appropriate relationships between technological innovation and product manufacturability and marketability; and
- evaluate unceasingly the positive and negative impacts of technologies in the work culture and environment, and provide corrective strategies and tactics as needed.

6. Evaluation of Technologies

At several points in the phased evolution of the technology development and transfer process, it will be necessary to assess and evaluate potential identified technologies and a hypothetical development and transfer process. Therefore, it is necessary to have criteria for evaluating relative appropriateness of various technology development or transfer strategies in order to determine appropriately meritorious technologies and associated development and transfer strategies. There are many factors that need to be considered in doing this. Large-scale technology development consumes financial and other resources, often for a significant time period. It is invariably necessary to recognize that the benefits of developing one particular technology strategy alternative must be weighed against the costs of foregoing other opportunities.

There are a number of issues to be resolved through the evaluation efforts accomplished as part of the systems management of technology development and transfer. These include:

- determining an appropriate *specific* process to use for the identification and evaluation of potential technologies for development and/or transfer;
- identifying the groups that should be involved in this identification and evaluation process;

- identifying the criteria that will be used to determine length and type of support;
- identification of appropriate criteria to determine transferability of the technology to full-scale operational deployment status or termination.

Each of these relates to the criteria that will actually be used for evaluating emerging technology development strategies. I have discussed many of these criteria in our efforts this far. Dutton and Crowe [12] provide an excellent summary of many appropriate evaluation attributes:

(1) Technological merit
   (a) Technological objectives and significance
   (b) Breadth of interest of strategy
   (c) Potential for new discoveries and understandings
   (d) Uniqueness of proposed development strategy
(2) Social benefits
   (a) Contribution to improvement of the human condition
   (b) Contribution to national pride and prestige
   (c) Contribution to international understanding
(3) Programmatic (management) issues
   (a) Feasibility and readiness for development
   (b) Technological logistics and infrastructure
   (c) Technological community commitment and readiness
   (d) Institutional infrastructure and implications
   (e) International involvement
   (f) Cost of the proposed strategy

I have rephrased these as attributes for a multiple attribute evaluation effort. These attributes may be viewed in several ways. For example, they can also be seen as ingredients at the various gateways for the development of emerging engineering technologies shown in figure 3.

In addition to obtaining an evaluation of proposed technologies and associated development and/or transfer strategies, an appropriate approach to evaluation should also allow for full exploration of the needed functions to insure satisfactory development of appropriate technologies. To do this, people are needed who fill the roles of [41]:

- *Idea generators* that contribute ideas from technology push or market pull considerations to extend the ultimate potential of the emerging technology under development;
- *Idea exploiters*, or innovators or technology development champions or research entrepreneurs, who take research ideas and attempt to get them fully explored, supported, and adopted;
- *Management leadership*, or business leaders, who see to it that the various planning, scheduling, monitoring, and control functions are carried out effectively;
- *Information (and knowledge) gatekeepers* who provide informed wisdom to the parties at interest as to the emerging engineering technology development relative to contemporary realities that effect technology, capital, manufacturing, standards, and market potential; and

– *Sponsors* who are not directly involved with the development strategy, in order to insure objectivity, and who provide leadership and resources from the very highest levels to enable development of the technology or to restrict it when it proves cost-ineffective.

These needs should also be translated into attributes that can be used to measure the success of a particular technology and development strategy.

The assessment and evaluation of alternative technologies proposed for development may be approached through the application of formal decision theoretic methods. A major objective in technology assessment and evaluation is forecasting the potential costs and the resulting effectiveness of technology development or transfer. It is desirable to minimize the error associated with this prediction. If a set of inductive beliefs do not conform to those associated with the use of the probability calculus, then the expected error can always be reduced by modifying those belief values to conform to the calculus. Multi-attribute utility models are generally additive. They calculate utilities of an alternative or outcome by a weighted addition of the utility values of the alternative or outcome across the multiple attributes. The reasons that support using these MAUT models are that the necessary and sufficient conditions for these models to hold are well known. Further, the very important notion of *value independence* results. A cost effectiveness model [53, 58] for evaluating alternatives may be utilized.

The broad goals of cost-benefit analysis are to provide procedures for the estimation and evaluation of the benefits and costs associated with alternative courses of action. In many cases, it will not be possible or appropriate to obtain a completely economic evaluation of the benefits of proposed courses of action. In this case, the word "benefit" is replaced by the multi-attribute term "effectiveness."

Identification and quantification of the benefits and costs of possible alternative courses of action for technology development and/or transfer is a difficult task. It is generally not as difficult perhaps as formulation of the issue and identification of the alternatives themselves, but it is still not easy. Here I use the word benefits to mean the possible effects of a project. These include the totality of both positive and negative benefits, or disbenefits. We must first identify benefits, and then we should quantify them by assigning a value to them. Many benefits (and disbenefits) will be intangible and will occur to differing groups or individuals in differing amounts. Problems with intangibles may be especially difficult in the public sector, where agencies are designed primarily to deliver services or public goods rather than products for individual consumption. A major goal of a private sector organization is profit maximization, and it is relatively easier to measure profit as a benefit. Often there will be a variety of reasons why people will be uncomfortable with providing a strict economic measure for benefits. The word "effectiveness" is often used instead of benefit when a strictly economic valuation is not needed. When effectiveness is substituted for benefit we obtain a cost-effectiveness analysis. The benefits of a public service are much more difficult to define because they are intangible or indivisible (or both).

The political environment of many public-sector efforts further complicates the task, and variables other than those associated with efficiency, economy, and equity should be measured. This suggests a multiple perspective approach to effectiveness. Among the many possible perspectives that need to be considered are economic, technical, legal, social, and political.

In cost-effectiveness analysis, we desire to rank projects in terms of economic costs, and in terms of effectiveness. The reason for this is that there are non-commensurate attributes of a project. Certainly we would wish to eliminate conspicuously inferior projects, that is to say projects that are more expensive and less effective than other projects, from consideration for selection. Beyond this, a cost-effectiveness analysis does not specify which of several projects is "best." This can be accomplished if one is willing to trade off cost for effectiveness, such as to obtain a "scaler performance index." It can be done by considering cost as one of the attributes in the effectiveness evaluation approach selected.

The effectiveness of an alternative is the degree to which that alternative is perceived as satisfying identified objectives. The effectiveness assessment approach described here provides an explicit procedure for the translation of quantitative evaluation of alternatives when the impacts of the alternatives are described by multiple attributes. This is accomplished by identifying and organizing the attributes of event outcomes (or alternatives, if there are no probabilistic uncertainties that influence the outcome that will result from alternative selection) into a tree-type hierarchy of attributes that is used together with measures of effectiveness to compare alternative technologies and development strategies as a basis for choice-making. Generally, an effectiveness assessment study involves a number of major analytical steps and is illustrated in figure 12. The final results of a cost-effectiveness assessment are used for comparison, ranking, and prioritization of the identified technology development alternatives according to effectiveness. This effectiveness assessment can be very useful for interpreting and evaluating the results of an analysis effort. In order to use the approach, we need a set of attribute of objective measures, information on the relative importance of attributes or objectives, and sufficient knowledge about project alternative scores and their outcomes to be able to assign effectiveness scores to the attribute measures that characterize the impacts of each outcome.

7. Information Technology Perspectives

While information technology does indeed enable better design of systems and existing organizations, it also enables the design of fundamentally new organizations and systems. Thus, efforts in this area include proactivity in the sense of being aware of future technological, organizational, and human concerns so as to support graceful evolution over time to new information technology–based services. Among these would be improved access to knowledge of all types.

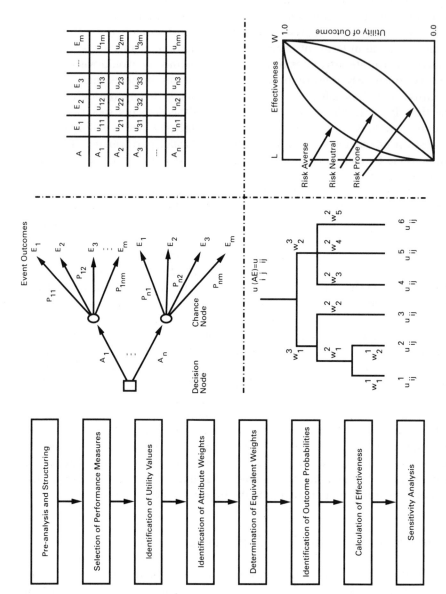

Figure 12 Steps in formal evaluation of technology cost and operational effectiveness

The initial efforts at provision of information technology–based systems concerned support through more advanced information technology–based hardware and software. Some time ago it was recognized that support could be provided not only to individuals in accomplishing such tasks as report preparation but that the ubiquitous computer could provide support for groups in answering queries of a *what if* nature with an *if then* response. This led, two or three decades ago, to the development of *support through management information systems* (MIS). These systems have become quite powerful today and are used for a variety of purposes, such as scheduling airplane flights and booking passenger seats on them, and registering university students in classes.

As management information systems began to proliferate, it soon became recognized that at least two difficulties remained. While the MIS was very capable of providing support for organizing data and information, it did not necessarily provide much support for human judgement and choice activities. Many such activities need support. They range from providing support in assessing situations, such as to better detect issues or faults, and to support diagnosis in order to enable the identification of likely causative or influencing factors. Nor did the classical MIS provide support for decision-related issues that involve selection of alternatives that have multiple and non-commensurate attributes. This capability was provided by *support through judgement and decision support systems*. These systems involved linking the database management systems (DBMS) so common in the MIS era with the model base management system (MBMS) capability made possible through advances in operations research and artificial intelligence with the visualization and interactive presentation capability made possible through dialogue generation and management systems (DGMS). The resulting systems are generally known as decision support systems (DSS) [57]. These systems provide needed support for information processing by individuals and organizations.

An additional difficulty is that it has become essentially impossible to cope with the plethora of new information technology–based support systems. The major reason for this is the lack of systems integration across the large variety of such products and services. This has led to the identification of an additional role for information technology professionals, one involving *support through information systems integration engineering*. An information systems integration engineer is responsible for overall systems management, including configuration management, to insure that diverse products and services are identified and assembled into total and integrated solutions to information systems issues of large scale and scope. There are many contemporary technological issues here. There is a need for what are often called "open systems architectures," or open systems environments, that provide for such needs as inter-operability of applications software across a variety of heterogeneous hardware and software platforms. The key idea here is the notion of *open*, or public, a notion that is intended to produce consensus-

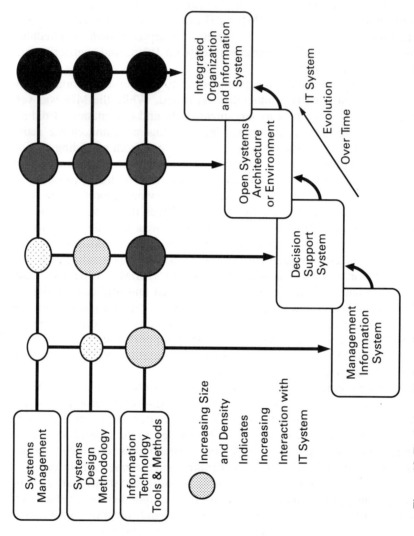

Figure 13 Evolution of information technology systems over time and development effort

based developments that will ameliorate difficulties associated with lack of standards and the presence of proprietary interfaces, services, and protocols.

When brought to fruition, these information systems integration developments, and the associated open systems environments, will enable more efficient and effective configuration managements. This will result in the development of information technology solutions in enabling existing organizations to function "better." However, there is also a need for "better" organizational designs. These needs exist in a variety of areas that range from more efficient and effective enterprise management, to more efficient and effective education of students in universities, to more efficient and effective manufacturing processes. It is in this area that contemporary *proactive* information technology developments promise the greatest pay-off. This is what we attempt to illustrate in figure 13. Through appropriate development efforts, and aided by effective approaches to operational and strategic quality assurance and management, systems integration, and standards, it should be possible to achieve the many objectives for development through systems engineering illustrated in figure 14.

It is important to consider the many impacts that various usages of contemporary information technology have on the environment. Huber [23], in an especially insightful article, identifies 13 propositions that relate to the effect of information technologies on organizations and on associated organizational situation assessment and decision-making.

The first three of the propositions deal with the effects on subunit structure and processes. Huber states that the use of information technology will lead to: a larger number and variety of people participating as information sources in the making of decisions; decreases in the number and variety of people comprising the traditional face-to-face decision unit; and less organizational time being absorbed by decision-related meetings. Six propositions deal with the organization as a whole. He indicates that the use of information technology in a given organization will lead to: a more uniform distribution, across organizational levels, of the probability that a specific organizational level will make a particular decision; a greater variation across organizations in the levels at which a particular type of decision is made; a reduction in the number of organizational levels involved in authorizing proposed organizational actions; fewer intermediate nodes within organizational information processing networks; more frequent development and use of computerized databases as components of organizational memory; and more frequent development and use of in-house expert systems as components of organizational memories.

Huber presents two propositions that deal with situation assessment. He states that the use of information technology will lead to more rapid and more accurate identification of problems and opportunities and to organizational situation assessment that is more accurate, comprehensive, timely, and available. Finally, he presents three propositions that deal with information technology effects on decision-making. It is postulated that the use of information technology will lead to higher quality decisions, to a reduction

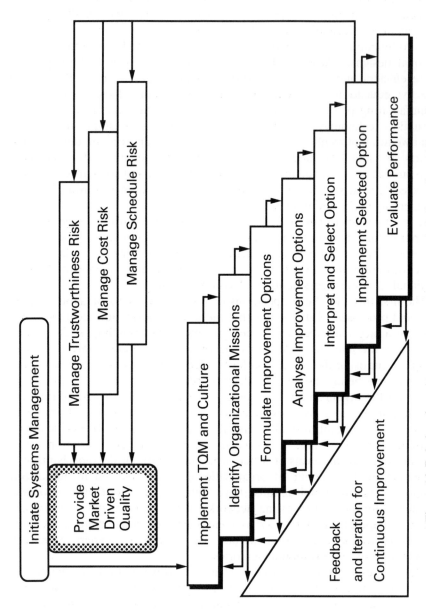

Figure 14 Relations between systems management and total quality achievement

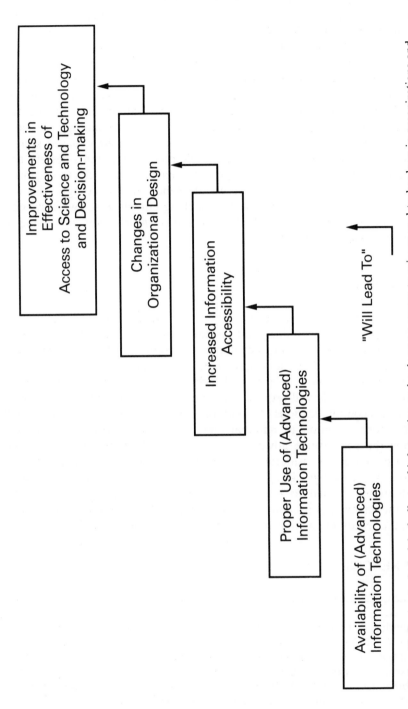

Figure 15 Conceptual model of effects of information technology on access to science and technology in organizations and nations

in the time required to authorize proposed organizational actions, and that the time required to make decisions will be reduced. On the basis of these 14 propositions, Huber identifies four elements or constructs and obtains a causal structural model as shown in figure 15. This is a fitting concluding picture for this paper, as it does indicate the considerable role that modern information technology–based systems can be expected to have in present and future organizational environments. This will involve: generating the context for information technology through integrating and aligning the strategies for the organization *and* for information technology developments; engineering the design of an information technology–based system through involvement of users in all life-cycle phases of systems design; and fielding the system in a manner that provides maximum benefits to organizations and humans, including their access to science and technology. This will require very careful attention to such critical ingredients as software systems engineering [59] and related efforts that involve information processing in systems and organizations [56].

8. Summary

This paper provides a framework for the use of the methods and methodologies of information technology and system engineering, and the application of these to enhance ultimate development and/or transfer of information technologies, especially in developing nations. A major potential use for this is in enhancing access to science and technology for societal improvement. To do this effectively will require much attention to international issues relating to telecommunications [35] and other information technologies. It will require much attention to national and international prosperity concerns affecting the competitive advantage of nations [37]. It will necessarily involve a plethora of considerations concerning the international transfer, or infusion, of technology [44]. I have discussed the many systems engineering considerations needed to beneficially address several problem areas that affect the many groups and issues involved. I have outlined, in general terms, the requirements that such a support system should fulfil.

REFERENCES

1. Allen, T.J. (1977). *Managing the Flow of Technology*. Cambridge, Mass.: MIT Press.
2. Andriole, S.J., and S.M. Halpin, eds. (1991). *Information Technology for Command and Control*. New York: IEEE Press.
3. Anthony, R.N. (1988). *The Management Control Function*. Cambridge, Mass.: Harvard Business School Press.
4. Bainbridge, L., and S.A. Ruiz-Quintanilla, eds. (1989). *Developing Skills with Information Technology*. Chichester, UK: John Wiley and Sons.

5. Beutel, R.A. (1991). *Contracting for Computer Systems Integration*. Charlottes-ville, Va.: Michie Co.
6. Cargill, C.F. (1989). *Information Technology Standardizaton: Theory, Processes and Organizations*. Bedford, Mass.: Digital Press.
7. Clemons, E.K. "Evaluation of Strategic Investments in Information Technology" (1991). *Communications of the ACM* 34 (1): 22–36.
8. Cohen, M.D., and J.G. March, eds. (1991). "Organizational Learning: Papers in Honor of and by James G. March" *Organization Science* 2 (1): 1–147.
9. Cohen, S.S., and J. Zysman (1988). "Manufacturing Innovation and American Industrial Competitiveness." *Science* 239 (4844): 1110–1115.
10. Computer Science and Telecommunications Board of the USA National Academy of Engineering (1991). *Keeping the U.S. Computer Industry Competitive: Systems Integration*. Washington, D.C.: National Academy Press.
11. Detrouzos, M.L., R.K. Lester, and R.M. Solow (1989). *Made in America: Regaining the Productive Edge*. Cambridge, Mass.: MIT Press.
12. Dutton, J.A., and L. Crowe (1988). "Setting Priorities among Scientific Intiatives." *American Scientist* 76 (November): 599–603.
13. Edosomwan, J.A. (1989). *Integrating Innovation and Technology Management* New York: John Wiley and Sons.
14. Fasser, Y., and D. Brettner (1992). *Process Improvements in the Electronics Industry*. New York: John Wiley and Sons.
15. Florida, R., and M. Kenney (1990). *The Break-through Illusion: Corporate America's Failure to Move from Innovation to Mass Production*. New York: Basic Books.
16. Frey, D.N. (1989). "R&D to the Marketplace: A New Paradigm?" *The Bridge* 19: 16–20.
17. Gallo, T.E. (1988). *Strategic Information Management Planning*. Englewood Cliffs, N.J.: Prentice Hall.
18. Goodman, P.S., and L.S. Sproull, eds. (1990). *Technology and Organizations*. San Francisco, Calif.: Jossey Bass Inc.
19. Harvard Business Review (1990). *Revolution in Real Time: Managing Information Technology in the 1990s*. Cambridge, Mass.: Harvard Business School Press.
20. Hayes, R.H., S.C. Wheelwright, and K.B. Clark (1988). *Dynamic Manufacturing: Creating the Learning Organization*. New York: Free Press.
21. Heylighen, F. (1992). "A Cognitive Systemic Reconstruction of Maslow's Theory of Self-Actualization." *Behavioral Science* 37 (1): 39–58.
22. Horowitch, M., and C.K. Prahalad (1976). "Managing Technolgical Innovation: Three Idea Modes." *Sloan Management Review* 17 (2).
23. Huber, G.P. (1990). "A Theory of the Effects of Advanced Information Technologies on Organizational Design, Intelligence, and Decision Making." *Academy of Management Review* 15 (1): 47–71.
24. Kaplan, R.S. (1989). "Management Accounting for Advanced Technological Environments." *Science* 245 (25 August 1989): 819–823.
25. Keen, P.G.W. (1991). *Shaping the Future: Business Design through Information Technology*. Cambridge, Mass.: Harvard Business School Press.
26. Lehr, L.W. (1979). "Stimulating Technological Innovation: The Role of Top Management." *Research Management* 22 (6).
27. Madnick, S.E. (1987). *The Strategic Use of Information Technology*. New York: Oxford University Press.

28. Malone, T.W., and S.A. Smith (1988). "Modeling the Performance of Organizational Structues." *Operations Research* 36 (3): 421–436.
29. March, J.G., and J.P. Olsen (1989). *Rediscovering Institutions: The Organizational Basis of Politics*. New York: Free Press.
30. National Research Council (1987). *Management of Technology: The Hidden Competitive Advantage*. Washington, D.C.: National Academy Press.
31. Nilsson, E.G., E.K. Nordhagen, and G. Oftedal (1990). "Aspects of Systems Integration." In: *Proceedings of the First International Conference on Systems Integration* (April 1990). IEEE Computer Society Press, pp. 434–443.
32. Orlikowski, W.J. (1991). "Integrated Information Environment or Matrix of Control? The Contradictory Implications of Information Technology." *Accounting, Management and Information Technologies* 1 (1): 9–42.
33. Orlikowski, W.J., and J.J. Baroudi (1991). "Studying Information Technology in Organizations: Research Approaches and Assumptions." *Information Systems Research* 2 (1): 1–28.
34. Parker, M.M., R.J. Benson, and H.E. Trainor (1988). *Information Economics: Linking Business Performance to Information Technology*. Englewood Cliffs, N.J.: Prentice Hall.
35. Pool, I.S. (1990). *Telecommunications Without Boundaries: On Telecommunications in a Global Age*. Cambridge Mass.: Harvard University Press.
36. Porter, A.L., A.T. Roper, T.W. Mason, F.A. Rossini, and J. Banks (1991). *Forecasting and Management of Technology*. New York: John Wiley and Sons.
37. Porter, M.E. (1990). *The Competitive Advantage of Nations*. New York: Free Press.
38. Raelin, J.A. (1991). *The Clash of Cultures: Managers Managing Professionals*. Cambridge, Mass.: Harvard Business School Press.
39. Rasmussen, J., K. Duncan, and J. Leplant, eds. (1987, 1991). *New Technology and Human Error*. Chichester, UK: John Wiley and Sons. Vol. 1 – 1987, Vol. 2 – 1991.
40. Repo, A.J. (1989). "The Value of Information: Approaches in Economics, Accounting, and Management Science." *Journal of the American Society for Information Science* 40 (2): 68–85.
41. Roberts, E.B., ed. (1987). *Generating Technological Innovation*. Oxford: Oxford University Press.
42. Roberts, E.B. (1988). "Managing Invention and Innovation." *Research and Technology Management* (January 1988): 11–29.
43. Roberts, E.B., and C.A. Berry (1985). "Entering New Businesses: Selecting Strategies for Success." *Sloan Management Review* 26 (3).
44. Robinson, R.D. (1988). *The International Transfer of Technology*. Cambridge, Mass.: Ballinger.
45. Rockart, J.F., and C.V. Bullen, eds. (1986). *The Rise of Managerial Computing*. Homewood, Ill.: Dow Jones-Irwin.
46. Rockart, J.F., and D.W. DeLong (1988). *Executive Support Systems: The Emergence of Top Management Computer Use*. Homewood, Ill.: Dow Jones-Irwin.
47. Rossak, W., and P.A. Ng (1991). "Some Thoughts on Systems Integration: A Conceptual Framework." *Journal of Systems Integration* 1 (1): 97–114.
48. Rossak, W., and S.M. Prasad (1991). "Integration Architectures: A Framework for System Integration Decisions." In: *Proceedings 1991 IEEE Systems, Man and Cybernetics Conference*, Charlottesville, Va.: October 1991, pp. 545–550.
49. Rouse, W.B. (1991). *Design for Success: A Human-Centered Approach to Designing Successful Products and Systems*. New York: John Wiley and Sons.

50. Rouse, W.B. (1992). *Strategies for Innovation: Creating Successful Products, Systems, and Organizations*. New York: John Wiley and Sons.
51. Roussel, P.A., K.N. Saad, and T.J. Erickson (1991). *Third Generation R&D: Managing the Link to Corporate Strategy*. Cambridge, Mass.: Harvard Business School Press.
52. Rubenstein, A.H., A.K. Chakrabarti, R.D. O'Keefe, W.E. Souder, and H.C. Young (1976). "Factors Influencing Innovation Success at the Project Level." *Research Management* 19 (3): 15–20.
53. Sage, A.P. (1983). *Economic Systems Analysis: Microeconomics for Systems Engineering, Engineering Management, and Project Selection*. North Holland: Elsevier.
54. Sage, A.P. (1987). "Knowledge Transfer: Innovative Roles for Information Engineering Education." *IEEE Transactions on Systems, Man and Cybernetics* 17 (5): 725–728.
55. Sage, A.P. (1989). "Systems Management of Emerging Technologies." *Information and Decision Technologies* 15 (4): 307–325.
56. Sage, A.P., ed. (1990). *Information Processing in Systems and Organizations*. Oxford: Pergamon Press.
57. Sage, A.P. (1991). *Decision Support Systems Engineering*. New York: John Wiley and Sons.
58. Sage, A.P. (1992). *Systems Engineering*. New York: John Wiley and Sons.
59. Sage, A.P., and J.D. Palmer (1990). *Software Systems Engineering*. New York: John Wiley and Sons.
60. Souder, W.E. (1989). "Improving Productivity Through Technology Push." *Research/Technology Management* 32 (2): 19–24.
61. Strassmann, P.A. (1985). *Information Payoff: The Transformation of Work in the Electronic Age*. New York: Free Press.
62. Strassmann, P.A. (1990). *The Business Value of Computers*. New Canaan, Conn.: Information Economics Press.
63. Thurow, L. (1992). *Head to Head: The Coming Economic Battle Among Japan, Europe, and America*. New York: William Morrow and Company.
64. Tinnitello, P.C., ed. (1989). *Handbook of Systems Management, Development and Support*. Boston, Mass.: Auerbach.
65. Walton, R.E. (1988). *Up and Running: Integrating Information Technology and the Organization*. Cambridge, Mass.: Harvard Business School Press.
66. Zuboff, S. (1988). *In the Age of the Smart Machine: The Future of Work and Power*. New York: Basic Books Inc.

A Role for the UNU/IIST: Developing Countries' Access to New Information Technologies

Dines Bjørner, Presented by Zhou Chao Chen

ABSTRACT

The paper has three parts: first I present the UNU/IIST, then I relate its work to the issues raised during the Kyoto Symposium, and finally I illustrate some additional possibilities. The three parts define complementary access approaches to information technology – axes that the UNU/IIST wishes to offer and pursue. The UNU/IIST offers state-of-the-art training and dissemination, projects and research in the software technology aspects of information technology.

1. PART 1: UNU/IIST

1.1 The Mission

1.1.1 Motivation

The United Nations University, International Institute for Software Technology (UNU/IIST) is fundamentally concerned with the software technology needs of the developing world. It is the first international institute devoted to this subject. The Institute has been set up by the UNU at a time when, although the growth of computer usage in the developing nations is quite high, most of them urgently need the establishment or expansion of both basic and advanced training, as well as development and research facilities. Thus, there is at once a lack of available software professionals for industrial growth and a chronic shortage of educators and trainers of professionals of the future.

Moreover, most developing countries have few software companies and little experience in industrial software development. Local software development is essential in the developing world not only to establish and strengthen local industry but also to provide software in local languages and with cultural features adapted to the specific needs and conditions of each country.

Although the knowledge and competence gaps between the industrialized and developing nations may be substantial, the UNU/IIST can still help specialists reach state-of-the-art levels in software technology. The discipline is in its infancy.

1.1.2 Intellectual vs. Material Technology Transfer

The UNU conducted, during the period 1987–1989, two studies into the modalities and the need for an institute like the UNU/IIST.

The first, for which I prepared a personal report [1], basically established the subject areas in which the UNU/IIST, if need were found, could engage. The second, the feasibility study [7], successfully argued the need.

The UNU/IIST is deemed a *stable body* that can carry out the expectations raised in the feasibility study and that can *persist* in doing so. Since the identified need must be alleviated through international cooperation, and since a significant academic component is necessary, the UNU is strongly believed to be the appropriate institution.

When in the following I refer to *technology transfer*, I mean primarily the transfer of *intellectual technology*.

The UNU/IIST will exclusively focus on the transfer of knowledge, prototyped products, and awareness of software tools to support intellectual techniques and software.

The UNU/IIST is not concerned with the important issue of transfer of material technology. The UNU/IIST hopes to be able, with various funding and loan agencies of the UN System (UNDP, World Bank, UNIDO, ADB, IDB, etc.) as the primary agents, to assist, in a secondary role, in furthering material technology transfer where needed.

1.1.3 Foundation

The UNU/IIST is the most recently established Research and Training Centre (RTC) of the UNU. It formally came into being on 12 March 1991, with the signing in Macau of agreements between the UNU, the Governor of Macau, and the Governments of Portugal and the People's Republic of China. It is basically financed by an initial fund of US$20 million contributed to the UNU/IIST Endowment Fund, and it is pledged that this capital fund will be increased to US$30 million through contributions from other sources.

1.2 Aims and Objectives

The UNU/IIST aims at assisting the developing world in meeting needs and strengthening capabilities in five activity areas:
(1) usage of as sophisticated a variety of advanced software as reasonable,
(2) software technology management,
(3) development of its own and exportable software,
(4) university education curriculum development, and
(5) participation in international research.
The UNU/IIST will help bridge gaps between the industrial and developing worlds, theory and practice, university and industry, and consumer and producer.

The UNU/IIST will be a *cradle* for educating young university computing scientists and industry programmers and software engineers above post-graduate level. The UNU/IIST will be a *showroom* for demonstrating paradigmatic approaches to requirements engineering, programming, software engineering, and software project plus product management techniques and tools. The UNU/IIST will create a *bridge* for enhancing cooperation between the industrial and the developing world in the areas of software. The UNU/IIST will be a *channel* for bringing to international attention the achievements of developing countries.

The above five points define the main technology and application-independent, that is the methodology-dependent, components by means of which the UNU/IIST believes it can help to further improve the developing countries' access to information technology.

In parts 2 and 3 of this paper, I shall outline respectively "soft" applications and "hard" technology-specific components, which, if offered by the UNU/IIST, are believed to further improve access.

1.3 The UNU/IIST Constituency

The UNU/IIST, roughly covering the five areas mentioned above, targets the following audiences: public administrations and small businesses and industries in the least developed countries, budding software industry management, industry software developers, university lecturers, and research institute scientists.
– Software users: National and local government planners and administrators are faced with increased demands for computer-supported gathering and evaluation of socio-economic data. Local, national, and regional professionals in, for example, public health, agriculture, fisheries, and transport are expected to apply computers in forecasting, planning, and decision-making. Small business and industry management must increasingly apply computers in day-to-day operations: budgeting, accounting, forecasting, inventory control, order-processing, invoicing, product planning, monitoring and control, marketing, and sales. The UNU/IIST

intends to offer such users 2–4 week training courses in software installation, operation, data preparation, and output interpretation. These training courses will typically be aimed at the least developed countries – and are expected to be externally funded.

- Technology management: Large-scale software users, and software house, utility, and information technology management need to apply the latest techniques related to technology management. Users need to know techniques for tendering and procuring software, and for managing computing facilities. Developers need to know techniques for bidding for software development contracts, and for managing projects and products. The UNU/IIST intends to offer such managers 2–4 week awareness courses in applicable techniques.
- Software developers: The software industry is rapidly accelerating the sophistication of its production techniques and tools. To their providers of software, the most advanced of the industrial nations demand software qualities only attainable through the use of rather advanced development techniques. The UNU/IIST will offer to leading-edge developers, as well as to university lecturers, three-month education courses in software development techniques, and will offer them active participation in 9–15 month feasibility, demonstrator, and prototype-product-development projects.
- University lecturers: With the advent of modern techniques for software development, existing computation science and engineering courses are rapidly becoming obsolete. Universities of the developing world also need to revise their curricula. Through the three-month software development education courses mentioned above, and through special 2–3 week events, the UNU/IIST will offer the contents of a modern computing systems curriculum and conduct in-depth workshops analysing such curricula.
- Computing scientists: The research communities in the industrial countries are becoming increasingly internationalized. They generally have easy access to electronic networks, participate in internationally sponsored joint research projects, and have reasonable funds for journals, books, and equipment, including software tools. These researchers travel regularly, often three or four times annually, enabling participation in meetings of various kinds. The UNU/IIST will help towards globalizing the research of developing world scientists through joint research, through the Organic UNU/IIST Network, by conducting international workshops and seminars, and thus by communicating developing world research worldwide.

1.4 Programme Activities

The UNU/IIST Progamme Activities centre around the organization of projects, training, consultancies, research, dissemination, and events.

1.4.1 Projects

The UNU/IIST intends to engage in feasibility, demonstrator, and technology transfer projects. All projects develop software using advanced techniques and are expected to last from 9 to 15 months. The UNU/IIST staff, visiting experts, and project fellows will conduct these Macau-based projects, which are expected to be externally funded.

- Feasibility projects *formally*, but experimentally develop small but difficult subsets of innovative software applications – and may lead to follow-on demonstrator projects. These projects may be prompted by, or lead to, research done at the UNU/IIST or within the UNU/IIST Organic Network.
- Demonstrator projects *rigorously* apply scalable, state-of-the-art techniques to applications that can serve as the basis for software development education courses – and may lead to follow-on technology transfer projects.
- Technology transfer projects *systematically* develop core, prototype parts of planned products and shall lead to detailed plans and technical directions that are then transferred to a developing-world company for concluding full-scale development.

The software developed may apply to *operational* systems such as railway monitoring and control, river monitoring and flood control, or cargo and customs clearance, via *analytic* systems such as traffic or crop fertilization planning, disease monitoring, or disaster management, to *conceptual* systems such as expert or knowledge-based systems for decision support, or university administration and management information. The UNU/IIST expects to assist in the creation of a local software industry.

1.4.2 Training: Courses and Seminars

The UNU/IIST offers training through fellows participating in carefully supervised projects and through courses and seminars.

The UNU/IIST will conduct three kinds of courses for *training participants* from the developing world.

- Two to four week *training* courses, usually located away from the Institute, will instruct software users and computer centre operators to install and operate large-scale software systems, and to prepare data for and evaluate results of their computations.
- Two to four week *awareness* courses, usually offered away from the Institute, will expose management to the intricacies of software technology management – in how to procure software, put out tender or bid for the development of software, and manage software development projects, software products, and computing facilities. Common aspects include quality assurance, risk analysis, resource estimation, planning, allocation and scheduling, process modelling, and simulation.
- Three-month *education* courses, based in Macau, will teach software de-

velopers to develop application-specific requirements, abstract and concrete programs, and the engineering of large-scale software systems, fit for use and purpose, correct, fault tolerant and safety critical, efficient, maintainable, and portable. Subset education courses may be given, normally away from Macau, as two-week seminars.

1.4.3 Consultancy

The UN system as well as governmental and non-governmental agencies today typically employ industrial world consultants to obtain in-depth studies of policy, usage, industry and technical issues of software technology. The UNU/IIST, with its experienced staff of software technologists, including visiting experts, offers to perform such studies – thereby bringing them "closer to home," while at the same time training a new generation of developing-world graduates to themselves perform such work in the future.

The UN system should look to the UNU/IIST, with its objective background in European, US, and Japanese practices, as its natural first, and independent consultants when it conducts its inquiries.

The UNU/IIST will typically link up with other UNU RTCs and Programmes when consulting on cross-disciplinary issues.

1.4.4 Research

Leading information technology countries and regions of the industrial world have long been and are still engaged in coordinated multi-billion dollar information technology research and pre-competitive projects involving sizable software technology components. Many years of research have brought these countries up to a very sophisticated level of software technology techniques that increasingly are being applied industrially under controlled supervision. The UNU/IIST will bring results of this research to scientists of the developing world through frequent research seminars, usually offered away from Macau. The UNU/IIST will itself conduct research, this research being conducted by UNU/IIST staff, visiting experts, and research fellows. The UNU/IIST will also co-sponsor joint research collaboration elsewhere.

Young *research fellows* may thus consider their time at the UNU/IIST as an internship part of their Ph.D. studies.

1.4.5 Dissemination

To reach growing software technology segments of the developing world, the UNU/IIST plans to regularly issue a *Software Technology Monitor* free of cost to managers, programmers, and software engineers of the developing world. The monitor will track international public domain as well as commercial software products, events, and developments – in both the industrial and the developing world. Eventually it is planned to establish an electronically accessible *Software Catalogue* that gives technical details on prerequisites, functions and operating characteristics of software products of interest to the developing world. A low-cost UNU/IIST *NewsLetter* additionally will provide survey, technical, and research articles on software technology issues and otherwise inform of UNU/IIST and developing-world activities. Finally the UNU/IIST plans to publish low-cost case-study books that record actual development of advanced, application-specific software.

1.4.6 Events

In addition to projects, courses and seminars, consultancy, research, and physical and electronic document dissemination, the UNU/IIST will arrange or co-sponsor task forces, panels, workshops, and symposia. Task forces and workshops may explore such issues as university curricula, professional accreditation, and software product and process standards. Task forces will start out by identifying issues worthy of study at panel events, will work in decentralized fashion over extended periods, offer workshops at regular intervals, and will present their findings at final panel events. These events will involve middle to low-level management staff from the developing world, together with the UNU/IIST staff and visiting experts. Additionally, scientific and technical workshops, aimed at leading-edge scientists and developers from the developing world, will explore new research directions and software development techniques. *IMaS²T* – The International Macau Symposium on Software Technology – is to be held regularly. It will typically feature two days of tutorials, three days of keynote lectures by internationally renowned experts, by invited and contributed-paper presentations, poster sessions and tool exhibits, and demonstrations – all within the area of the software technology profession as outlined in 1.6.

1.5 Fellowship Programme

The main purpose of the UNU/IIST fellowship programme is intellectual and technological redistribution. The UNU/IIST will focus on the best and the brightest, seeking out new talent in young people.

The UNU/IIST will seek funding in order to offer three kinds of fellowships: 9–15 month *projects*, from 2 to 4 week to 3-month *courses*, and 6–9 month *research fellowships*. Fellows will come from the developing world to work with the UNU/IIST scientific and technical staff and visiting experts. Fellows shall return to their place of origin and be committed to the transfer of their experience to home-country affairs: software development, university teaching, and research. The UNU/IIST will monitor the effectiveness of this transfer and will assign priority in its cooperation accordingly.

Young students will be invited to spend part of their studies at the UNU/IIST in the form of Ph.D. internships, for example in the form of repeated three-month visits. The UNU/IIST is itself not a degree-awarding institute, but will endeavour to profile its research so as to become an internationally recognized centre. As such, the UNU/IIST will also be open to developing world post-doctoral research fellows and visiting experts.

1.6 The Profession

In this section I shall examine one technical approach to software development. Another approach – together with a comparative analysis – will be sketched in the interlude section, "Two Worlds of Computing," after section 1.10.

To develop software involves several disciplines: requirements development, programming, software engineering, management, and the support of resident computation scientists. Although the UNU/IIST is also engaged in other, non-developmental issues, the above disciplines will be a major focus.

The UNU/IIST is dedicated to the professionalization of software development.

A professional is a developer, a resident computation scientist, or a manager who is scientifically and technically interested in these fields; who knows whether a product can be developed in a trustworthy manner; who will develop software in such a manner, performing as a natural part of development calculations that reveal properties of requirements models, abstract functional and behavioural specifications, and of designs long before costly implementation has taken place; who is aware of the forefront of the field and is able to stay abreast; and is a person who is interested in further developing the science and techniques of these fields.

As is well established in other branches of engineering, the use of mathematical modelling and reasoning about contemplated designs is at the centre of any of the development activities that the UNU/IIST will propagate.

1.6.1 Requirements Development

In order to ensure that the customer, the user, gets software adapted to its intended purpose, a careful study must be made of the domain to which the software is to be applied. Thus, requirements development may involve the mathematical modelling of concepts and facilities of the problem domain, and of functional and behavioural characteristics; together these form the conceptual model. Additionally, the physical model involves modelling interfaces between components implementing the conceptual model, as well as safety criticality, dependability, performance, and human-computer interface facets. Finally, simulation models may allow experimental model executions before final requirements are captured – and in preparation for the programming and engineering of the software.

Requirements development may, depending on application characteristics, also, or instead, involve establishing knowledge-based models that allow system reasoning during execution.

1.6.2 Programming

In order to secure efficiency and correctness of implementations, the latter with respect to requirements, programming now traditionally employs the precise, logical techniques that allow objective validation and the controlled, carefully motivated design decision introduction of efficiency. Typically programming thus includes abstract functional and behavioural specifications, their stepwise design transformation into efficient implementation, followed eventually by the coding in some executable programming language. Proofs of properties of individual stages as well as correctness of stage transformations intersperse this somewhat simplified process picture.

The UNU/IIST will, in its projects and education courses, propagate programming techniques of the kind represented by the *B, Estelle, Esterel, HP-SL, Larch, Lotos, RAISE, VDM* and *Z* methods, will feature proof assistant and theorem proving systems of the kind represented by *B, Boyer-Moore, Isabelle, Gypsy, mural* and *RAISE* tools, and will otherwise rely on such programming languages as *Modula-3, Oberon II, occam, Prolog* and *Standard ML*.

1.6.3 Software Engineering

In order to insure conformable, maintainable, and portable software systems, software engineering employs such subsidiary techniques as ongoing – in the field – conformance testing; version control and configuration management; change request identification; monitoring and control; test-case generation and validation; requirements and design decision tracking; and hypermedia-supported documentation.

The UNU/IIST will feature, in its projects and its education courses, *Computer Aided Software Development Environments* that will assist the software engineer in carrying out the above tasks.

1.6.4 Computation Science

Software developments usually entail new application domains, new functions and behaviours, and new software designs – for which state-of-the-art development techniques only partially apply. A resident computation scientist is hence a person who can check the case-by-case validity of the techniques applied and, in cases, provide justification for the use of novel techniques. The UNU/IIST will exemplify, in its projects and its education courses, instances of such problems and examples of their remedy.

1.6.5 Management

In order to secure timely, economic, and manageable development projects, managers must apply resource planning: estimation, risk analysis, allocation and scheduling, resource monitoring and control techniques, and overall quality assurance: planning, monitoring, and control. The UNU/IIST will, in its projects and courses, propagate state-of-the-art management techniques.

1.7 Organic Network

An Organic Network will be linked to the UNU/IIST. It will be an expanding circle of affiliated, cooperating software technology development centres, university computation science and engineering departments and research institutes. The network will focus on the developing world, but industrial world centres are expected to help secure the objectives of the network.

Aims of the network are to strengthen identity, stability, quality, and productivity in developing-world centres, with respect to development projects, university education, and research in the areas covered by the UNU/IIST.

Emphasis within the cooperation areas will be put on formulation of professional accreditation criteria, including university curriculum development, affinity of software to its intended use – thereby helping to close the gap between consumer and producer – and correctness of software with respect to requirements definitions – thereby aiding the developing world in competitively producing highest quality software.

Cooperation means are based on the UNU/IIST fellowship programme, rotation of scholars, joint projects, joint courses and seminars, joint events, joint publications, and extensive, timely circulation of the planned *Software Technology Monitor*, the UNU/IIST *NewsLetter*, and other, low-cost or free UNU/IIST publications.

The UNU/IIST will assist affiliated centres in becoming connected to an electronic mail service. This electronic aspect of the Organic Network shall eventually include ability for general file transfer, log-in to remote computers, and general access to international academic and technological (commercial) e-mail networks. This form of on-line connection will form the backbone of the UNU/IIST Organic Network. The UNU/IIST will endeavour to secure uniform electronic network access so as to avoid a degradation into "first" and "second"-class affiliates.

1.7.1 Network Projects

In 1.4.1, I briefly outlined project classes and project topics. In 1.6, I briefly outlined components of professional project development methods.

Organic Network institutions are expected to be involved in projects in two ways: by 9–15-month secondment of project fellows from affiliated software development centres to Macau, and by some of these centres eventually technology transferring prototyped products for final completion. Only by actual hands-on experience under careful supervision can newest methodologies and technologies be efficiently transferred into industry.

1.7.2 Professional Accreditation

The UNU/IIST assumes a heretofore unprecedented high level of professionalism to be achieved in its projects, to be conveyed in its courses, to be applied in its consultancy, and to otherwise be further developed in its research. In cooperation, typically with professional (computer science and information technology) societies – also affiliated with the Organic Network – the UNU/IIST wishes to develop criteria and standards for industry-oriented professional accreditation.

1.7.3 Curriculum Development

The UNU/IIST is introducing the latest, but proven "didactics" and techniques in its course programme and projects. In cooperation with Organic Network–affiliated national and regional universities, the UNU/IIST wishes to help these upgrade local computation science and engineering course curricula.

1.7.4 Network Research

The UNU/IIST software development model represents a rich, multifaceted picture – for which not all aspects are yet in a state where matured courses can be taught or techniques applied using conventional management principles. The UNU/IIST projects will illuminate specific methodological and technological facets in need of further research. The UNU/IIST intends to cooperate with Organic Network–affiliated research centres in pursuing this research – thereby bringing many of the positive aspects encountered in advanced university-industry projects (within for example the European ESPRIT programme) to bear also on universities and industries of the developing world.

The UNU/IIST offers advice to information technology agencies of regions of the developing world with respect to possible multilateral research and pre-competitive development programmes.

1.8 The UNU/IIST: Novel Facets

The UNU/IIST will approach its task by novel compositions of actual, live feasibility, demonstrator, and prototype product technology transfer projects with on-the-job training in the form of advanced education courses. The UNU/IIST is to be geared explicitly to narrowing gaps between theory and practice, and thus between universities and industries. The UNU/IIST will thus differ substantially from most conventional university, educational, and research systems, including those in the industrial world.

The UNU/IIST personnel – administrative and technical/scientific, and staff, visiting experts, and fellows – will likewise present novel aspects of cooperation. Over the build-up, initial years of the UNU/IIST, staff and visiting experts will gradually shift from coming from the industrial to representing the developing world.

1.9 Funding and Donation

The UNU/IIST depends, in carrying out a majority of its programme activities (notably projects and courses), on funding primarily from the UN system: the UNDP, Unesco, UNIDO, World Bank (WB), and regional banks such as the Asian, African, and Inter-American Development Banks. All such funding requests will be coordinated by the UNU Centre in Tokyo, Japan.

The UNU/IIST expects external funding for the production and low-cost, or even free, distribution of UNU/IIST publications, including the *Software Technology Monitor* and the UNU/IIST *NewsLetter*.

The UNU/IIST shall additionally seek operational donations thereby making endowment income available for fellowships. Donations will thus be sought to cover such day-to-day base facets as library, soft and hard computing, and networking. Finally the UNU/IIST invites private companies to establish Distinguished UNU/IIST Research Chairs in their names.

1.10 UNU/IIST Interfaces

The diagram attempts to illustrate operational interfaces between components of the UNU/IIST and the Organic Network.

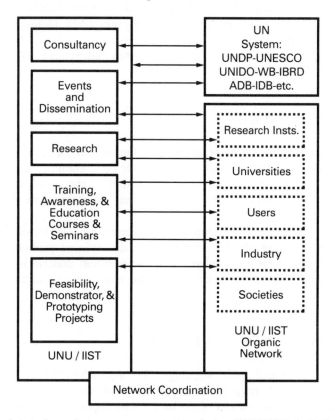

Operational interfaces between components of the UNU/IIST and the Organic Network

Two Worlds of Computing

In preparation for parts 2 and 3 of this paper, I discuss issues relating to so-called "intelligent access" – mentioned in the Announcement of this Kyoto Symposium.

Part 2 assumes some familiarity with the field of computing science. It is introduced in order to make reasonably qualified statements on one of the roles that the UNU/IIST might play. This role relates to those topics of the Kyoto Symposium that are based on advances in Computer-Human Interfaces (CHI), Cognition, and Natural Language Interfaces (Computational Linguistics). (These terms were found in the Symposium Announcement.)

In section 1.6, I have outlined one approach to software development. For technical problems and facets, see [5, 6, 2, 3]. That approach is currently the predominant one, but perhaps not the best. The approach outlined in section 1.6 can, with a few terminological changes, be brought to cover also the next one.

This at times and in certain circles somewhat controversial approach is represented by so-called Knowledge-based Systems (KBS) Engineering (KE). Derivative terms are: Artificial Intelligence (AI), Heuristic or Expert System (ES) approaches.

Novel applications, as we shall see, are often "advertised" as being of the latter kind (KBS, AI, ES).

Algorithmics and KBS

The "classical" approach more or less assumed in section 1.6 is based on the Fortran, Algol 60, Pascal, and C tradition of programming. In this style of programming, one transforms some abstraction of the problem domain, in several stages of development, into compiled, execution time, and storage efficient machine code.

The KBS/AI/ES approach, briefly characterized here, is based on the Lisp and Prolog tradition of programming. In this style of programming, one attempts to represent knowledge about the problem domain in such a way that it is this representation that forms the basis for execution time interpretation.

Algorithmic vs. KBS Development

We call the "classical" programming style "algorithmic" (and by ALG we mean Algorithmic Software Development), while the other style is referred to as "knowledge engineered" (and by KBS we mean Knowledge-based Systems Development). The choice of terms is not quite pragmatic – as will be argued below.

Let us contrast the two styles of programming and programs in order better to identify differences and complementarities.

A.a. In ALG program computations, all reasoning has been deeply "compiled" (away) into efficient, but not very general, not very adaptable, code.

K.a. In KBS programs, computations take the form of reasoning based on generally adaptable knowledge representations.

A.b. In ALG specification, we use logic at the specification level,

K.b. versus as an object language in KBS.

A.c. In ALG development, we refine by steps,

K.c. versus remaining faithful to the problem domain in KBS.

A.d. In ALG, we aim at efficiency through specialization.

K.d. versus through general methods in KBS.

A.e. In ALG, our units of specification are functions *in extenso*,

K.e. versus rules or facts in KBS.

A.f. The composition principle in ALG is functional composition,

K.f. versus catenation of assertions and object classification with multiple inheritance – however including non-monotonicity – in KBS.

A.g. In ALG, we speak of computation,

K.g. while in KBS we speak of deduction, induction, and abduction.

A.h. In ALG, we operate with partial evaluation,

K.h. versus partial deduction in KBS.

A.i. We speak of compilation in ALG,

K.i. versus intelligent backtracking in KBS.

For a more thorough analysis, see [4].

2. Part 2: Advanced Applications

By Information Technology (IT) we mean the combination, first of software with computing equipment: computers and input/output peripherals (visual display units, sensors, actuators, etc.), and then this software/hardware computing system with human operators and users (as in administrative and end-user systems) and/or electro-mechanical gear (as in computer-integrated manufacturing, railway train traffic regulation, etc.).

By Advanced Applications we mean computing systems that apply state-of-the-art information technology to novel end-user problems. In this part, I will review state-of-the-art applications, while part 3 reviews state-of-the-art technology.

2.1 Background

The UNU/IIST seeks to be a *bridge* between the industrial and the developing world for new applications.

Several presentations at the Kyoto Symposium have outlined specific application technological means such as: databases, data banks, communications networks, electronic libraries, information retrieval, multi-media, cognitive (expert) systems, natural language processing, etc.

It can be foreseen that the UNU/IIST will attempt to organize advanced courses on: Database and Information Retrieval Systems; Knowledge- and

Expert-based Systems and Data Engineering; Visualization of Data, Hyper-media, Data Communication Networks and ISDBN, etc.

2.2 Specific Application Disciplines

Each of the following subsections addresses one of the topics covered by our colleague speakers. From these you may get the erroneous impression that the UNU/IIST will assist in transfer of material technology. The UNU/IIST will not be so occupied. The UNU/IIST will not advise the developing world on which hard, material technologies to adopt. Instead, the UNU/IIST will consult on the software technological consequences of most such hard tech-nologies, will conduct technology management awareness courses, will train in their software components, will help develop these, and will educate in their development.

2.2.1 Database and Knowledge Systems

We speak here of a spectrum: from conventional "old-fashioned" databases – in which data ("knowledge") are represented in a form that does not readi-ly permit the kind of reasoning usually experienced in deductive, inductive, and "abductive" systems – to knowledge bases where such reasoning is readily provided for.

In creating a database, one often imperceptibly transforms the knowledge gained, when analysing the application, into some efficient physical storage representation – thereby "compiling" away those knowledge facets that might be useful in heretofore unforeseen contexts.

In creating a knowledge base, such generally adaptable knowledge is pre-served. Execution time is longer, but the time it takes for humans to formu-late a new application and get it "running" is generally very short.

In either case (of information-base creation), the requirements developer spends a very long time analysing the domain of interest. In the former case, the knowledge gained is specialized with reference to the specific, usually narrow, foreseen application – and hence most is lost with reference to un-foreseen, future, related applications.

Take an example: one can write a software system for the regulation (monitoring and control) of a railway system in either of three ways.

In the first, using the ALG approach, all knowledge about the specifics of railways is hidden in the code of the programs, in their compiled con-stants, and in the structure of the code itself, in the specific composition of statements.

In the second case, one represents the railway system implementation as a database: all the tracks, the trains, the schedule, etc., are data structures, and computation proceeds through interpretation. The railway system spe-cific interpreter is, however, general in that the same interpreter, written once, can be used to handle any country's railway system. Execution is slow-er than in the former case, but one can easily introduce changes to the rail-way network, trains, schedule, etc.

The compiled, first version represents a so-called partially evaluated version of the second. In either case, the interpreter does not reason about train regulation (based on logic) but, typically, arithmetically computes based on classical mathematical laws of kinematics and operations research.

Instead of either of these two approaches, one could represent these laws in logical form – in addition to, but in the same form as, the representation of the railway system components. "Computation" now proceeds by means of a general meta-interpreter, also called an inference machine, the same for all kinds of systems and not only railway systems.

This "tri-partite" view is novel, not scientifically, but in being accepted in everyday industrial practice.

UNU/IIST wishes to propagate awareness of possibilities and of techniques relevant for all three cases. And the UNU/IIST will attempt to arrive at some kind of "didactics" and "pragmatics"-based classification criterion that advises the designer on which of the three design approaches to apply for given applications.

2.2.2 Communication Networks and Distributed Systems

Techniques for implementing secure and fault-tolerant communication networks and distributed systems include cryptographics and verified protocol designs. It seems inescapable to have any computing system being developed today involve networking and distribution.

Typical networks are either Local Area Networks (LANs) or Wide Area Networks (WANs), or are heterogeneous combinations thereof. LANs are typically based on fibre optics, whereas WANs typically involve radio telescope communication via satellites.

There is a tendency to vastly over-sophisticate, especially the WAN networks. Too many optional features and too general communication capabilities, for example, between any two ground stations involving only one intermediate satellite call usually take such systems economically out of the reach of most nations.

The National Informatics Centre (NIC), part of the Indian Government's Department of Electronics, has developed a low-cost WAN called NIC-NET. NIC-NET claims to connect about 500 computer centres all over India. There is a central, "expensive" computer centre, in New Delhi, and each of some 500 ground stations communicate via one satellite (India-Sat) directly with this centre. If the communication is intended for some other ground station, then the centre passes it on. NIC-NET then combines this network communication with seemingly reasonably homogeneous, conventional databases at each ground site. Together the network communication plus the databases form a typical distributed system.

It seems that NIC-NET offers a very interesting solution to a global problem, one that is usually frowned upon in the industrial world for being "too simple" (industrialists would claim), but one that might appeal to countries and regions of the developing world by being low cost.

The tendency to over-design networked, distributed systems is a typical result of engineering – of electronic engineers putting too much emphasis on getting maximum functionality and flexibility out of the hardware. The result is oftentimes that the software part of these systems is hard to get right and that the end-user is offered too many options.

The UNU/IIST wishes to propagate state-of-the-art techniques both for the design of sophisticated, not necessarily low-cost, as well as of simple, low-cost, secure, and fault-tolerant (distributed) network systems. The UNU/IIST will emphasize that systems that reflect user requirements, as it seems the NIC-NET does, are provably correct, secure, and fault tolerant, and that such systems offer flexibility through implementing user concepts, not by implementing computing systems concepts that the users must then combine, often in tricky ways, to achieve user functions!

We finally remind the reader of the desire of the UNU/IIST to provide for an electronic mail service network connecting (at least) all the UNU/IIST Organic Network affiliates (see last paragraph of the introduction to section 1.7). We have such "e-mail" nets to greatly enhance international cooperation – and I personally believe it extremely important that our colleagues, in academia and in industry in the developing world, be integral parts of such "gateway interconnected" networks.

2.2.3 Electronic Libraries and ISDN

Here we are talking about a special application area within information base systems – and hence the question arises as to whether such systems should be databased or knowledge oriented.

ISDN stands for "integrated services digital network." With the ISDN the possibility exists of freely mixing structured (tables) and unstructured (text) data, structured (drawings, diagrams) and unstructured (photos), still and live (video) pictures, and voice in any one communication.

Whereas the "narrower" issue of electronic databases may primarily be application driven, ISDN is a classical case of a technology- and monopoly-driven development. There still seems to be some international uncertainty with respect to the introduction and spread of ISDN.

The focus that the UNU/IIST could emphasize is that of software implications of the ISDN with respect to end-user applications. Once the ISDN becomes popular there will be a great need for a richly faceted software industry that can provide a multitude of applications without which the ISDN would not be economically feasible.

Again I remind the reader that the UNU/IIST eventually would like to establish a form of electronically accessible database in connection with its *Software Catalogue* (see subsection 1.4.5.). Establishing such a facility could eventually become a "carrier" for a UNU/IIST engagement in the ISDN.

2.2.4 Higher Education Management Information Systems

There seems to be a demand, voiced through Unesco, that the UN system help, specifically African universities, in jointly developing and installing

Management Information Systems (MISs) for Higher Education (colleges and universities).

The application is a typical (LAN, and possibly WAN-based) distributed information planning, control, and base system. Typically such systems can principally be databased, rather than knowledge based, with no loss in generality.

The UNU/IIST is currently in the process of formulating for Unesco a proposal for the requirements development and software development of such a system. The UNU/IIST will propose that it coordinate a feasibility study, to be followed by a demonstrator project, and hopefully a prototype technology transfer product development project.

A university MIS usually enables the following functions: (i) budgeting (planning and approval) and accounting; (ii) personnel; (iii) student registration; (iv) course/class planning, scheduling, and registration; (iv) examination planning, monitoring, and control (including student and grade registration, external reviewer liaison, etc.); (v) research project planning, monitoring, and control; (vi) central and distributed library database; (vii) services management (purchasing, invoicing, facilities management, etc.); and so forth. A computerized university management system thus requires software that helps in planning, regulatory control, and information dissemination – hence it may contain both classically, i.e. formally developed, as well as conceptually developed components – all integrated.

The ambitious aim is to create, in one or more developing countries, on each of two or three continents (Africa, Asia, Latin America), specialized software houses whose expertise is exactly in such college and university MISs. The long-range hope is that such software houses, by being specialists in a very application-specific domain, can become competitive in the global market.

2.2.5 Education Technology

Much is being said about computing system learning and instruction technology: primary and secondary schools, high school education, trade school training, profession-oriented instruction, etc.

Here we are dealing, in general, with systems that require computing components that reflect national language translation and cognition and hence will typically be knowledge based. Usually it is believed that the computer-human interfaces should additionally include multimedia technology.

It is not clear to me – I am not an "expert" in the field – whether the requirements issues are well understood:
– There are up to four "players" in any such educational technology system: (i) there is the computer system with all its "gadgets"; (ii) there is the student; (iii) there is the teacher; and (iv) there is the subject – the topic being taught or learned.

The sequence in which I listed these four "actors" is usually reflected in many proposals and actual products. That is: often the multimedia components of the computer hardware "swamp" the intellectual content of the subject field. It also seems that we do not always expose, let alone exploit, all the possible interfaces between the four "players" consistently.

- In this connection, there are different emphases to be made: either the teacher is supported in the education process *vis-à-vis* the students, or the students are supported in their learning process. Thus there are times in an expert system like educational technology when it is unclear who is being the expert!

- When an educational technology system is offered it naturally appears that definitive means of "didactic" and "pedagogic" nature are being provided. But is that the case? Do there exist generally accepted criteria for good "didactics" and good "pedagogics" at the operational level required by educational technology?

Thus the role of cognition seems far from settled.

- The few systems that we have seen demonstrated have usually been developed by one person. Most often, this results in less than professionally engineered and programmed systems, which in addition emphasize only one of the four cornerstones of the educational technology–based teaching/learning process. By less than professionally engineered systems I mean for example: systems featuring disturbing, ad hoc interfaces by the users and technology. By less than professionally programmed systems I mean for example: obvious, generally adaptable knowledge-based techniques are usually omitted in favour of standard, "hard-wired," non-flexible coding tricks.

The sum total of the above suggests that this exciting and undoubtedly much-needed educational technology has not been launched as the result of commercially sound or publicly high priority endeavours. Industries and businesses are not embarking on the production and marketing of educational technology. If it exists, yes; then such companies will willingly offer it via their marketing channels, as something that will enhance their other, commercially viable products. Ministries of education in several countries are likewise "holding their breath" – regarding, it seems, educational technology as an "add-on luxury." Contributing to the attitude of the commercial field is that of reduced government funds for education, and contributing to the attitude of education ministries is that of widespread resistance among teachers to the new technology. It seems, however, that, for example, efforts in Bulgaria and Russia are exceptions that offer promising possibilities.

The UNU/IIST is not interested in becoming involved in controversial issues, but the UNU/IIST is potentially interested in establishing, possibly as a Unesco-sponsored project, a series of increasingly ambitious feasibility, demonstrator, and technology transfer product development software projects, combined with training courses about existing technology.

It is most likely an over-ambitious aim that such studies can bring together, on a global scale, the best expertise in order that some "standardizable" consensus can be obtained, and such that this pedagogics, didactics, and cognition consensus could form the basis for a distributed software "village" industry in the developing world, where each "cottage" specializes in some field and subject. To wit: secondary school mathematics, university biochemistry, trade school auto mechanics, etc.

The UNU/IIST is interested in offering its good offices in an international, cooperative effort that critically examines the problem of educational technology and proposes portable structures for implementation.

2.2.6 Multimedia Technology

It has been implied that multimedia technology may divert the developer and the user from the "real thing"; namely, that of reflecting, respectively utilizing, by means of computer software, deep knowledge about the problem domain.

The UNU/IIST will attempt not to forget issues of multimedia technology and will be ready to undertake projects and train in multimedia issues. Should the UNU/IIST be so involved, then it is likely that it will emphasize: (i) how to choose such technology based on proper requirements studies, (ii) how to let its functions be determined by considered use of cognition, (iii) how such technology influences technical solutions, and (iv) techniques for their programming.

It seems that many small software houses could potentially base their project services or their marketable products on specialization in the use of specific media in specific application areas. It seems that one does not need vast armies of programmers in order to develop appropriate software.

2.3 Material versus Intellectual IT Worlds

It is normally the material quantities, the physical components (including the "smart, gimmicky, colourful" user-interface equipment), the high-speed execution, the large-volume storage, and the low price that enable ergonomically and economically feasible access to IT.

And often that is naïvely all that IT journalists see and report on!

The real advances in our improved understanding of problems and in our algorithmic or deductive ability to solve this problem using computers are harder to communicate and digest.

The UNU/IIST will try to present a balanced picture.

3. Part 3: Advanced Technologies

In the final part of the talk I express the desire of the UNU/IIST to become a *showroom* for demonstrating advanced computer-based technologies.

In addition to the "soft technologies" of methodologies and the "less soft" applications (parts 1–2), the UNU/IIST also aims at featuring, as an integral part of its operations, some of the underlying technologies implied by the presentations at the Kyoto Symposium. Thus, the UNU/IIST is expected to become a *showroom* where *awareness* courses and "hands-on" experience can illustrate front-of-the-wave technology in a context not readily available elsewhere in the developing world.

This then represents the third axis by means of which the UNU/IIST hopes to be able to help the developing world improve access to information technology.

3.1 Specific Material Technologies and Tools

From the next subsections you may get the erroneous impression that the UNU/IIST will assist in the transfer of material technology. Remarks made in subsection 2.2 apply. The UNU/IIST, however, may, in Macau, feature some of the hard, material technologies – some of which are now mentioned.

3.1.1 Supercomputing

This subsection is somewhat speculative! The UN system has a great need for gathering and evaluating socio-economic data – but lacks facilities for the kind of supercomputing sometimes called for when performing extensive regional and global modelling and simulation (based on such data and with large-scale mathematical [operations research analytic and statistical] models). The developing world itself has no such well-identified supercomputer centre.

Modern technology now provides dramatically new, and, in some cases, rather lower priced supercomputers than heretofore experienced. The industrial world – the United States, Japan, and Europe – now has several such supercomputer centres established outside the nuclear physics sector.

The UNU/IIST would like to take the initiative to establish such a centre, for example as part of the UNU/IIST, and located in some developing country, for example, but not necessarily, in Macau.

The supercomputer centre could be a joint effort with a supercomputer manufacturer (XXX) and be announced as the "XXX/UN Supercomputer centre."

That centre could be the place where the entire UN system could satisfy its demand for supercomputing. Its presence in a developing country and close to many other developing countries could stimulate local build-up of expertise in supercomputing. The centre could also be a place where universities of the developing world fill their demand for supercomputing facilities. Its presence in one of their countries should provide a more sensitive affinity to the needs of the developing world.

It is envisaged that each user of the global (or Asia-Pacific regional) centre link up with that centre via satellite communication – much like the NIC-NET mentioned in section 2.2.2.

The UNU/IIST can train users in installing, operating, and applying existing software for the modelling and simulation of large-scale models, and for visualization of socio-economic and scientific data. The UNU/IIST can make policy planners and university administrators aware of the many possibilities supercomputing offers. The UNU/IIST can educate operations researchers, experimental scientists, and policy planning staff to develop their own large-scale (simulation) models. The UNU/IIST can finally coordinate regional and global efforts in large-scale modelling and simulation.

The above are "loose" plans, "dreams" perhaps "in the eyes of the beholder"!

3.1.2 Multimedia Laboratory

The UNU/IIST is interested in establishing, and many clients of the UNU/IIST expect it to establish, a Multimedia Laboratory – a showroom featuring the latest multimedia technology: hardware plus software, including tools and applications. Various conditions in the developing world prevent the ready availability of such technology. The UNU/IIST could repair such a situation.

The UNU/IIST can then, within the multimedia laboratory, give training, awareness, and education courses, in usage, product management, and actual software development of applications that are based on multimedia technology. The idea is to have easy access to this technology, provide a means for software companies of the developing world to skip a product generation or two, and step right into one of the commercially most viable forefronts.

3.1.3 Other Technologies

It might be useful to feature several other technologies at the UNU/IIST Telematics, CIM (Computer Integrated Manufacturing, incl. Robotics), etc.

3.2 Conclusion

There is imminent danger that in focusing on the "hard" technologies, applications often become artificially technology motivated and driven. On the other hand, the newest versions of these technologies often enable applications that were previously not feasible.

4. Conclusion

In part 1 I emphasized the methodology aspects: techniques for achieving applications. In part 2 I emphasized developing-world needs: applications that solve problems. And in part 3, I have emphasized the hard technology: ultimate "carriers" of such applications – carriers often requiring special development techniques, novel technologies that often remove constraints on previous usage.

I have thus sketched three axes of possible UNU/IIST offerings – each aimed at helping developing countries improve their access to information (including computing) science and technology. The first axis stressed "soft-soft" methodologies, the second stressed "soft" applications, and the third stressed "hard" technologies. The UNU/IIST emphasizes the intellectual possibilities afforded through any of these axes.

Obviously the UNU/IIST cannot feature a complete range of access provisions – as outlined in parts 1, 2, and 3 of this paper. Which will actually be pursued depends on policy priorities within the UNU/IIST and on funding opportunities originating outside the UNU/IIST.

The UNU/IIST is, on the one hand, to be an institute located in and offering many activities in Macau. On the other, the UNU/IIST will strive to see the establishment, if needed, of similar institutes, perhaps specializing in specific IT areas elsewhere in the developing world. The UNU/IIST will be ready to advise, on the basis of many years of experience with industrial world IT (including software technology) programmes, regions of the developing world should they wish to embark on similarly successful R&D programmes.

REFERENCES

1. Bjørner, D. (1991). *Proposed Strategy, Tactics and Operational Plans: The United Nations University (UNU), International Institute of Software Techology (IIST), "The Macau Institute"*. Dept. of Comp. Sci., Techn. Univ. of Denmark, 13 and 31 July, 5 Aug. 1991.
2. Bjørner, D. (1992) "Trustworthy Computing Systems: The ProCoS Experience." In: *14th ICSE: Intl. Conf. on Software Eng., Melbourne, Australia*. ACM Press, 11–15 May 1992.
3. Bjørner, D., C.A.R. Hoare, and H. Langmaack, eds. (1992). *The ProCoS Project*, volume 5XX of *Lecture Notes in Computer Science*. Heidelberg: Springer-Verlag.
4. Bjørner, D., and J.F. Nilsson. (1992). "Algorithmic and Knowledge Based Methods—Can They Be Unified?" In: *International Conference on Fifth Generation Computer Systems: FGCS '92* ICOT, 1–5 June 1992.
5. Brooks, F.P. (1986). "No Silver Bullet: Essence and Accidents of Software Engineering." In H.-J. Kugler, ed. *Information Processing '86*. Amsterdam: North-Holland Publ. Co. IFIP World Computer Congress, pp 1069–1076.
6. Harel, D. (1992). "Biting the Silver Bullet." *IEEE Computer* 25 (1): 8–20.
7. Ramani, S. et al. (1989). *International Institute for Software Technology, A Proposed UNU Research and Training Centre*. UNU, June 1989.

The Potential of Information Technologies for International Cooperation

Blagovest H. Sendov

ABSTRACT

The paper emphasizes information technologies as an object of international cooperation. The activities conducted by a number of international organizations active in this area are outlined. Reflections follow on the identification of appropriate thrusts for international education and training activities concerned with new information technology, in particular with regard to the developing countries.

It is impossible to work together without exchanging information. In this respect, information technologies may be considered as basic instruments for cooperation. But when we discuss the potential of information technologies for international cooperation, it is necessary to distinguish two different aspects: (1) information technologies as a tool for, and (2) information technologies as an object of international cooperation. In this paper I emphasize the latter.

When considering information technologies as an object (and objective) of international cooperation, I shall briefly discuss the notion of "information" as a substance to be processed through these technologies.

As examples of successful ongoing international cooperation in information technologies, I review some activities of the International Federation for Information Processing (IFIP), the Intergovernmental Informatics Programme (IIP) of Unesco, the Committee on Data for Science and Technology (CODATA) of the International Council of Scientific Unions (ICSU), and the project of the International Association of Universities (IAU) called the "University-based Critical Mass System for Information Technology" (USIT). This selection is based on my personal acquaintance with these organizations. There are so many other governmental and non-governmental, international and regional organizations active in information technology that it would be difficult even to list their names.

Education is an important objective for international cooperation in information technologies. There are many important problems in devising strategies of education for the information age that concern all countries. I will discuss some of these problems with emphasis on needs for radical change.

Collecting and using different types of information (in the form of data) are also important problems of international cooperation, especially in science. Among the areas needing attention are standardization, accessibility, reliability, and, last but not least, the cost of obtaining and processing different kinds of scientific data.

1. The New Technologies

When speaking about "new technologies" we usually imply information technologies. From the vantage point of the approach to information, the history and foreseeable future of mankind can be viewed as consisting of the following eras [7, 19]:

(1) *Neolithic era.* Man acted on the basis of learning how to make use of the potential provided by Nature to have food available in sufficient amount and whenever needed.

(2) *Industrial era.* Man acted on the basis of learning how to use the laws of Nature to have energy available in sufficient amount and whenever needed.

(3) *Information era.* Man acts and will act on the basis of how to use Nature and its laws to have information available in sufficient amount and whenever needed.

The technologies of the industrial era, or the old technologies, are based on the transformation (processing) of energy. More generally speaking, the technologies of the industrial era related to the transformation of physical matter. The technologies of the information era, or the new technologies, are based on the processing (transformation) of intellectual product.

There is one substantial difference between material products and information products. To multiply a material product, one needs comparable material resources. But multiplication of an informational product does not require a proportional expense. This correlation is significant in planning strategies for international cooperation in information technologies.

The information technologies comprise basically the storage, processing, and transmission of information. These branches of information technology have roots in the history of knowledge and learning. The book, for example, represents an accumulation of knowledge and culture. Instruments for the automatization of arithmetic operation have been used for centuries. But only after information technologies were based on electronics and new physical principles were the new information technologies born.

These three types of information technologies have their specifics. Historically, they were developed as independent technologies. The electronic computer became the integration medium for these three branches of information technology. Today it is impossible to imagine communications without computers. But while the computer has a central role in the new information technologies, these are not to be identified with computers only.

2. Information and Knowledge

When using information, very often we talk about the information as knowledge. But it would be wrong to equate them. According to the *Random House Dictionary of the English language* (1971), "information" is:

> 1. Knowledge communicated or received concerning a particular fact or circumstance.
> 2. Information, knowledge, wisdom are terms for human acquirements through reading, study, and practical experience. Information applies to facts told, read, or communicated which may be unorganized and even unrelated . . . . Knowledge is an organized body of information; or the comprehension and understanding consequent on having acquired and organized a body of facts. . . . Wisdom is a knowledge of people, life, and conduct, with the facts so thoroughly assimilated as to have produced sagacity, judgment, and insight.

In the first approximation there is no substantial difference between information and knowledge. It is accepted that knowledge is "structured information." Today, the information industry is producing so-called "knowledge based systems." A point I would like to make here is the difference between "knowledge" stored in a computer and "knowledge" possessed by a human being. The "knowledge" in a computer is information structured according to given rules that are precisely defined and well known. The "knowledge" possessed by a human being is information structured through the learning process, and the rules for this structuring are not known. It is natural to expect that the learning process, or the rules for structuring that transform acquired information into human knowledge, is individualized. That means that we have to differentiate between "knowledge" of a human being and "knowledge" stored in the memory of a computer [14]. The first may be called simply knowledge (or natural knowledge), and the second artificial knowledge.

On the other hand, the notion of knowledge is closely related to the notion of "intelligence," which means the capacity for reasoning, understanding, and for similar forms of mental activity. Knowledge is a necessary condition for every manifestation of intelligence. I have to stress here that intelligence has been considered as a unique characteristic of a human being. That is why when the idea of imitating man's intelligence was born, the term "Artificial Intelligene" was accepted as the most correct description.

The successes and shortcomings of Artificial Intelligence are well-known. Critics like Hubert Dreyfus [4] many years ago called attention to its difficulties and restrictions. But in one aspect the enthusiasts of Artificial Intelligence are on the safe side. They have addressed the fundamental task of imitating intelligence, and this imitation, good or bad, they called Artificial Intelligence. Their aim is not to create intelligence artificially, but to create Artificial Intelligence. It is quite natural that artificial intelligence be something different from natural intelligence. Let us mention that in the material world we are happy with a lot of artificial materials such as artificial wool, artificial silk, artificial caviar, etc.

3. Activities of Some International Bodies in Information Technologies Cooperation

3.1 The International Federation for Information Processing (IFIP)

The IFIP came into official existence in January 1960. It was founded under the auspices of Unesco to meet a need identified at the first International Conference on Information Processing held in Paris in June 1959 under the sponsorship of this UN organization.

The IFIP is a non-profit organization with the following basic aims:
- To promote information science and technology;
- To advance international cooperation in the field of information processing;
- To stimulate research, development, and the application of information processing in science and in practical activity;
- To further the dissemination and exchange of information on information processing;
- To encourage education in information processing.

The IFIP is devoted to improving worldwide communication and increased understanding among practitioners of all nations about the role information processing can play in all walks of life.

Information, science, and technology, together with informatics, are potent instruments in today's world and affect people's lives, in everything from their education and work to their leisure and their homes. They are powerful tools in science and engineering, in commerce and industry, in education and administration. They are truly international in scope and offer a significant opportunity for the developing countries. The IFIP helps to bring together professionals in all spheres of practice, research, and education to share their knowledge and experience and acts as a catalyst to advance the state of the art.

The IFIP's mission is to be the leading, truly international, apolitical organization that encourages and assists in the development, exploitation, and application of information technology for the benefit of all people.

The IFIP aims to assist developing countries in their application of information processing and cooperates with Unesco to achieve this. Responsibility for planning and overseeing the work rests with the Developing Countries Support Committee (DCSC).

The aim of the DCSC is to promote the cooperation of the IFIP with the developing countries through the sharing of IFIP experience, technical information, and knowledge and to help developing countries and areas with their specific needs and requests. The DCSC's programme includes distribution of IFIP publications, regional activities, and training courses and seminars. Support has also been offered to the developing countries to participate in IFIP conferences.

3.2 The Intergovernmental Informatics Programme (IIP) of Unesco

The IIP is the youngest intergovernmental programme of Unesco [5]. The first meeting of the Committee was held in October 1986 in Sofia, Bulgaria, and the activities began in 1987. In the short intervening time, the IIP made considerable progress.

The programme for action of the IIP is based on the following priorities:
- Training of specialists, researchers, instructors, users, and maintenance technicians;
- Information exchange networks between institutions specialized in informatics with a view to carrying out joint or complementary activities;
- Production of software for informatics and computer-assisted teaching, and for the management of education systems or other applications in various fields of activity;
- Studies on the implementation of integrated policies and strategies for sectorial computerization;
- Research and development in informatics with a view to making the best use of the human resources available (so as to create, innovate, or adapt in the fields of theoretical and applied informatics).

These priorities, which concern in varying degrees both the industrialized and the developing countries, form a good basis for regional and international cooperation.

The importance of the IIP for member countries of Unesco may be judged by the quality and the number of projects submitted to the Bureau. The projects were evaluated on the basis of the selection criteria established by the Committee. The requests for financial support usually exceed by tenfold the money available. To prevent dissipation of the limited resources, the Bureau was obliged to make a strict selection and resort to partial financing, despite the high calibre of the projects.

In principle, training is one of the main priorities of the IIP as it has very often been the lack of skilled manpower that slowed development. The

Committee considered that training should extend beyond the training of specialists in informatics. It was felt a priority matter to train instructors, users, and maintenance technicians, and also to train researchers so that they could use informatic techniques to speed up their research and improve economic performance in their countries.

3.3 The Committee on Data for Science and Technology (CODATA) of the International Council of Scientific Unions (ICSU).

CODATA is an interdisciplinary ICSU body founded in 1966 [10]. CODATA is concerned with all types of quantitative data resulting from experimental measurement or observations in the physical, biological, geological, and astronomical sciences. Particular emphasis is given to data management problems common to different scientific disciplines and to data used outside the field in which they were generated. The general objectives are the improvement of the quality and accessibility of data, as well as the methods by which data are aquired, managed, and analysed; facilitation of international cooperation among those collecting, organizing, and using data; and the promotion of an increased awareness in the scientific and technical communities of the importance of these activities.

The activities of CODATA are extremely important for international scientific cooperation in astronomy, geology, ecology, biology, and many other sciences, where a large amount of data has to be used in a standardized way.

Factual data constitute the essence of scientific knowledge. Easy access to data is a basic condition of the evolution of science. The colossal increase in the generation and collection of data poses major financial, technical, and ethical problems to the potential user. The gap between producers and consumers of data is expected to widen throughout the world in the future.

There are two more interdisciplinary groups in the ICSU that deal in different ways with data activities. They are: GAGS, the Federation of Astronomical and Geophysical Data Analysis, and the WDC Panel, or World Data Centers Panel.

3.4 University-based Critical Mass System for Information Technology (USIT) of the International Association of Universities (IAU)

The International Association of Universities has introduced, in cooperation with Unesco's Intergovernmental Informatics Programme (IIP), the University-based Critical Mass System for Information Technology (USIT) [9]. The purpose of the USIT is to improve and unify the access, adaptation, and use of computer technology in higher education on a world scale. The USIT provides a comprehensive strategy for higher education institutions. It addresses management, research, and teaching needs, duly accounting for the particular economic, social, and cultural conditions in different national settings.

The USIT approach is based on the following assumptions:

- Higher educational institutions, as agents of change, are the best vehicles for creating a critical mass in the utilization of a new information technology – a "critical mass" consisting of levels of knowledge and usage that allow the activity to be self-sustained.
- The critical mass is decisive because, in industrialized and developing countries, the availability of computers and the utilization of information technology is neither uniform nor sufficient to allow the creation of effective knowledge and consumer bases.
- The higher education system, with its different types of institutions, should be a major developer and dispenser of technology. These institutions are in a unique position to lead in informatics development and to work as partners with the governmental and private sectors. By focusing on higher education, the USIT addresses an indispensable component of the larger national informatics effort.
- Optimization of informatics development and utilization within available resources is badly needed.

The USIT aims to promote optimization by assisting in such matters as original purchase and delivery, assuring hardware and software compatibility, and constant upgrading to meet changing needs and conditions.

Central to the rationale for the "critical mass" is the fact that the utilization of information technologies tends to be dispersed, of low intensity, and lacking support structures that permit either optimal or efficient applications. The critical mass system is designed to achieve long-term goals by promoting the self-sufficiency of users. Educational, human resource development, and support programmes are developed with the aim of reducing the involvement of outside personnel and the need for outside assistance.

The implementation of the USIT started in Africa. A Research Committee, consisting of 20 persons who indicated an interest, was organized. Six specialists with a special expertise in informatics will serve on a Project Steering Committee. A proposal to support the USIT was submitted to the International Development Research Centre and a grant to fund data collection was received. The study sample will be drawn from the 62 IAU member universities in Africa.

4. Educational Strategies

From an analysis of the activities of the international organizations just reviewed, it is obvious that the problems of education and training dominate. That is only natural, and a very clear view of concepts and strategies for education in the information age is essential.

If one regards the computer as an extension of the human mind, it follows that the aim of education in the future should be to educate human beings with an extended mind.

An analysis of the character of computer use in education [12, 13, 16] shows three main tendencies:

(1) the computer as an object of study;

(2) the computer as a training device;

(3) the computer as an extension of the human mind.

All three have their place in education, but the third will lead to the most profound and essential, since it changes the very object of education.

It is natural to expect changes in the methodology, as well as in the content, of education.

The change in school curricula is particularly significant. But this change cannot be carried out easily because of the traditional conservatism of educational systems, as well as the fairly rapid improvement of computers. I do not mean that the object of education has already changed; it is still undergoing a change. An adequate school training model will develop and crystallize only as the result of a long and difficult process.

The expected changes in the contents of education are measured against Bloom's taxonomy concerning the hierarchic aims of education [6]. These can be grouped in three main domains:

(1) the cognitive domain, concerned with knowledge and its use;

(2) the affective domain, concerned with emotional responses and values that are taught;

(3) the psychomotor domain, concerned with physical and manipulative skills.

The changes in education will affect most substantially the first domain of Bloom's taxonomy. The second and third domains will undergo minor changes. Manipulative skills will probably change considerably.

4.1. Education Curricula and Knowledge Structure

I shall share my view of some principles for education in the information era that have been implemented in an educational project in Bulgaria during the last decade [13, 15, 16, 17, 18].

The basic principle is the *integrity of knowledge*. The principle of integrity, or the *wholeness* of education, is not a new one. The idea of studying objects and phenomena from different points of view, and combining knowledge of different school subjects, is a well-known approach, particularly in primary education. The very integrity of knowledge itself can have different characteristics. Specialization is possible within the different subjects, while at the same time examining and emphasizing their interconnections remains the focus of the learning objectives.

Another approach is integration on the basis of fundamental ideas from different fields of knowledge. Methods using projects are well-known, and achieved great popularity in the 1920s and 1930s in the United States.

The integrated approach for the information era is of a special nature and differs in principle from other methods of education defined as integral. The basic difference lies in the emphasis placed on the need for integration of knowledge as a consequence of a qualitatively new situation – the emergence of new information technologies. This new situation can be characterized by the following embroilments:

- the school will no longer be the sole nor the most attractive source of knowledge;
- quick and unhindered acquisition of knowledge in a pleasant atmosphere will be provided through TV, radio, and computer networks.

These sources therefore provide strong competition for the school. However, there is one aspect of school education that cannot be rivalled by other sources of information. This is the unique commitment and capacity of the school to provide *systematized and well-structured* knowledge. Hence, in the era of highly advanced information technologies, the main preoccupation of the school should be the systematization and structuring of knowledge, whereby emphasis is laid on fundamental and universally valid principles. In this sense, the integration of knowledge acquires a special significance. The purpose of the integrated approach to education in the information era is not to limit the learning in different spheres to a certain body of facts needed to carry out a definite practical job, or to develop a project. What I have in mind is just the opposite. It is to concentrate the attention of the student on the basic and valid principles from the viewpoint of a large number of scientific subjects. That will encourage further independent study and the utilization of specific information through information technologies.

It is easy enough to formulate this requirement in principle, of course, but it is rather more difficult to implement it as a particular learning process incorporating curricula, textbooks, and study aids. The school, even the university, is not in a position to provide sufficient knowledge for the entire range of human working life. It can be said that man's life-span will continue to increase in the future, if by this we mean the degree of change man will experience in his environment throughout his life. Because of rapid change, it is no longer possible for a school to equip the future citizen with the knowledge and skills to serve for a lifetime. No doubt citizens will be compelled to study all their lives. So the chief task facing the school of today is *to teach pupils how to learn*. In this respect, the integrated approach has indisputable advantages; it enables the student to observe natural and social phenomena from different angles and stimulates the need for a constant search for new relationships and facts.

5. Developing Countries

Special attention in the area of international cooperation in information technology is given to the developing countries, as shown in the activities of Unesco [3] and its International Informatics Programme (IIP), as well as the International Federation for Information Processing (IFIP) and the International Council of Scientific Unions (ICSU). The importance of information technologies for developing countries is emphasized first of all in the field of education. To prepare for the coming new technologies, it is natural first to educate the new users.

Strategies for education in the information era may differ in different countries, but the main principles are more or less universal. The special difficulties in the developing countries are connected with the lack of hardware and communication.

It is very important for the developing countries to set their priorities in information technologies [8, 11]. It is strongly advised to develop first powerful communication facilities, even if this slows down, in the beginning, the general process of development. Investing in communication is decisive for all other activities. Reliable communications are necessary for effective international cooperation in all fields and for using international information services.

6. Negative Tendencies and Illusions

Some have said that in an information era, with powerful computers, excellent communications, and access to knowledge bases, the pupil will need less education. If we use the analogy between energy and information, the argument may go as follows: In the industrial era, men learned how to make use of the laws of Nature to have energy available in sufficient amount and whenever needed. That made human beings physically very powerful, having at hand energy from many different processors (transformers) of energy. The natural human physical energy lost its value as a productive force. The most physically powerful men and women in the industrial era often found their vocations in sports and entertainment, but not in industry. In the information era, man will learn how to use Nature and its laws to have information available in sufficient amount and whenever available. Following the analogy, we may expect that in the information era, natural human knowledge as capacity to produce information will lose its value. It is difficult to deny this possibility. But it would be wrong to conclude that because of this, in the information era we shall need less education. It is completely true, though, that we shall need a different type of education.

Let us mention also the opinion of Brauer and Brauer [1], leading specialists in computer education:

> There are different opinions on what the implications of computers to education really are. A number of people even think that the better the computerized tools (and in particular their man/machine interfaces and their knowledge bases) are, the less important becomes education. The opposite opinion (and this is ours) is that everybody needs a thorough education (adequate to its level of knowledge) on how a computer and its software work in order to be able to use the computer in an optimal way which includes knowing about its limitations and disadvantages. This

education should make clear that a computer is merely a tool which cannot assume any responsibility for what it does, but a tool which is of a completely new type: it amplifies our mental capabilities, in contrast to the traditional tools which serve for physical work.

The supposition, that in the information era we need less education, is wrong. On the contrary, education will become much more necessary and a lifelong process. International cooperation in education at all levels is very important to eliminate this dangerous illusion.

The enthusiasm born of the first use of computers, for processing highly structured information, or for processing of knowledge, as it is usually called, creates another illusion. It is believed that it will be possible to use knowledge processing and Artificial Intelligence to build fully automatic factories and eliminate human beings from participation in industry and industrialized agriculture. The opinion in this respect of Coy and Bonsiepen [2] is as follows:

> The general idea of the "fully automatic" factory which is supported by artificial intelligence products like expert systems is wrong and leads to the wreckage of many Computer Aided Manufacturing projects. It is dangerously wrong because of its consequences for all participants in this adventure. . . . The use of expert systems in risky environments – where quick decisions are demanded, must be considered irresponsible. The control of nuclear, chemical or petrochemical factories by expert systems leads to irresponsible actions and must therefore be prohibited. . . . Though we probably cannot avoid the transfer of responsibility from humans to machine systems, it should be clear that this responsibility transfer must stay transparent to the users.

It is important to realize the difference between knowledge as structured information in a human brain and that in a computer. The difference is in the rules defining the structure. We still do not know how the educational process structures the acquired information. This structure, as mentioned above, is most probably individualized and dynamic. The obvious difference between the rules producing these two types of knowledge has to convince us that they are different breeds of knowledge. To distinguish them, we use the notions "natural knowledge" and "artificial knowledge." The illusion that these two types of knowledge are identical and mutually exchangeable is wrong and may be dangerous.

REFERENCES

1. Brauer, W., and U. Brauer (1989). "Better Tools–Less Education?" In: *Information Processing '89*, IFIP Congr. 1989. Amsterdam: Elsevier, North-Holland, pp. 101–106.

2. Coy, W., and L. Bonsiepen (1989). "Expert Systems: Before the Flood?" In: *Information Processing '89*, IFIP Congr. 1989. Amsterdam: Elsevier, North-Holland, pp. 1167–1172.

3. Douglas, A.S., and A.L. Olver (1988). "The Fourth Annual Conference on Information Technology for Developing Countries." *Information Technology for Development* 3(3): 249–258.

4. Dreyfus, H. (1972). *What Computers Can't Do*. New York: Harper and Row.

5. General Conference of Unesco, Twenty-sixth Session, Paris 1991, Document 26 C/99.

6. Gronlund, N.E. (1976). *Measurement and Evaluation in Teaching*. New York: Collier-Macmillan.

7. Gruska, J. "Why We Should Not Only Repair, Polish and Iron Current Computer Science Education." *Proc. IFIP Workshop "Informatics at the University Level: Teaching Advanced Subjects in the Future."* Forthcoming.

8. Gupta, P.P. (1986). "Can Developing Countries Compete in Information Technology?" In: *Information Processing 86*. Proc. IFIP Congress, Dublin, Ireland, 1–5 Sept. 1986. Amsterdam: North-Holland, pp. 77–78.

9. Hayman, J. (1991). "LAU's USIT Informatics and Research Programme: Informatics Research in Africa." *Higher Education Policy* 4(3): 49–51.

10. International Council of Scientific Unions. *Year Book 1992*.

11. Kohly, F.C. (1986). "Information Policy Issues in Developing Countries." In: *Information Processing 86*. Proc. IFIP Congress, Dublin, Ireland, 1–5 Sept. 1986. Amsterdam: North-Holland, pp. 591–597.

12. Sendov, B. (1978). "Informatique, ordinateurs et education." *Impact: science et société* 28 (3): 287–292.

13. Sendov, B. (1992). "Une education adaptée à l'ere de l'information. *Impact: science et société* 146: 203–212.

14. Sendov, B. (1985). "Information and Knowledge." In: *Formal Models in Programming*. Amsterdam: North-Holland, IFIP, pp. 97–101.

15. Sendov, B. (1986). "Children in an Information Age, Tomorrow's Problems Today." In: *Children in an Information Age*. Oxford: Pergamon Press, pp. 195–200.

16. Sendov, B. (1988). "Children in an Information Age." *Education & Computing* 4: 21–26.

17. Sendov, B. (1989). "A Broad View of Informatics." *Higher Education Policy* 2 (4): 73–76.

18. Sendov, B., and A. Eskenasy (1991). "Quality of Education with the Application of New Information Technologies." In: D. Bjørner and V. Kofov, eds. *Images of Programming*. Amsterdam: North-Holland, IFIP, 1.71–1.78.

19. Toffoly, T. (1989). "Position Statement for Panel 1: Frontiers in Computing." In: *Information Processing '89*. Proc. IFIP 11th World Computer Congress, San Francisco, Calif. Amsterdam: North-Holland, pp. 591–597.

Discussion

Starting the discussion, I. Wesley-Tanaskovic noted that the initial assumptions of the Symposium were that the new information technologies would offer new modalities for cooperation in general, and especially for international cooperation, but this could be examined from two aspects: information technologies as an object (and objective) of cooperation, and as a tool for cooperation. The first aspect was the main concern of the papers presented in Session 5, while the information technologies used as tools for international cooperation, particularly those enhancing access to science and technology, were to be the central topic of the Panel 2 discussion.

Answering the question of G. Johannsen on "measures of effectiveness," A. Sage stressed that the "likelihood for success" when applying systems management in development of information technologies depends on a set of multi-attribute effectiveness facets, of which market share and profit, but also safety and environmental issues, are part.

There were two principal topics in the ensuing discussions: education for systems engineering, and transfer of technologies to the developing countries.

G. Johannsen remarked that while systems engineering is currently a major focus of industrial efforts, even in the most industrially developed countries, there is still little attention being paid to this subject at universities and technical colleges. A. Sage shared this view, explaining that it is simply easier to deal with the micro-level details associated with conventional engineering than with the broad scope of macro-level efforts that comprise systems engineering. He further commented that the increased emphasis now placed on global economics and the need for industry-government-university interaction will ameliorate this situation. Finally, he expressed the thought

441

that systems engineering is fundamentally an information-based endeavour, and it is only now that the new information and other meta-level concerns are beginning to be incorporated in engineering education. J. Alty mentioned that some advances have been made recently, in Great Britain (the University of Loughborough) for instance, in cooperation with large manufacturers. All agreed that education for the "information era" in general may be claimed as the key to development in the future for all societies and nations. International cooperation is much needed for the delivery of new scientific and technological modalities that generate enhanced productivity for all, while information technology itself can potentially do much to support needed cooperative efforts.

The old controversy of "technology push" and "societal pull" in the development and transfer of technology was raised again by several speakers (M. Lundu, C. Correa, A. Sage, W. Rouse) during discussion of specific needs and conditions in the developing countries. D. Torrijos underlined that it is a fallacy to consider the so-called "developing countries" as completely blank when it comes to transfer of technologies. They have their culture, practices, and traditions, which do very strongly influence the success of the new technologies being introduced: this has been clearly shown in the past, especially in the case of failures.

Comments were made by M. Stone and D. Torrijos in relation to the programme of the UNU/IIST, advocating that it should find its place within the UNU and in collaboration with other relevant UN organizations. M. Almada de Ascencio stressed that the IIST should endeavour to assist the developing countries to create their own software facilities and industries, while D. Torrijos added that it should start by providing them with an "objective" assessment of the appropriateness of commercially available products. The answer of Zhou Chao Chen was that while the UNU/IIST plans to establish an electronic software product catalogue, it is realized that such an assessment is not easy and may be misleading. He further stated that the UNU/IIST will cooperate with Unesco and other organizations active in the field of information technology. In this cooperation, he said, the IIST could play the role of consultant/partner/trainer in software technology.

Responding to the relevant concerns expressed by some participants, I. Wesley-Tanaskovic stated that within the UNU, the IIST, INTECH, and all the projects of the Micro-electronics and Informatics Programme, including the Microprocessor Laboratory at the International Centre for Theoretical Physics (ICTP) in Trieste, Italy, were cooperating closely. Their programmes are considered as being complementary within the framework of the UNU General Research and Training Programme.

Panel Discussion 2:
Towards New Modalities of International Cooperation

Coordinator: Ines Wesley-Tanaskovic

The Panel, chaired by Ines Wesley-Tanaskovic as moderator, was composed of the following members: Charles Cooper, UNU Institute for New Technologies, The Netherlands; Carlos Correa, Programa Regional de Cooperación en Informática y Microelectrónica, Argentina; Meinolf Dierkes, Science Center Berlin, Germany; Martha Stone, International Development Research Centre, Canada; and Hisao Yamada, National Center for Science Information Systems, Japan.

I. Wesley-Tanaskovic introduced the discussion by stating that the Symposium started with at least two underlying assumptions: (1) that "expanding access to science" is for the "benefit of mankind," and (2) that in this process the new information technologies, opening new perspectives, have an important "positive" or enhancing role to play.

Consequently, two questions have been raised from the beginning, she said, and the objective of the Symposium has been to try to answer them: (1) what effect have the new information technologies had so far on the access to science (and technology to a certain degree), and (2) what developments can be expected in this respect in the future?

On the basis of this analysis, she proposed that one might also forecast what kind of impact the information technologies will exert on international cooperation for expanding access to science, which is an international endeavour *per se*, and to technology, which is increasingly "science dependent."

She recalled that this last aspect, i.e. international cooperation, had been emphasized throughout the Symposium, at Panel 1 and in several papers, especially those in Session 5. There, the potential of information technologies was examined in the light of cooperation in ensuring access to science

443

and technology, but more particularly in the development of information technologies and the design of information systems. A synthesis of the views on international cooperation, presented so far during the Symposium, was to be given at Panel 2, which would add the views expressed by members of the Panel. But first, she stressed, two considerations should be kept in mind:

First, "access to science" is not necessarily the same as "access to technology," which often comprises the so-called "transfer of technology." Therefore, when dealing here with international cooperation in access to science and technology, those aspects of technology are dealt with that are related to the application of scientific results and that are of free access, in the "precompetitive phase" of R&D. The legal and other issues of property rights, restricted access, contractual conditions, and the like are not of concern here, notwithstanding their importance.

Second, "international cooperation" as perceived at the UNU, the "academic arm of the UN system," is different from the approach of the other UN and intergovernmental organizations. This is clearly stated in the Charter of this University, where it is said that the UNU should enhance communication among scholars and between scholars and the other communities, with the intention to disseminate worldwide and without any discrimination the results of scholarly work. In order to achieve this goal at the global level, particular attention must be given to the needs and conditions in developing countries, with the purpose to ensure their full involvement and partnership.

At the same time, it is said in the UNU Charter that in propagating ideas and results, the UNU enjoys complete academic freedom in the selection of topics and the expression of ideas. Therefore, the patterns or the modalities of international cooperation used by the UNU are different, and must be different, from those practiced by others, e.g. Unesco or UNIDO or other UN organizations that are also active in scientific and technical information transfer "for the benefit of mankind."

Developing this basic concept, I. Wesley-Tanaskovic said that innovative models of international cooperation, using new information technologies, introducing "intelligence" in man-machine systems and in communications networks should be promoted by the UNU. This would enable the transition from old applications to new ones, even without knowing exactly today what tomorrow's technical possibilities will be, since advanced systems could be readily integrated/upgraded as they become available. Technology offers vast new opportunities, but only if people have the imagination to grasp them. "User-oriented" or "human centred" design of information systems should be "technology-knowledgeable," as already stated at the Symposium. This means, she said, that systems design should take into account the constraints vs. the potentials of existing and future technologies, and progress one step ahead of the existing technologies – setting requirements, demands, and challenges to technological development.

She concluded that one should not forget that this Symposium is an activity in the UNU series "Frontiers of Science and Technology." It is, therefore, its mandate to look ahead – to the "real new world of computing."

It was stressed from the start of the Panel 2 discussion that generation, dissemination, and application of knowledge are critically important and will become even more so in the development process. This process is dependent on international cooperation, collaboration, and sharing to foster equity of access to scientific endeavours and to technology for social and economic benefit. Perhaps the most vital difference between developed and developing, rich and poor, is the knowledge gap – the capacity to generate, acquire, and use scientific and technological knowledge.

In relation to this topic, the mission of the International Development Research Centre (IDRC), Canada, was described as encapsulated into one phrase, "empowerment through knowledge," which is considered the key element in the development of nations, peoples, communities, and individuals:

> The capacity to conduct research is therefore a necessary condition for development. IDRC has dedicated its resources to creating, maintaining, and enhancing research capacity in developing regions, in response to needs that are determined by the people of those regions in the interest of equity and social justice.
>
> As part of its new strategic framework, IDRC continues to build upon its global perspective on mobilizing science and technology for development objectives, building bridges across continents and putting development-country researchers and policy-makers in contact with each other.
>
> A main component is the identification of communalities in development problems and solutions, fostering comparative research across regions, countries, and cultures, thereby allowing widely different developing countries to learn from one another.
>
> We have witnessed the beginning of several major groups of change in this decade which will influence our future in ways that are still unimaginable.
>
> Of these groups, which include the rapidly shifting political environment, explosive growth in social demands in developing regions, major transformation in the patterns of world economic interdependence, of principal interest to us, here today, is the group of change concerned with the plethora of technological advances that, while opening up new opportunities for some countries, will likely create deeper and more intractable problems for others.
>
> The new science based technologies have emerged in the last two decades at a pace which can be described as an *explosion*. Many of these technologies, which we have discussed in the past two days, are highly flexible and mobile, allowing for rapid and continuous modifications and improvements. As such, they are fast changing the way in which the inter-

national marketplace has functioned since 1945. Individuals, groups, and nations actively participating in the generation and exchange of these new technologies will prosper in the emerging new order; those left behind will become increasingly marginalized, and the risk of marginalization is particularly severe for the least developed countries. (M. Stone)

As an example, an introductory programme statement was made concerning the work of the Information Sciences and Systems Division (ISSD) of the IDRC, which is directed towards improving the management and use of scientific and technical information in support of development:

General objectives influence the program of work for the ISSD, including better access and use of knowledge required for development research; improved collaboration by exchanging information and experience to stimulate cooperation and coordination in development research; capacity-building within developing countries for better management of information and effective application of knowledge. Principal among these objectives are: information innovation – the enabling of developing countries to benefit from applied research into problems of sharing and using knowledge for development, and on ways to improve and adapt information systems, methods, and technologies.
To achieve these objectives, increased emphasis has been placed on supporting applied research on information and communications issues.
The research agenda has and will continue to be defined in partnership with developing countries and regions, enabling them to play a more substantial role in a field that to date has largely excluded them. The program includes innovation and adaptation of modern information technologies, as well as research on policy, economic, and other issues that influence the successful introduction of information technologies and systems.
The program focuses on electronics-based technologies which can be used to collect, store, process, package and communicate information and provide access to knowledge. These are technologies which clearly can be integrated with computing capacity, and are at the "leading edge" of development to ensure that developing regions can have some experience with their design, adaptation, and use before the development/ introduction cycle fully solidifies and excludes their interest. (M. Stone)

The information technologies that are included in the ISSD/IDRC programme were enumerated, exemplifying the broad scope of this programme dedicated to the developing countries:

Technologies for information acquisition and management; and effective systems design, development and integration.
 – remote sensing via new systems – radar satellites, automated surveys, various computer-related input methods

- expert systems techniques for data analysis; image analysis and pattern recognition
- spatial information handling via Geographic Information Systems (GIS)
- software and system design development.

Technologies for knowledge access: to add value to the information base and help turn it into knowledge; or to provide access to already-packaged knowledge.

- communication technologies for networking, the introduction, transfer, and use of information technologies in developing countries
- expert systems and other knowledge-engineering methods
- artificial intelligence applications
- computer assisted training.

Knowledge-communicating technologies – i.e. hypertext, multimedia integration with computing, and video-based techniques. (M. Stone)

In addition, the place in the ISSD/IDRC programme of future research on information technology policies and policy instruments (informatics and telecommunications) was indicated:

- social, economic and political impact studies dealing with the introduction, transfer, and use of information technologies in developing countries
- technology methods, information research, networking of information technology research, dissemination of research results
- cooperation with institutions supporting information technology research
- providing selective support to prototype activities in capacity development with other development assistance agencies with common goals and interests
- highly selective infrastructure development
- centres of excellence, model technology programs, in developing regions fostering South/South transfer, etc. (M. Stone)

In concluding the description of the IDRC, cited as an example of ongoing work, it was underlined that the Centre has just emerged from a massive restructuring exercise that has resulted in a leaner information science programme, but one with an extremely high profile, responding to the mission statement of "empowerment through knowledge." What is new in this restructuring for the ISSD is the new focus on *research and impact evaluation*. It is said that one must be able to demonstrate concretely, in ways that are understood and accepted by resource allocators, that information technologies are critical tools in the decision-making process at all levels: from the institutions/communities to the highest country/regional levels.

The IDRC is currently involved in such a long-term research project that will explore the possibility of assessing the impact/relevance of information technologies on the development process. It is stressed that the "indicators" must be built into a project or system at the time of design, and not after it has been completed, to ensure the focus of its output or impact:

> This [the IDRC research project] is divided into three phases. A theoretical discussion via a closed computer conference with a group of international experts in the field of information science. Following the report of this conference, a meeting of highly specialized information practitioners and users, concerned with the developing countries issues, will consider this theoretical framework and they will attempt to devise the criteria for creating assessment indicators – not input indicators. Following this, two or three new projects will be selected to field test the results from the earlier activities. . . . Already a project in health policy formulation for Uganda has been identified as a likely field test technology. Inexpensive satellite and PC linkage is already being used with medical practitioners, researchers, and extension workers in this country to disseminate requested information. What is needed now is to be able to answer with clarity what difference does it make. (M. Stone)

In respect to the Symposium discussion, it was pointed out that the challenge for the future is to cooperate internationally on evaluating the benefits from information and communications systems and determining for whose benefit they are being developed.

The point was raised repeatedly that it is inability to demonstrate relevance to the users or impact on the development process that prevents the allocation of adequate financial resources for the provision of even very important information services, especially in/for the developing countries.

There was a report on one case of setting up an information system adapted to users' needs in a highly industrialized country that showed how the National Center for Science Information Systems (NACSIS) of Japan contributes to academia through its database (DB) and communication technologies:

> NACSIS is one of the National Inter-University Research Institutes under the Ministry of Education, Science and Culture of Japan. Its mission is to gather, organize and provide scholarly information, as well as to carry out research and development of scholarly information DBs and their service system.
>
> NACSIS presently provides services on 40 academic DBs of various origins to the academic community of Japan and abroad. These DBs include (1) secondary and primary information DBs imported from abroad, (2) secondary information DBs created by NACSIS, (3) DBs created in collaboration with academic societies and the like (including full-text DBs), (4) repository DBs of academic research teams and other organizations, and (5) catalogue DBs (including those constructed in NACSIS).
>
> They are presently serviced on-line via NACSIS's own communication

network to libraries of about 200 universities and organizations and to several thousand individual academic researchers both in Japan and, on trial-basis, to some universities and other institutions abroad.

One unique category of DB is on academic conference papers titles and abstracts within three months of the conferences. This will keep researchers globally aware on who is doing what and where in Japan. It serves to keep the Japanese researchers aware of the utmost front-line of research in Japan. This will make interdisciplinary cross-fertilization active as well as make Japanese contribution visible to the world. (H. Yamada)

Illustrating the decisive importance of strong motivation for access to scientific information, in finding ways and means to get to information sources, even under most unfavourable conditions, an interesting example was described relating to Japan:

Developing countries ask for information. But I'll tell you an episode. During the Occupation by Allied Forces after World War II, we were not allowed to import foreign books. So I went to what is now the American Center Library in Tokyo and took out a book and hand-copied the entire book, and it was used by my fellow students and even my professor. That is the way Japan achieved its present status. I am not suggesting that this is the way to do it now. But when there is will, there should be a way. (H. Yamada)

In connection with this, a provocative question was raised:

Sometimes we are wrongly accused of hiding information. But when we publish journals in English, besides journals in our language, very few copies are sold. Do they really need information [those who accuse us]? (H. Yamada)

In the light of experiences with the use of information technologies to improve access to scientific knowledge in general and the results of international cooperation in this field aimed to improve the situation in the developing countries (discussed in Panel 1), it was stated that the rapid technological progress in the area of telecommunications and informatics does open new opportunities for developing countries to improve their information systems. However, caution was advocated:

Technology by itself cannot solve all problems. While the cost of equipment has declined substantially, telecommunications costs are unaffordable for many countries. Training of personnel to use the technology is also required. In many cases, the most suited solutions may be based on the use of old technologies or in their blending in various forms with the new ones. Due attention should be paid, moreover, to peculiarities of demand for information in developing countries, where there is no tradition to use databases, and to the problems of existing information retrieval systems still present there. . . . The obstacles are numerous and important, including language difficulties, high cost of acquiring primary literature, low relevance of available information to local problems and lack of recognition of information role and value. Decisions at the public and private level are often taken on the basis of incomplete and/or outdated information, which obviously affects the *quality* of decisions. Therefore, the role of public agencies should not be limited to satisfying existing demands: they should also help to develop and expand such demands. (C. Correa)

The current trend towards privatization of scientific results and the expanding scope of intellectual property rights in industrialized countries may further restrict flows of scientific knowledge to developing countries:

The trends towards a growing protectionism and privatization of scientific knowledge need to be mentioned. Four main factors can be mentioned in this regard: (a) there are new restrictions in some countries for the participation of foreign researchers or students in certain activities or courses; (b) universities and scientists have become more conservative in the publication of research results, whenever there is a likely application thereof for commercial purposes; (c) new forms of government intervention have emerged in this field; and (d) the expansion of intellectual property laws is blurring the distinction between *discovery* and *invention*. These trends may have substantial impact not only on the access by developing countries to science but also on scientific development in the long term. (C. Correa)

In spite of the considerable scope and important achievements of international cooperation in providing access to science and technology all over the world, especially in the past two decades, it has been said that there appears to be an urgent need to expand and strengthen actions in this direction:

If not done [international cooperation], developing countries may be growing left out of the main contemporary scientific development. . . . Such a cooperation should be based on the self-identification of problems by the developing countries concerned. It should emphasize the creation of absorptive capabilities – projects based only on the provision of equipment are clearly insufficient –, focus on the demand side via an active promotion of and education on the use of the information systems, and develop cost-effective solutions. International cooperation should create awareness on available technological alternatives and consider the establishment or enhancement of information systems as a process rather than as a single action. [International] cooperative activities should also take into account current changes in economic approaches in some developing countries and encourage the participation of the private sector in the building up and exploitation of information capabilities. (C. Correa)

In further reflection on the problem of scientific and technical information adapted to users' needs and peculiarities, which was emphasized many times during the discussion from different points of view, the importance of the broad vision of the closely integrated cultural diversity in the "global village" was stressed. This aspect is considered as essential for the success of international cooperation:

On the basis of the discussions here in Tokyo, it may be desirable . . . to advocate a large-scale international and multidisciplinary collaborative effort to enhance the relevant [information] technologies – as outlined by our contributors – in a culturally sensitive, fair, and economically sound manner. This includes a careful assessment of the possible social, economic, political, and cultural consequences of such a joint global effort to convert into reality what today seems to be a desirable and feasible vision for technological development. As with other technologies, one should bear in mind that even such a convincing vision may have indirect and unforeseen side-effects that some, or even most, people may consider undesirable. Those effects should therefore be avoided when building international cooperation to bring the vision of flexible information-processing closer to reality. (M. Dierkes)

Other comments on international cooperation advanced by several discussants were oriented towards three main topics:
- Scarcity of available funding for developing new modalities of international cooperation that would, on the one hand, satisfy needs and, on the other, exploit the full potential of new information technologies (J. Tocatlian, M. Stone, G. Johannsen, N. Dusoulier);
- Most "appropriate" information technologies, especially for use in the developing countries (J. Alty, D. Lide, M. Almada de Ascencio, C. Correa);
- Task force for follow-up action after the Symposium (Almada de Ascencio, M. Stone, N. Dusoulier, M. Dierkes).

Considering the funds available, especially with regard to support for the South, it was said that the industrialized countries have many economic problems to solve and might divert funds from development programmes, while the "West" is increasingly turning its attention to Eastern Europe. Though recognizing this tendency, it was further said that nowadays priorities have changed and that large, expensive projects get support, e.g. in the European Community (EC), to the detriment of smaller ones. It was therefore important to persuade the high levels of decision-making to re-allocate more funds for science and technology, and the relevant information systems, by demonstrating their overall social and economic impact.

One response to the disturbing trends mentioned should be to strengthen the close collaboration and pooling of resources among all the agents involved in international cooperation for development, bringing together both public and private sources. Another condition for attracting funds is the preparation of proposals of high quality, corresponding to the real needs, that would create systems for the benefit of large user communities.

In further comment on the last point, the habitual question was raised about the "appropriate" technologies for the less developed countries. While some voices were heard in favour of less sophisticated systems for the South, most participants stressed the fact that for the so-called "developing" countries, the new information technologies are indeed the most "appropriate." However, it is important to have the projects for systems defined by the concerned users, i.e. countries. Some of them might need basics, while others require advanced solutions, but they themselves must decide. One should be reminded that it is a characteristic of the latest advances in information technologies that they are usually the most efficient, powerful, easy to implement, and hence the cheapest for the user.

In this connection N. Streitz made the observation that this Symposium had offered a unique opportunity for a meeting between two distinct groups of practitioners: the researchers presenting the state of the art of information technologies, and the specialists involved in international cooperation of information and communication systems. Their encounter, a rare occasion, was judged as inspiring and fruitful: both groups have learned from each other. The question was raised about what kind of interaction would be envisaged in the future.

There was a general feeling that the Symposium, notwithstanding the publication of its proceedings, should not constitute a single event without follow-up. The goals of the Symposium require continuous action. Many questions had been raised, but only a few answers could be given in the short time of three days. The request was made and strongly supported that the United Nations University should find ways to continue the process initiated by this Symposium.

In the future, the UNU, by initiating and co-sponsoring studies and pilot projects, which would show how an innovative idea really "works," should continue its activities in this field, closely linking its efforts to those of other organizations that conduct similar regional and international projects.

As a possible mechanism, the establishment of a task group was proposed that would, taking advantage of the excellent papers and the rich discussions heard here, identify the most promising areas for international cooperative activities in the future.

Closing the discussion, I. Wesley-Tanaskovic stated that:
1. One major conclusion could be drawn from the Panel discussions: new information technologies offer manifold opportunities for opening an era of more flexible global communications.

The use of new information technologies may exacerbate, instead of reduce, the gap that separates the industrialized societies from those in development. However, thanks to their potential ubiquitous availability, when used in an appropriate way, they are able to combat many obstacles more efficiently than before – among others, the trend towards privatization of scientific outputs – and thus foster their applications for social and economic development. The new information technologies, facilitating communication in the "global village," may help to preserve linguistic diversity and eliminate the cultural problems of access to knowledge, intellectual experience, and indigenous expression.

In access to knowledge (science and technology), the era of rigid centralized information systems has irrevocably gone. Fortunately, *technology* can be directed to create user-oriented and user-defined networks that will outperform (and have already outperformed in many instances) the centralized models of information systems, which create "isolation" by their limited views and not by the limited access to the data in their files. At the same time, technology enables the interfacing to a variety of networks through the modern multi-task (multimedia) terminal of the user, while the universal "intelligent" networks will ensure global communications among this variety of local networks.

The countries called "developing" are those that would benefit most from the creation of local networks, adapted to their needs, connected in a flexible "network of networks," provided adequate planning and funding for an efficient communications infrastructure become available. This is considered a priority for the developing world: building modern communication networks that could support all the local value-added information systems, comprising those dealing with access to science and technology.

Thus, "mastering new networking technologies" might become a priority area for research and training oriented mainly to responding to the present needs of the developing countries. This could follow up the UNU project of "mastering microprocessor technologies" that was initiated in the early 1980s and at present involves several universities in the South.

2. The other conclusion, which obtained full consensus during the Panel session, is that international cooperation in this field is of critical importance for everybody – for the developing countries first of all, but for the industrialized societies as well.

Global development is interdependent and the lagging behind of the major part of the world is jeopardizing the welfare of all. Moreover, in this field of rapid technological advances that require many and varied human resources, the denial to a large part of humanity of the possibility to face this challenge and to actively contribute to scientific and technological progress is inequitable but also counter-productive. At the global level, much of the human potential is being irreparably lost.

Therefore, a commensurate effort is needed to propagate these views in order to obtain overall cooperation and the funding required to include those parts of the world that cannot afford technological development using their own resources.

3. At the end, the panel recommended that the United Nations University

should find ways and means to follow up on the deliberations of the Symposium, continuing the interface between all the various relevant professional groups, and the development funding agencies, public and private, interested in international cooperation for enhancing access to science and technology.

Closing Remarks

Roland J. Fuchs, Vice-Rector, The United Nations University

Friends and colleagues,

The hour is late and, after three very full days, I trust you will forgive me for making my closing remarks very brief ones. There is in any case very little to say beyond what has already been said.

Our conference was based on several assumptions that I believe were confirmed in the course of our meetings:

(1) That the telecommunication, computer, and related information technologies together represent a fundamental technological revolution with profound and far-reaching implications;

(2) That increasingly the wealth of nations will rest on access to knowledge and development of the so-called knowledge-based industries;

(3) That the emerging information technologies, rather than "broadening the community of learning and knowledge," as we may hope, could serve instead to widen the economic and knowledge gaps between nations.

Our meetings gave us cause for both hope and concern:

- We were provided with an overview of the various exciting frontiers of information technology;
- While not every prospect necessarily pleases, even a novice (as I most assuredly am) could not help but be excited by the vistas opening not only for science and technology, but for humankind more generally;
- This is indeed a testament to the imagination and dedication of many of you in this room and the communities of scientists and technologists you represent;
- However, one could also not help but notice that this technological progress, not unexpectedly, is being driven by the internal imperatives of the

research communities and the perceived market opportunities for the private sector;
- In reviewing this progress, one senses little explicit concern with the applications of these technologies to increasing the access of LDCs to science and technology, and their benefits;
- The nature of the gap between developed countries and underdeveloped countries was dramatically illustrated by the example of Japan on the one hand planning a nationwide fibre optic network to be completed within 20–30 years at a cost of 200–300 billion dollars, while many developing countries, on the other hand, cannot even look forward to a reliable phone system in that span of time. We received also many examples of the lopsided international distribution of databases, computers, and information technology generally.

Therefore, there was a particular interest in those sessions, papers, and discussions that dealt explicitly with this problem of expanding access to science and technology and the role of international cooperation. Progress in the development of science and technology will certainly go on with or without those of us in this room, even without the eminent researchers present, but international cooperation at this point in time is very much dependent on several organizations represented at this conference – the UN agencies, selected scientific and scholarly organizations, several national information centres, and organizations such as the IDRC. But, however important, these organizations can only do a part of what is required; we desperately need the help of leading researchers and the private sector. One result of this conference we hope is that some of you will turn your interest and talents to the problems of developing countries and imaginative ways of overcoming the technological gap.

In conclusion, on behalf of the UNU, I wish to extend our sincere thanks to those who made this conference possible – our co-organizer Kyoto University, especially President Imura and past President Nishijima, our financial supporter Fujitsu Corporation, our scientific organizers Dr. Araki, Dr. Tocatlian, and Ms. Wesley-Tanaskovic. I would also like to acknowledge the assistance of Sir John Kendrew from the time the meeting was first considered some two years ago. Finally, our warmest thanks to each of you who took the trouble to participate, despite the many competing demands on your time.

We are grateful to you all and we hope to involve you in our future activities. We remain open to your suggestions and ideas at any time.

Except for those who must meet tomorrow morning, I now declare the conference closed.

Domo Arigato Gozaimashita – Sayonara.

Contributors

Margarita Almada de Ascencio graduated in chemistry at the National University of Mexico (UNAM) and attended courses in information and library systems planning in Mexico and the United Kingdom. Currently Director of the Science and Humanities Centre (CICH) and Professor of Library and Information Science at UNAM, she has undertaken research projects in informatics, information planning and policy, and her works have been published extensively.

James L. Alty has a degree in physics and a Ph.D. in nuclear physics from the University of Liverpool. He was Professor of Computer Science at the University of Strathclyde, Glasgow, and Executive Director of the Turing Institute, which specializes in research and exploitation of Artificial Intelligence. He heads the Computer Studies Department of Loughborough University.

Michael Keeble Buckland is Professor and former Dean at the School of Library and Information Studies, University of California, Berkeley. He has a degree in modern history (Oxford University) and a Ph.D. in economic analysis and librarianship from Sheffield University. He has published several books on library services and information systems.

Charles Morrison Cooper was Professor of Development Economics, the Institute of Social Studies, The Hague, and a member of the Royal Economic Society. He worked as Senior Fellow at the Science Policy Research Unit, University of Sussex, and the Directorate for Scientific Affairs, the OECD,

Paris. He is presently Director of the United Nations University's Institute for New Technologies (INTECH) in Maastricht, Netherlands.

Carlos Maria Correa earned his degrees in law and economics at the Universidad Nacional de Buenos Aires. Among his positions were Director of the Research Project on Technological and Economic Trends in Informatics and Electronics, Undersecretary of State for Informatics and Development, and Member of the Technical Advisory Group for the GATT negotiations.

Meinolf Dierkes holds a Ph.D. in economics and social science from the University of Cologne. He was Founding Director of the International Institute for Environment and Society at the Science Centre, Berlin, and was its first President. At present, he is Director of the research unit, "Organisation and Technology," Science Centre, Berlin, and Professor of Sociology of Science and Technology at the Technische Universität, Berlin.

Nathalie Dusoulier is Director-General of the Institut de l'Information Scientifique et Technique of the National Scientific Research Centre (CNRS) in Nancy, France. She has a Ph.D. in pharmacology from the University of Paris and a degree in business administration from the same university. She is former Director of the UN Information Services at the Dag Hammarskjöld Library in New York as well as in Geneva.

Masaru Harada is Associate Professor of Library and Information Science at the Faculty of Education, Kyoto University. He earned his M.Sc. in applied science from the University of Tokyo and worked several years at Unesco, Paris. He has authored and co-authored five books.

Gunnar Johannsen is Professor of Systems Engineering and Man-Machine Systems in the Department of Mechanical Engineering, University of Kassel, Germany. He received his Ph.D. in electrical engineering and guidance and control from the Technical University of Berlin. He co-authors and co-edits several scientific publications.

Takahiko Kamae is Executive Manager of the Human Interface Laboratories of the Nippon Telegraph and Telephone Corporation (NTT). He has a B.E. and Master of Electronics Engineering from Kyoto University. He obtained his Ph.D. in engineering from the University of Illinois, USA. He has co-authored two books in Japanese – "Local Area Networks" and "New Media in Telecommunications."

Sir John Cowdery Kendrew was awarded the Nobel Prize for Chemistry in 1962. He received his Ph.D. from Cambridge University. He was involved with the founding of the European Molecular Biology Organisation (EMBO), and is former Director-General of the EMBO laboratory in

Heidelberg. He has been elected Secretary-General, Vice-President, and President of the International Council of Scientific Unions (ICSU). He is former Member and Chairman, the UNU Council, Tokyo, and was President of St John's College, Oxford.

Lian Yachun is Chief, the Computer Centre, the Institute of Scientific and Technical Information of China (ISTIC), Beijing. He graduated from the Physics Department, Zhong Shan University. He trained in computer hardware maintenance in Tokyo and software programming at the Royal Institute of Technology, Stockholm.

David R. Lide was educated at the Carnegie Institute of Technology and obtained his Ph.D. in chemical physics from Harvard. He worked at the National Bureau of Standards as Head of the Molecular Spectroscopy Section, then as Director, Standard Reference Data Program. As President of CODATA, he coordinated data programmes in areas of the physical, biological, and geosciences. He is Editor of the *Journal of Physical and Chemical Reference Data* and the *Handbook of Chemistry and Physics*.

Maurice Chinfwembe Lundu is University Librarian at Copperbelt University, Kitwe, Zambia. A graduate in sociology with a Master's Degree from Case Western Reserve University, Cleveland, USA, he also has a Ph.D. from the University of Sheffield. He is former Regional Vice-President for East, Central, and Southern Africa of the Commonwealth Library Association (COMLA).

Makoto Nagao has been full Professor of Electrical Engineering at Kyoto University from 1973, obtaining his Master's Degree and Ph.D. from the same. Acting editor of various international journals including *Computer Graphics, Vision and Image Processing*, and *Artificial Intelligence*, and Editor-in-Chief of *IECE Transactions on Information Systems*.

Nobuyuki Otsu received his degree in mathematical engineering from the University of Tokyo. He has been with the Electrotechnical Laboratory (ETL), Agency of Industrial Science and Technology (AIST), and the Ministry of International Trade and Industry (MITI) of Japan. He is currently Director of the Machine Understanding Division of the ETL, and Professor, Graduate School, Department of Electronics and Information of Tsukuba University.

Stephen E. Robertson is Head of the Department of Information Science at City University, London and directs a research centre within the department. Originally a mathematician, he received his Ph.D. in information studies at University College, London. He is co-author with Karen Sparc-Jones of a book on the theory of relevance weighting of search terms.

William B. Rouse completed his Ph.D. at the Massachusetts Institute of Technology. He has served on the faculties of four American universities. Currently he is Chief Executive Officer of Search Technology, which specializes in software products, contract R&D, and engineering services in decision-support and training systems. He is editor of the research annual *Advances in Man-Machine Systems Research*.

Andrew P. Sage received his engineering degree from the Massachusetts Institute of Technology, his Ph.D. from Purdue, and a doctor of engineering *honoris causa* from the University of Waterloo. He is First American Bank Professor and Dean of the School of Information Technology and Engineering at George Mason University, and Editor of *IEEE Transactions on Systems, Man and Cybernetics, AUTOMATICA, Large Scale Systems*, and a textbook series on systems engineering.

Blagovest Hristov Sendov is a Member and former President of the Bulgarian Academy of Sciences. He earned his Ph.D. at Steklov Mathematics Institute of the CIS Academy of Sciences, Moscow, specializing in numerical methods. He studied computer science at the Imperial College, London. He is Professor of Mathematics, former Dean of Faculty and Rector of Sofia University; President, International Federation for Information Processing (IFIP); Director, Centre for Informatics and Computer Technology, Bulgarian Academy of Sciences, Sofia.

Linda C. Smith is Associate Professor, Graduate School of Library and Information Science, University of Illinois at Urbana-Champaign (UIUC). She received her Ph.D. from the School of Information Studies, Syracuse University, an M.Sc. in computer science, Georgia Institute of Technology, and a B.S. in physics and mathematics. She is co-editor of the textbook *Reference and Information Services* and served as Consultant to Unesco's General Information Programme.

Martha B. Stone graduated in philosophy, political science, and sociology and earned an M.Sc. degree in library/information science at Drexel University, Philadelphia. She was involved in library services with the Canadian government. As Director-General of the Information Science and Systems Division, International Development Research Centre (IDRC) in Ottawa, Canada, she has assisted developing countries in the use of information technologies to enhance their capacities for addressing development problems.

Norbert A. Streitz received his Ph.D. in cognitive psychology from the Technical University, Aachen, and a Ph.D. in theoretical physics from the University of Kiel. He is currently Vice-Director of the Integrated Publication and Information Systems Institute (IPSI) of the Gesellschaft für Mathematik und Datenverarbeitung (GMD) (National Research Centre for Computer Science and Technology) and Manager of the research division "Cooperative